IUTAM Symposium on Designing for Quietness

SOLID MECHANICS AND ITS APPLICATIONS
Volume 102

Aims and Scope of the Series

The fundamental questions arising in mechanics are: *Why?*, *How?*, and *How much?* The aim of this series is to provide lucid accounts written by authoritative researchers giving vision and insight in answering these questions on the subject of mechanics as it relates to solids.

The scope of the series covers the entire spectrum of solid mechanics. Thus it includes the foundation of mechanics; variational formulations; computational mechanics; statics, kinematics and dynamics of rigid and elastic bodies: vibrations of solids and structures; dynamical systems and chaos; the theories of elasticity, plasticity and viscoelasticity; composite materials; rods, beams, shells and membranes; structural control and stability; soils, rocks and geomechanics; fracture; tribology; experimental mechanics; biomechanics and machine design.

The median level of presentation is the first year graduate student. Some texts are monographs defining the current state of the field; others are accessible to final year undergraduates; but essentially the emphasis is on readability and clarity.

For a list of related mechanics titles, see final pages.

IUTAM Symposium on

Designing for Quietness

Proceedings of the IUTAM Symposium held in Bangalore, India, 12-14 December 2000

Edited by

M.L. MUNJAL

Facility for Research in Technical Acoustics (FRITA),
Department of Mechanical Engineering,
Indian Institute of Science,
Bangalore, India

Springer-Science+Business Media, B.V.

A C.I.P. Catalogue record for this book is available from the Library of Congress.

Printed on acid-free paper

ISBN 978-90-481-6081-5 ISBN 978-94-017-0095-5 (eBook)
DOI 10.1007/978-94-017-0095-5

Originally published by Kluwer Academic Publishers in 2002.
Softcover reprint of the hardcover 1st edition 2002

Contents

Preface

It is being increasingly recognized that noise is not only a nuisance but also is often a measure of wasted energy, or an indication of a badly designed, fabricated or maintained machine, or a symptom of a malfunction. The most cost-effective way for noise control is to control it at the source, or better still to design for quietness. Design engineers need to be taught how to design for quietness through continuing education courses. Quieter technologies need to be adopted where available and developed where necessary. Towards this end, an International Symposium on Designing for Quietness was held at the Facility for Research in Technical Acoustics (FRITA), Department of Mechanical Engineering, Indian Institute of Science, Bangalore in December 2000. It was sponsored by the International Union for Theoretical and Applied Mechanics (IUTAM).

There is hardly any book or monograph on the subject of Designing for Quietness or Noise Control at the Source. This was probably the first symposium on this important subject. There was in fact a strong suggestion at the end of the symposium that we should try to repeat such symposia elsewhere and eventually make them periodical. Therefore, the proceedings of this symposium presented in this volume after appropriate review, revision and editing, should be a welcome edition on the desk of all engineers and designers. In particular, this volume will be of great use to all engineers and consultants dealing with noise control of the existing machinery and design of new machinery and industrial lay-outs. Teachers of technical acoustics will find useful supplementary material in this volume.

The papers presented here are invited from the well known experts in the field, hailing from different countries in the east as well as the west. There are several case studies illustrating successful design for quietness or noise control at the source. Further reading will be facilitated by the references listed at the end of every paper. The particular applications covered in this volume are diesel engines, automobiles including trucks, exhaust systems, ventilation systems, ships, air conditioners, refrigerators, helicopters and space structures. Besides, there are a few papers dealing with basic concepts like statistical energy analysis, structural optimization, acoustical materials, acoustic transmission in ducts and mufflers, active noise control of acoustic cavities with flexible walls, etc. There are also a couple of papers dealing with the related issue of the effect of noise on human hearing and health.

As indicated above, the emerging area of designing for quietness falls in the general area of technical acoustics. However, as is obvious from the list of applications covered in this volume, this is an inherently multidisciplinary area encompassing duct acoustics, vibro acoustics, computational acoustics (FEM, BEM, etc), mechanical engineering, instrumentation and applied mathematics.

Over the last one year, subsequent to the symposium, the Editor has been helped by members of the scientific committee and other reviewers, who found time from their busy schedule to review the manuscripts and offer their valuable comments. In the course of editing this volume, the computer work by Mr. B. Venkatesham and Mr. Avuluri Sreenivasa Rao have been very crucial indeed. Communication with authors and reviewers has been carried on by Prabha Raju with great patience and perseverance. In the organization of the symposium in December 2000, the Editor got the whole hearted support of all the faculty and supporting staff associated with the Facility for Research in Technical Acoustics. In particular, the patronage of the Chairman of the Department, Prof. S.K. Biswas, is acknowledged with thanks. The technical advice of the Series Editor, Prof. Graham Gladwell, is gratefully acknowledged. Dr. Arno Schouwenburg and Ms. Nathalie Jacobs of Kluwer Academic Publishers have indeed been very patient and helpful with their suggestions from time to time. Finally, the Editor would like to acknowledge the International Union for Theoretical and Applied Mechanics for sponsoring the symposium, and Department of Science and Technology of the Government of India for supporting the Facility for Research in Technical Acoustics at the Indian Institute of Science.

M. L. MUNJAL

Contributing Authors

Jorge P. Arenas Mechanical Engineering Department, Auburn University, Auburn, AL 36849, U.S.A.

Hans-Joachim Beer, Institut fr Festkrpermechanik, Technische Universitt, 01062 Dresden, Germany.

V. I. Bujakas, Astro Space Center of P. N. Lebedev Physical Institute, Russian Academy of Science, Profsoiuznaya st. 84/32, Moscow, 117810, Russia.

Malcolm J. Crocker, Mechanical Engineering Department, Auburn University, Auburn, AL 36849, U.S.A.

Yash K. Dungerpuria, Facility for Research in Technical Acoustics, Department of Mechanical Engineering, Indian Institute of Science, Bangalore 560 012, India.

Albrecht Eiber, Institute B of Mechanics, University of Stuttgart, Germany.

Joachim Gier, Institut fr Festkrpermechanik, Technische Universitt, 01602 Dresden, Germany.

S. Gopalakrishnan, Department of Aerospace Engineering, Indian Institute of Science, Bangalore 560 012, India.

Hans-Jurgen Hardtke, Institut fr Festkrpermechanik, Technische Universitt, 01062 Dresden, Germany.

R. J. Hooker, Department of Mechanical Engineering, The University of Queensland, Brisbane, Australia.

C.S. Jog, Facility for Research in Technical Acoustics, Department of Mechanical Engineering, Indian Institute of Science, Bangalore 560 012, India.

K. Konishi, Ken-on Engineering Inc., 1-17-1 Shofudai, Tarumi, Kobe 655-0871, Japan.

Z. Maekawa, Environmental Acoustics Laboratory, Kobe University Faculty of Engineering, Rokko, Nada, Kobe, 657-8501, Japan.

D. Roy Mahapatra, Department of Aerospace Engineering, Indian Institute of Science, Bangalore 560 012, India.

Dhanesh N. Manik, Department of Mechanical Engineering, Indian Institute of Technology Bombay, Mumbai 400 076, India.

S. Manivasagam, Product Engineering Department, Kirloskar Copeland Ltd., Karad- 415 110, India.

Steffen Marburg, Institut fr Festkrpermechanik, Technische Universitt, 01062, Dresden, Germany.

Parag H. Mathuria, Section Engineer, Mahindra (R& D), Nasik, India.

A. R. Mohanty, Department of Mechanical Engineering, Indian Institute of Technology, Kharagpur 721 302, India.

M.L. Munjal, Facility for Research in Technical Acoustics, Department of Mechanical Engineering, Indian Institute of Science, Bangalore 560 012, India.

S. Narayanan, Department of Applied Mechanics, Indian Institute of Technology Madras, Chennai 600 036, India.

C. Padmanabhan, Department of Applied Mechanics, Indian Institute of Technology Madras, Chennai 600 036, India.

Avinash R. Patil, Mechanical Engineering Department, Auburn University, Auburn, AL 36849, USA.

M. G. Prasad, Noise and Vibration Control Laboratory, Department of Mechanical Engineering, Stevens institute of Technology, Hoboken, New Jersey 07030, USA.

Deepak Prasher, Institute of Laryngology and Otology, University College of London, London, UK.

Rudra Pratap, Facility for Research in Technical Acoustics, Department of Mechanical Engineering, Indian Institute of Science, Bangalore 560 012, India.

V. Bhujanga Rao, Associate Director, Naval Science and Technological Laboratory, Visakhapatnam 530 027, India.

Werner Schiehlen, Institute B of mechanics, University of Stuttgart, Germany.

J. Senthilnathan, Product Engineering Department, Kirloskar Copeland Ltd., Karad - 415 110, India.

B. Venkatesham, Facility for Research in Technical Acoustics, Department of Mechanical Engineering, Indian Institute of Science, Bangalore 560 012, India.

ANALYSIS AND DESIGN OF AN ANNULAR AIRGAP LINED DUCT FOR HOT EXHAUST SYSTEMS

M. L. MUNJAL and B. VENKATESHAM
Facility for Research in Technical Acoustics
Department of Mechanical Engineering
Indian Institute of Science
Bangalore 560 012, India
munjal@mecheng.iisc.ernet.in

Abstract One of the reasons why exhaust systems of automobiles are designed as reactive mufflers is that high velocity, hot and impure exhaust gases would blow away fibers of dissipative material, thereby affecting the acoustic performance of automotive mufflers with time. On the other hand, absorptive mufflers have a distinct advantage over reactive ones in that their performance is wide band, without sharp troughs. Therefore, annular-gap lined duct is proposed as an alternative for dissipative mufflers.

Starting with the basic equations, wave equations are derived for the three interactive regions. For harmonic time dependence, these equations are reduced to Helmholtz equations, which in turn are converted to six first order equations. Eigen analysis of these equations leads to a 6×6 matrix. Use of the four boundary conditions reduces this to the required 2×2 transfer matrix between the upstream and the downstream point in the inner flow duct of the element. This has been validated for self-consistency and then used in a parametric study which has resulted in design guidelines for such like mufflers.

1. Introduction

Automotive exhaust systems are characterized by hot gas flows, often containing some unburnt carbon particles. In an acoustically lined duct shown in Fig. 1, there is a strong possibility of some fibers of a fibrous absorptive material like glass wool or mineral/rock wool being swept away continuously, and the perforated protective plate getting progressively clogged with unburnt carbon particles and possibly lubrication oil.

M.L. Munjal (ed.), IUTAM Symposium on Designing of Quietness, 1–19.

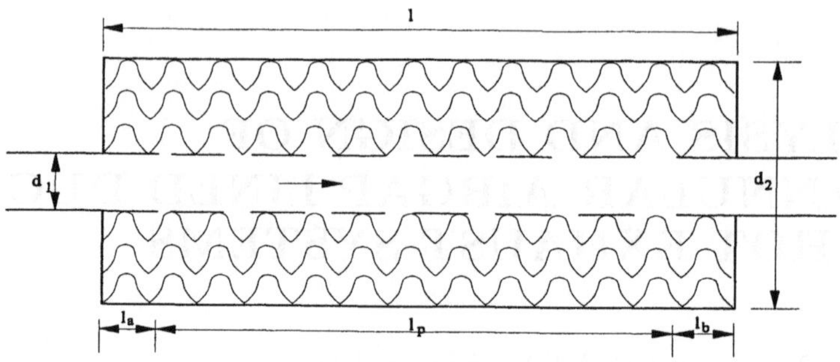

Figure 1. Schematic representation of lined duct

Therefore, most of the automotive exhaust mufflers are of the reactive or reflective or non-dissipative type that work on the principle of impedance mismatch [1]. Insertion loss curves of these mufflers, unfortunately, suffer from many sharp troughs, not all of which can be predicted or avoided. Dissipative mufflers or acoustically lined ducts shown in Fig. 1 do not suffer from this disadvantage. They have low back pressure, their transmission loss graphs are characterized by a relatively broadband dome which can be predicted quite well right at the design stage. It is desirable therefore to incorporate a dissipative stage in series with a main reactive muffler. But then, the problem of high temperature remains. Resin-bonded glass wool or mineral wool or ceramic wool cannot be used because resins would generally melt at such temperatures (typically exceeding 500 degrees centigrade). Loose or un-bonded fibrous material would need to be enclosed in thin impervious materials, specially designed for the purpose. These are generally proprietary in nature. Inertive impedance of the impervious covering lowers the high-frequency acoustic absorption considerably [2]. An alternative would be to make use of another perforated cylinder between the inner flow pipe and the absorptive layer with airgap in between, as shown in Fig. 2.

The annular airgap would not only ensure mechanical protection of the absorptive material, but also would have interesting acoustic properties, as investigated in this paper.

Starting with one-dimensional basic equations of mass continuity, momentum balances and isentropicity, coupled Helmholtz equations are derived for the three regions, viz., the inner flow duct, annular airgap and the absorptive layer. The basic logic is the same as in Refs. [3, 4]. Equivalent density and sound speed of the absorptive material are obtained

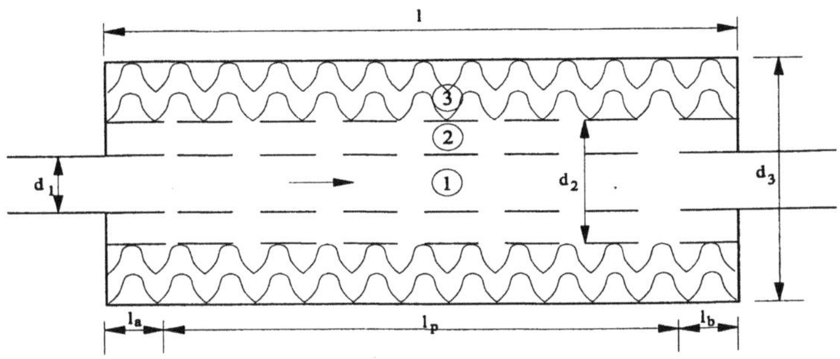

Figure 2. Schematic representation of annular airgap lined duct

from wave number and characteristic impedance, which in turn are got from the specific flow resistance. The coupled Helmholtz equations are solved as an eigenvalue problem, and appropriate boundary conditions are applied to obtain the transfer matrix between points u and d (denoting upstream and downstream, respectively)as shown in Fig. 2. Transfer matrix enables cascading of this element with other elements constituting the exhaust system. In this paper, however, transmission loss, TL, for the annular airgap configuration of Fig. 2 is computed, verified for self-consistency, and used for a parametric study, leading to the required design guidelines.

2. The Governing Equations

Following the physical concepts explained in Refs.[3, 4], one-dimensional continuity equations for the three interacting regions shown in Fig. 3 are

$$\rho_0 \frac{\partial u_1}{\partial z} + U_1 \frac{\partial \rho_1}{\partial z} + \frac{4}{d_1} \rho_0 u_{12} = -\frac{\partial \rho_1}{\partial t} \tag{1}$$

$$\rho_0 \frac{\partial u_2}{\partial z} - \frac{4 d_1 \rho_0 u_{12}}{(d_2^2 - d_1^2)} + \frac{4 d_2 \rho_0 u_{23}}{(d_2^2 - d_1^2)} = -\frac{\partial \rho_2}{\partial t} \tag{2}$$

$$\rho_w \frac{\partial u_3}{\partial z} - \frac{4 d_2 \rho_w u_{23}}{(d_3^2 - d_2^2)} = -\frac{\partial \rho_3}{\partial t} \tag{3}$$

where

ρ_0 is the mass density of the exhaust gases at ambient temperature

ρ_w is the equivalent mass density of the absorptive material, $\frac{Y_w k_w}{\omega}$

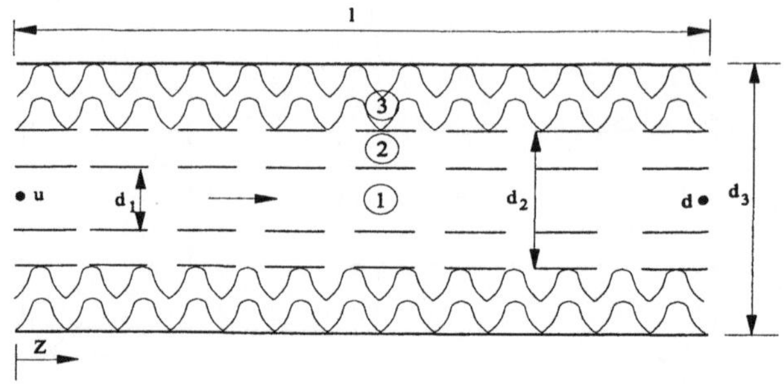

Figure 3. Interaction of waves in annular airgap lined duct (1. Main flow duct 2. Annular air gap 3. Absorbing material lining)

Y_w and k_w are the characteristic impedance and the wave number of the absorptive fibrous material respectively, described by the empirical formula of Delany and Bazley [5] as modified and improved by Mechel [6]:

$$\frac{Y_w}{\rho_0 a_0} = \left[\begin{array}{lll} 1 + 0.0485A^{0.754} - \jmath 0.087A^{0.73} & \text{for} & A < 60 \\ \frac{\frac{0.5A}{\pi} + \jmath 1.4}{(-1.466 + \jmath 0.212A)^{0.5}} & \text{for} & A > 60 \end{array} \right. \tag{4}$$

$$\frac{k_w}{k_0} = \left[\begin{array}{lll} 1 + 0.0978\ A^{0.6929} - \jmath 0.189\ A^{0.6185} & \text{for} & A < 60 \\ (1.466 - \jmath 0.212A)^{0.5} & \text{for} & A > 60 \end{array} \right. \tag{5}$$

A is the normalized flow resistance of a λ-deep layer; i.e.,

$$A = \frac{E\lambda}{\rho_0 a_0} \tag{6}$$

λ is the wave length of the gas at ambient temperature
a_0 is the speed of sound in the gas at ambient temperature
E is the flow resistance of the absorptive material
u_i is the acoustic particle velocity in region i; i=1, 2, 3.
ρ_i is the density perturbation in region i
U_1 is the mean flow velocity in the inner flow tube (region 1)
u_{12} is the radial velocity across the inner perforated tube, from region 1 to region 2:

$$u_{12} = \frac{p_1 - p_2}{\rho_0 a_0 \zeta_1} \tag{7}$$

u_{23} is the radial velocity across the intermediate perforated tube, from region 2 to region 3:

$$u_{23} = \frac{p_2 - p_3}{\rho_0 a_0 \zeta_2} \tag{8}$$

a_0 is the sound speed in the gas at ambient temperature
ζ_1 is the grazing flow impedance of the inner perforated tube [7]

$$\zeta_1 = \frac{[7.337 \times 10^{-3}(1 + 72.23\ M_1) + \jmath\ 2.2245{\times}10^{-3}\ (1 + 51t)(1 + 204\ d_h)\ f]}{\sigma} \tag{9}$$

M_1 is the mean flow mach number, $\frac{U_1}{a_0}$
t is thickness of the perforated tube
d_h is the hole diameter
σ is porosity
ζ_2 is the stationary medium impedance of the intermediate perforated tube [3]:

$$\zeta_2 = \frac{[6{\times}10^3 + \jmath\ k_0(t + 0.75\ d_h)]}{\sigma} \tag{10}$$

f is frequency in Hz, $\frac{\omega}{2\pi}$
k_0 is the wave number, $\frac{\omega}{a_0}$
Equations of momentum balance are :

$$\rho_0 \frac{Du_1}{Dt} = -\frac{\partial p_1}{\partial z} \tag{11}$$

$$\rho_0 \frac{\partial u_2}{\partial t} = -\frac{\partial p_2}{\partial z} \tag{12}$$

$$\rho_w \frac{\partial u_3}{\partial t} = -\frac{\partial p_3}{\partial z} \tag{13}$$

where

p_1,p_2 and p_3 are pressure perturbations in the three regions
$\frac{D}{Dt}$ is the total or substantive derivative:

$$\frac{D}{Dt} = \frac{\partial}{\partial t} + U_1 \frac{\partial}{\partial z} \tag{14}$$

It may be noted that radial mass transfer does not affect the one-dimensional momentum equations.
Finally, the isentropicity equations are:

$$dp = a_0^2 d\rho \qquad \text{for regions 1 and 2} \tag{15}$$

and

$$dp_3 = a_w^2 d\rho_3 \qquad \text{for region 3} \tag{16}$$

where a_w is the equivalent wave speed in the absorptive material:

$$a_w = \frac{\omega}{k_w} \tag{17}$$

For harmonic time dependence $e^{\jmath\omega t}$, combining Eqs. (1)-(3), and (11)-(14) with (15) and (16), so as to eliminate ρ_1, ρ_2, ρ_3, u_1, u_2 and u_3, yields the following three Helmholtz equations:

$$(1-M_1^2)\frac{d^2p_1}{dz^2} - (2\jmath k_0 M_1 + \frac{4M_1}{d_1\zeta_1})\frac{dp_1}{dz} + (k_0^2 - \frac{4\jmath k_0}{d_1\zeta_1})p_1 + \frac{4M_1}{d_1\zeta_1}\frac{dp_2}{dz} + \frac{4\jmath k_0}{d_1\zeta_1}p_2 = 0 \tag{18}$$

$$\frac{d^2p_2}{dz^2} + (k_0^2 - \frac{4\jmath d_1 k_0}{(d2^2 - d1^2)\zeta_1} - \frac{4\jmath d_2 k_0}{(d_2^2 - d_1^2)\zeta_2})p_2 + \frac{4\jmath d_1 k_0}{(d_2^2 - d_1^2)\zeta_1}p_1 + \frac{4\jmath d_2 k_0}{(d_2^2 - d_1^2)\zeta_2}p_3 = 0 \tag{19}$$

$$\frac{d^2p_3}{dz^2} + (k_w^2 - \frac{4\jmath d_2 Y_w k_w}{(d_3^2 - d_2^2)Y_0\zeta_2})p_3 + \frac{4\jmath d_2 Y_w k_w}{(d_3^2 - d_2^2)Y_0\zeta_2}p_2 = 0 \tag{20}$$

3. Solution of the Governing Equations

Eqs. (18)-(20) may be rearranged in the matrix form

$$\begin{bmatrix} D^2 + \alpha_1 D + \alpha_2 & \alpha_3 D + \alpha_4 & 0 \\ \alpha_5 D + \alpha_6 & D^2 + \alpha_7 D + \alpha_8 & \alpha_9 D + \alpha_{10} \\ 0 & \alpha_{11} D + \alpha_{12} & D^2 + \alpha_{13} + \alpha_{14} \end{bmatrix} \begin{bmatrix} p_1 \\ p_2 \\ p_3 \end{bmatrix} = 0 \tag{21}$$

where

$$\alpha_1 = -\frac{2\jmath k_0 M_1 + \frac{4M_1}{d_1\zeta_1}}{1-M_1^2} \qquad \alpha_2 = \frac{k_0^2 - \frac{4\jmath k_0}{d_1\zeta_1}}{1-M_1^2}$$

$$\alpha_3 = \frac{4M_1}{d_1\zeta_1(1-M_1^2)} \qquad \alpha_4 = \frac{4\jmath k_0}{d_1\zeta_1(1-M_1^2)}$$

$$\alpha_5 = \alpha_7 = \alpha_9 = \alpha_{11} = \alpha_{13} = 0$$

$$\alpha_6 = \frac{4\jmath d_1 k_0}{(d_2^2 - d_1^2)\zeta_1} \qquad \alpha_8 = k_0^2 - \frac{4\jmath d_1 k_0}{(d_2^2 - d_1^2)\zeta_1} - \frac{4\jmath d_2 k_0}{(d_2^2 - d_1^2)\zeta_2}$$

$$\alpha_{10} = \frac{4\jmath d_2 k_0}{(d_2^2 - d_1^2)\zeta_2} \qquad \alpha_{12} = \frac{4\jmath d_2 Y_w k_w}{(d_3^2 - d_2^2) Y_0 \zeta_2}$$

$$\alpha_{14} = k_w^2 - \frac{4\jmath d_2 Y_w k_w}{(d_3^2 - d_2^2) Y_0 \zeta_2} \tag{22}$$

Defining

$$\frac{dp_1}{dz} \equiv p_1' \equiv y_1$$

$$\frac{dp_2}{dz} \equiv p_2' \equiv y_2$$

$$\frac{dp_3}{dz} \equiv p_3' \equiv y_3$$

$$p_1 \equiv y_4, \qquad p_2 \equiv y_5 \quad \text{and} \quad p_3 \equiv y_6 \tag{23}$$

The second order Eqs. (21) may be rearranged in the canonical form as a set of six first order equations:

$$\begin{bmatrix} y_1' \\ y_2' \\ y_3' \\ y_4' \\ y_5' \\ y_6' \end{bmatrix} = \begin{bmatrix} -\alpha_1 & -\alpha_3 & 0 & -\alpha_2 & -\alpha_4 & 0 \\ -\alpha_5 & -\alpha_7 & -\alpha_9 & -\alpha_6 & -\alpha_8 & -\alpha_{10} \\ 0 & -\alpha_{11} & -\alpha_{13} & 0 & -\alpha_{12} & -\alpha_{14} \\ 1 & 0 & 0 & 0 & 0 & 0 \\ 0 & 1 & 0 & 0 & 0 & 0 \\ 0 & 0 & 1 & 0 & 0 & 0 \end{bmatrix} \begin{bmatrix} y_1 \\ y_2 \\ y_3 \\ y_4 \\ y_5 \\ y_6 \end{bmatrix} \tag{24}$$

Solving this as an eigenvalue problem by means of one of the standard subroutines, we can obtain eigenvalues β_i, $i = 1, 2, \ldots, 6$, and the eigenmatrix or modal matrix ψ. The generalized variables $\{y\}$ are related to the principal coordinates $[\Gamma]$ through the nodal matrix.Thus

$$\{y\} = [\psi][\Gamma] \tag{25}$$

where

$$\Gamma_i(z) = C_i e^{\beta_i z}, \qquad i = 1, 2, \ldots, 6 \tag{26}$$

C_i are arbitrary constants.
Next, Eqs (18)-(20) may be used to obtain expressions for $u_1(z)$,$u_2(z)$, and $u_3(z)$. Finally, one may write

$$\begin{bmatrix} p_1(z) \\ p_2(z) \\ p_3(z) \\ \rho_0 a_0 u_1(z) \\ \rho_0 a_0 u_2(z) \\ \rho_w a_w u_3(z) \end{bmatrix} = [\ A\ ] \begin{bmatrix} C_1 \\ C_2 \\ C_3 \\ C_4 \\ C_5 \\ C_6 \end{bmatrix} \tag{27}$$

where

$$A_{1,i} = \psi_{4,i} e^{\beta_i z}$$

$$A_{2,i} = \psi_{5,i} e^{\beta_i z}$$

$$A_{3,i} = \psi_{6,i} e^{\beta_i z}$$

$$A_{4,i} = -\frac{\psi_{1,i} e^{\beta_i z}}{(\jmath k_0 + M_1 \beta_i)}$$

$$A_{5,i} = -\frac{\psi_{2,i} e^{\beta_i z}}{(\jmath k_0)}$$

$$A_{6,i} = -\frac{\psi_{3,i} e^{\beta_i z}}{(\jmath k_w)}, \qquad i = 1, 2 \ldots, 6 \tag{28}$$

Now p's and u's at z=0 can be related to those at $z = l$ by eliminating $\{C\}$ from Eqs.(27). Thus,

$$\begin{bmatrix} p_1(0) \\ p_2(0) \\ p_3(0) \\ \rho_0 a_0 u_1(0) \\ \rho_0 a_0 u_2(0) \\ \rho_w a_w u_3(0) \end{bmatrix} = [\ T\ ] \begin{bmatrix} p_1(l) \\ p_2(l) \\ p_3(l) \\ \rho_0 a_0 u_1(l) \\ \rho_0 a_0 u_2(l) \\ \rho_w a_w u_3(l) \end{bmatrix} \tag{29}$$

where

$$[\ T\ ] = [\ A(0)\ ]\,[\ A(l)\ ]^{-1} \tag{30}$$

4. Derivation of the Transfer Matrix

In order to relate the state variables $p_1(0)$ and $v_1(0)$ at the upstream point u with $p_1(l_p)$ and $v_1(l_p)$ at the downstream point d in Fig. 2, we make use of the following four boundary equations:

$$\frac{p_2(0)}{-u_2(0)} = -\jmath\ \rho_0 a_0\ \cot k_0 l_a$$

$$\frac{p_2(l)}{u_2(l)} = -\jmath\ \rho_0 a_0\ \cot k_0 l_b$$

$$\frac{p_3(0)}{-u_3(0)} = -\jmath\ \rho_w a_w\ \cot k_w l_a$$

$$\frac{p_3(l)}{u_3(l)} = -\jmath\ \rho_w a_w\ \cot k_w l_b \tag{31}$$

There are 12 variables, six on either side of the matrix equation (29). Eqs. (29) and (31), total ten in number, may now be rearranged in the matrix form

$$[\ \mathrm{A}\ ]\,\{x\} = \{C_p\}\ p_1(l) + \{C_v\}\ \rho_0 a_0\ u_1(l) \tag{32}$$

where the state vector $\{x\}$ contains 10 of the 12 state variables, other than those two on the right hand side.
Eqs.(32) may be solved readily to obtain the matrix equation

$$\begin{bmatrix} p_1(0) \\ \rho_0 a_0 u_1(0) \end{bmatrix} = \begin{bmatrix} \mathrm{TM}_{11} & \mathrm{TM}_{12} \\ \mathrm{TM}_{21} & \mathrm{TM}_{22} \end{bmatrix} \begin{bmatrix} p_1(l) \\ \rho_0 a_0 u_1(l) \end{bmatrix} \tag{33}$$

Finally, writing

$$\rho_0 a_0 u_1 = \frac{\rho_0 a_0}{S_1} S_1 u_1 = Y_1 v_1 \tag{34}$$

Eq (33) may be rewritten in the form

$$\begin{bmatrix} p_1(0) \\ v_1(0) \end{bmatrix} = \begin{bmatrix} \mathrm{TM}_{11} & \mathrm{Y}_1\mathrm{TM}_{12} \\ \frac{\mathrm{TM}_{21}}{\mathrm{Y}_1} & \mathrm{TM}_{22} \end{bmatrix} \begin{bmatrix} p_1(l) \\ v_1(l) \end{bmatrix} \tag{35}$$

which yields the required transfer matrix between points u and d in Fig. 2
Transmission loss, if required , may readily be computed from the transfer matrix parameters [1].

5. Validation

In order to validate the foregoing analysis, certain self-consistency checks were run. Fig. 4 shows the TL spectra for an annular-airgap lined duct with the specific flow resistance SFR of the absorptive material as a parameter . The curve for SFR=0 coincides identically with that of an unlined duct, thereby providing a necessary check. Incidentally, the qualitative effect of increasing SFR on the TL troughs is exactly as would be expected.

Specific flow resistance, SFR, of typical fibrous materials lies between 10000 to 20000 $Pa.s/m^2$. Fig. 5 compares TL curves of air (SFR=0) and an absorptive layer of SFR-10000 $Pa.s/m^2$. While the air-curve behaves like a concentric-tube resonator (see Fig. 8.15 of Ref. 1), the absorptive material curve behaves like a normal lined duct (see Fig. 6.7 of Ref. 1).

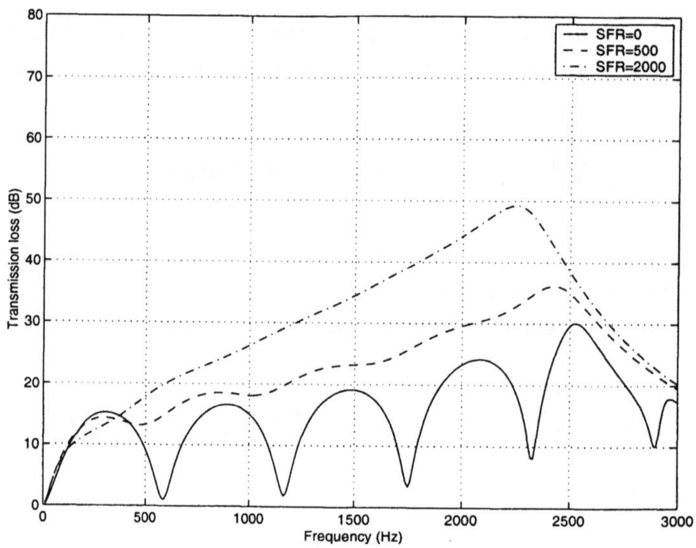

Figure 4. Effect of Specific flow resistance

Fig. 6 shows the effect of the annular airgap, keeping the overall shell diameter remaining the same ($d_3 = 0.15$ m). The pressure of an annular airgap of 10 mm, radial thickness of the absorptive layer got reduced from 50 mm to 40 mm. Therefore, there is considerable reduction in TL.

Incidentally, their is considerable numerical instability at lower frequencies despite double-precision arithmetic. This happens primarily when the normalized flow resistance parameter A exceeds 60; see Eqs. (4), (5) and (6). However, instability can also be observed at very high frequencies (at above 5000Hz), as shown in Fig. 7, where the effect of outer shell diameter d_3 is shown on a logarithmic scale for frequencies of on 100 to 10000 Hz.

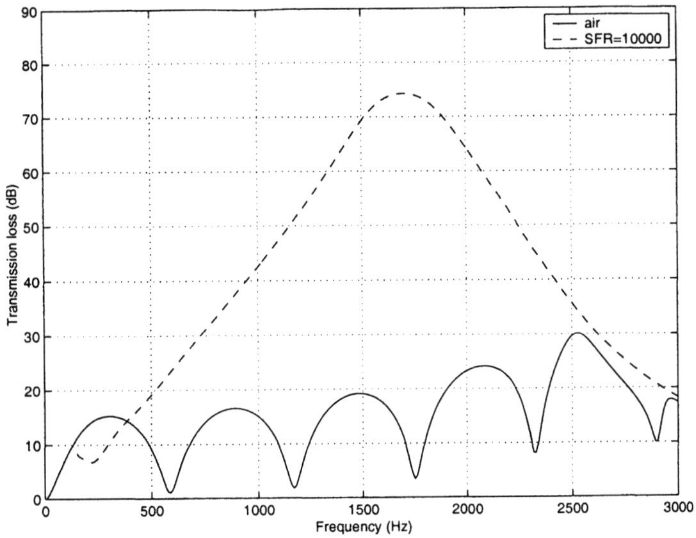

Figure 5. Comparison of air and absorptive material

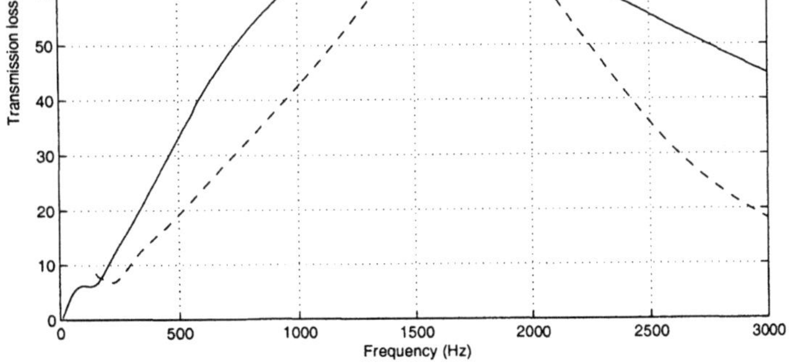

Figure 6. Effect of annular airgap

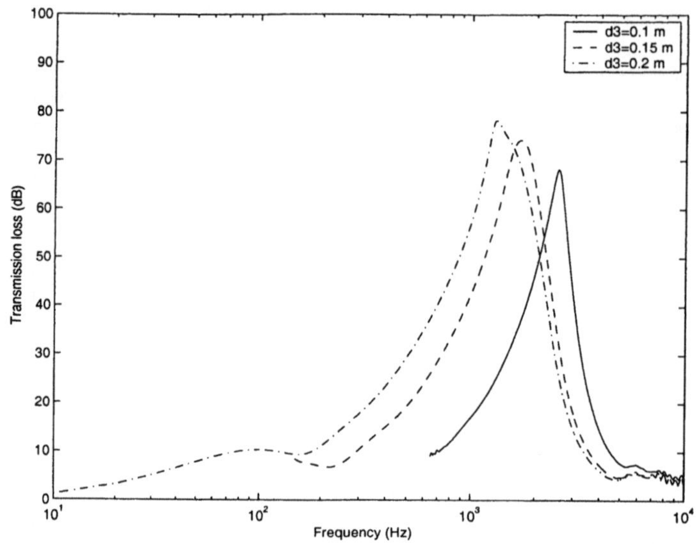

Figure 7. Effect of outer diameter on Transmission Loss

6. Parametric Studies

Referring to the configuration of Fig. 2, the following parameters have been kept fixed :

$l_p = 0.5$ m
$l_a = l_b = 0$
$d_1 = 0.05$ m
hole diameter $d_h = 3$ mm
thickness of the perforated plates, t=1 mm

The remaining parameters are varied in steps as follows :

Intermediate diameter, d_2 = 0.06, <u>0.07</u>, 0.08 m
Outer shell diameter, d_3 = 0.1, <u>0.15</u>, 0.2 m
SFR of the absorptive material =0, 5000, <u>10000</u>, 20000 Pas/m^2
Mean flow Mach number, M_1 = 0.0, <u>0.1</u>, 0.2
Porosity of both plates , σ = 0.02, <u>0</u>.05, 0.07

Default values of the parameters above are indicated by an underline. Both the perforated plates are assumed to be identical. As TL is known to vary linearly with axial length of the lined duct, l_p has not been varied in this parametric study. Fig.8 shows effect of a variation in the intermediate diameter d_2, keeping the outer shell diameter d_3 constant at 0.15 m .

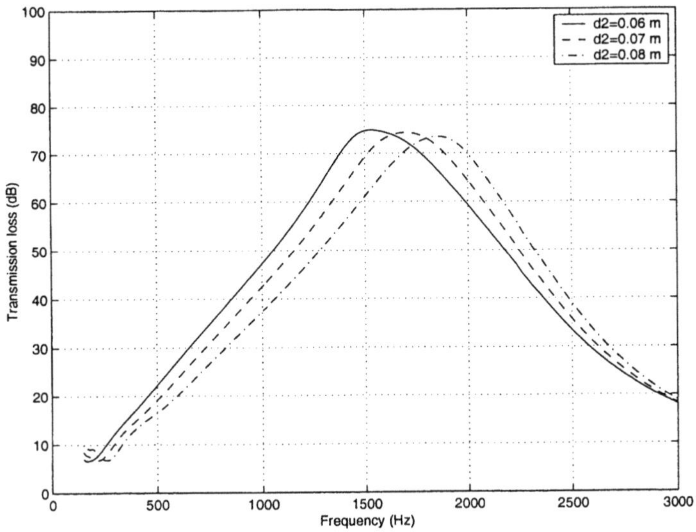

Figure 8. Effect of intermediate tube diameter

Therefore in reality, this figure shows the effect of variation in the radial thickness $\frac{(d_3-d_2)}{2}$ of the absorptive layer. The same effect is shown, rather magnified, in Fig. 9 where d_3 is varied in larger steps.

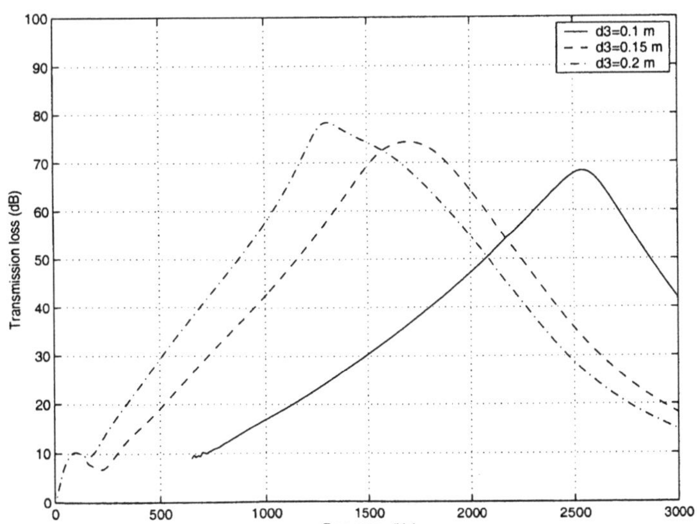

Figure 9. Effect of outer diameter

Fig. 10 shows the effect of the specific flow resistance SFR in $Pa.s/m^2$. It may be noted that on the whole increased SFR leads to higher transmission loss, TL . Interestingly, this effect is not so evident in rectangular lined ducts (see Fig. 6.8 of Ref. 1).

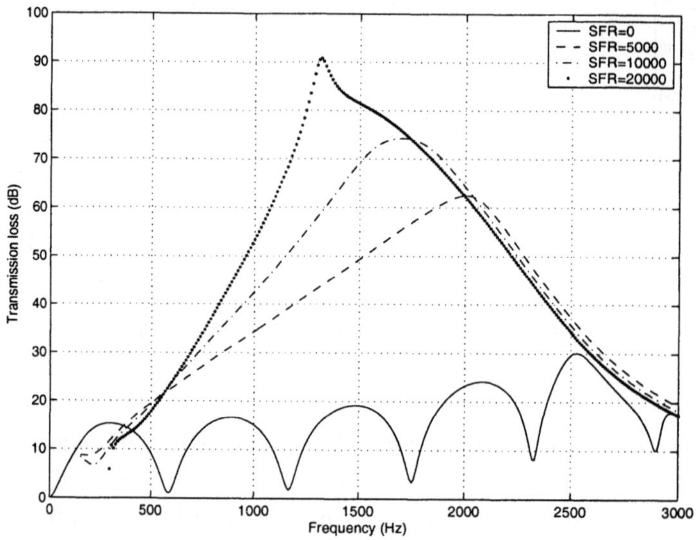

Figure 10. Effect of Specific flow resistance

The effect of mean flow Mach number on TL is shown in Fig. 11. This effect is not so pronounced as the mean flow changes impedance of the inner perforated plate only (see Eq. (9))

On the other hand, effect of porosity of the perforated plates is much more pronounced, as shown in Fig. 12. Higher porosity improves TL at the medium and high frequencies, while reduces it marginally at the lower end.

Incidentally, this trend is not shown by simple concentric tube resonators, while it is very pronounced for simple lined ducts [2]. Thus, an annular-airgap lined duct lies in between a concentric tube resonator and a lined duct, in shape as well as in acoustic behavior . Finally, Fig. 7 is reproduced as Fig. 13, with porosity of both the perforated plates being changed from the default values of 0.05 to 0.07, with the frequency axis being drawn to a logarithmic scale.

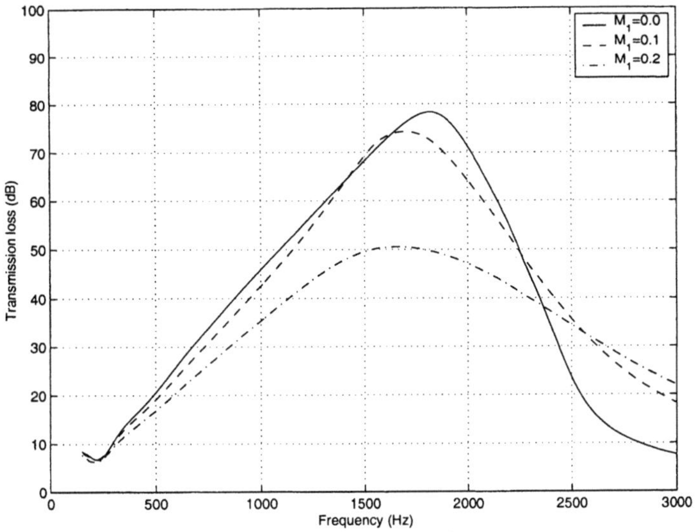

Figure 11. Effect of Mach number

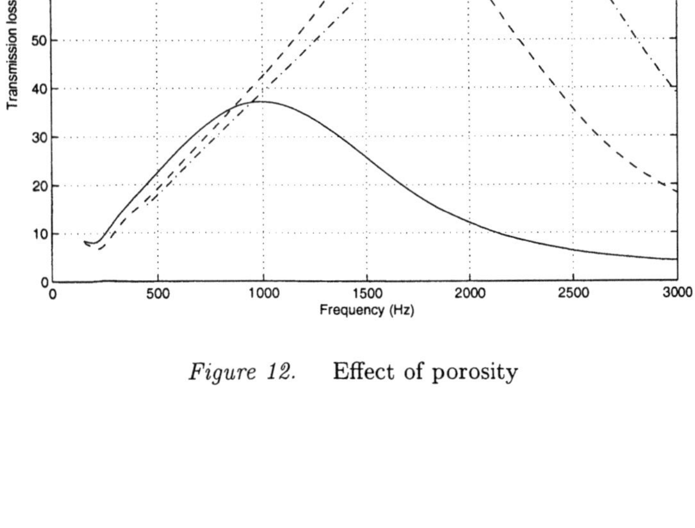

Figure 12. Effect of porosity

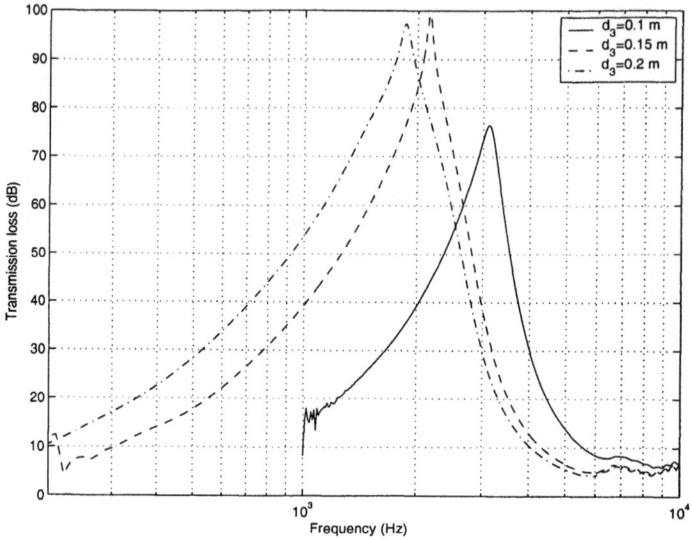

Figure 13. Effect of outer diameter on Transmission Loss

7. Concluding Remarks

The three-duct perforated element analysis is extended for the analysis of an annular airgap lined duct for hot, impure and high velocity flow exhaust systems. The resulting transfer matrix has been validated for self-consistency. The parametric studies have revealed that, in general the annular airgap lined duct is a combination of concentric-tube resonator and a simple lined pipe, both in shape and acoustical performance. However, the lined pipe character is dominates. Thus, an annular-airgap lined duct would act more or less like a normal lined duct for silencing purposes, while protecting the lining from high temperatures and the eroding effect of hot exhaust flows of the internal combustion engines.

Acknowledgements

The authors are grateful to the Department of Science and Technology of the Government of India for funding this research through the Facility for Research in Technical Acoustics.

Nomenclature

a_0	Speed of sound in air at room temperature
E	Flow resistance of the unit thickness bulk material
j	$\sqrt{-1}$
k_o	Wave number
M	Mach number at any section z.
M_{ex}	Mach number in exhaust pipe.
p	Acoustic pressure.
ρ	Acoustic perturbation on density of fluid medium.
ρ_w	Euivalent mass density of the absorptive material.
u_i	Acoustic particle velocity.
U	Mean flow velocity at any section.
ω	Circular frequency.
k_W	Wave number of the absorptive material.
SFR	Specific flow resistance.
σ	porosity.
d_i	Diameter of the pipe.
ζ	Specific Acoustic impedance.
v	Acoustic mass velocity
Y_o	Characteristic impedance.

References

Munjal, M.L.: *Acoustics of Ducts and Mufflers* , Wiley-Interscience, New York, 1987
Munjal, M.L. and Thawani, P.T.: Effect of protective layer on the performance of absorptive ducts, *Noise Control Engineering Journal* **45** (1997), 14-18
Sullivan, J.W. and Crocker, M.J.: Analysis of Concentric tube muffler components, *Journal of the Acoustical Society of America* **64** (1978), 207-215.
Munjal, M.L., Rao, K.N. and Sahasrabudhe, A.D.: Aeroacoustic analysis of perforated muffler components , *Journal of Sound and Vibration* **114** (1987), 173-188.
Delany, M.E. and Bazley, B.N.: Acoustical characteristics of fibrous absorbing materials, *Applied Acoustics* **3** (1970), 106-116.
Mechel, F.P.: Extension of low frequencies of the formulae of Delany and Bazely for absorbing material (in German), *Acustica* **35**(1976), 210-235.
Rao, K.N. and Munjal, M.L.: Experimental evaluation of impedance of perforates with grazing flow, *Journal of Sound and Vibration* **108**(1986), 283-295.

ACOUSTICAL MATERIALS FOR AUTOMOTIVE NVH REDUCTION

A. R. MOHANTY
Department of Mechanical Engineering
Indian Institute of Technology, Kharagpur 721 302, INDIA

Abstract

Acoustical materials are widely used by noise control engineers for NVH (Noise Vibration and Harshness) reduction. However, in order to effectively select and apply these materials which are commercially available in various types, the acoustical properties of these materials need to be known. Well established standards exist for the determination of such properties. In this paper we describe the details of two experimental rigs developed in house at IIT, Kharagpur based on available standards for the determination of the normal sound absorption co-efficient and the transmission loss of the sound absorbing materials and the sound barrier materials, respectively. Results from sample measurements are also provided. Some of the materials discussed in this paper are widely used in the automobile industry for NVH reduction in automobiles as well.

1. Introduction

There are three important sources of noise in an automobile, namely power-train/engine noise, road/tire interaction noise and wind noise. In general, noise control for any machinery is either done at its source, at the receiver or in the path between the source and the receiver. In an automobile to reduce the NVH levels inside the passenger cavity, usually noise control methods are applied in the path between the various sources and the passenger or the driver's ear.

For NVH reduction in an automobile interior, suitable acoustical materials are selected from the wide varieties commercially available and then applied. These acoustical materials can be classified into three categories, namely, sound absorbing materials, sound barrier materials and sound damping materials. The sound absorbing material like open cell polyurethane foam is usually porous and its sound attenuating property

M.L. Munjal (ed.), IUTAM Symposium on Designing of Quietness, 21–31.

is determined by the normal specific sound absorbing co-efficient. A good sound absorber will have a value of normal specific sound absorbing coefficient close to 1. The normal specific sound absorbing materials have traditionally also been rated, based on their noise reduction coefficient (NRC) values. The sound absorbing co-efficient of these materials can be found out for normal incidence as well as random incidence of the acoustical waves [1,2]. In small machinery enclosures, as well as in passenger cars, because of small dimensions as in comparison of a concert hall or auditorium the normal specific impedance will suffice with reasonable accuracy in noise level predictions. Numerical methods like FEM (Finite Element Method) and BEM (Boundary Element Method) have been used for the prediction and control of vehicle interior noise levels using sound absorbing materials[3].

The sound barrier materials are usually heavy and need not have pores in them. Below the coincident frequency of the material for every doubling of the mass of the barrier the transmission loss of the material improves by 6 dB [4]. A good sound barrier material will have a high value of transmission loss. Typical sound barrier materials used for noise control in automobiles are sheet metal, glass and loaded rubber sheets. Many standard test procedures are established to determine the transmission loss of such materials [5,6].

Damping materials are used to dampen the structural vibrations at resonance of the structure to which it is attached. Such reduction in vibration leads to the reduction in the radiated noise levels. The damping materials are characterized by their damping factor. Again well established standards exist to determine such damping factors [7].

In this paper we present the details of the experimental rigs developed at IIT, Kharagpur for determination of the normal specific sound absorbing co-efficient and transmission loss of sound absorbing and sound barrier materials, respectively. Results from sample measurements are also presented.

2. Sound Absorption Co-efficient

To determine the normal specific impedances of materials, the SWR (Standing Wave Ratio) technique has given way to the faster two- microphone, random excitation technique[2]. The technique based on the initial works of Chung & Blaser, and Seybert & Ross [8,9] have been used successfully in the past for acoustical property determination [10].

Here we describe the test setup based on the ASTM E1050 standard for measuring the acoustical impedances of sound absorbing materials. The setup along with the developed software for property computation has been checked for measured impedance of a closed tube of a finite length.

The impedance tube used for measurement was made of brass having an overall length of 900mm with precision machined end plate and side ports to accurately maintain microphone alignment. The spacing between the two microphones was kept at 35mm. The usable frequency range of the setup is from 200 Hz to 4500Hz. The lower limit is due to the frequency response of the acoustic driver. A PC based FFT analyzer based on the National Instruments AT-DSP2200 card have been used along with the digital spectrum analyzer software developed in the LabVIEW programming language[11]. Two Brüel & Kjaer (Type 4136) $\frac{1}{4}$" condenser microphones along with their preamplifiers are used for the sound pressure measurement, inside the tube. The microphone pair was calibrated as per the procedure given in [10].

Most of the materials whose acoustical absorbing properties were measured were fibrous in form, hence could not used as a sample in that state. Such materials were wrapped around by a Mylar sheet, thus forming a cylinder which were then inserted in the sample holder. For repeatability of measurements, four samples of each material were prepared and tested. The average value has been reported here.

The complex reflection coefficient of the system under test is given in equation (1), the notations for the symbols are given in the nomenclature.

$$R = \frac{H_{12} - e^{j2k(l+s)}}{e^{jks} - H_{12}} e^{jk2(l+s)} \tag{1}$$

The normal specific coefficient of absorption is given by,

$$\alpha = 1 - |R^2| \tag{2}$$

And the normal specific impedance is given by

$$\frac{Z}{\rho c} = \frac{1+R}{1-R} \tag{3}$$

Since the test setup was fabricated as per the ASTM E-1050 standard and the software for property computation was developed using the LabVIEW program, a validation test was performed, before using the developed system for determining the sound absorbing properties of the materials [1].

3. Transmission Loss

The developed experimental rig to determine the normal incidence transmission loss of the barrier material is shown in Fig.1. A random noise from the bandwidth of 200 Hz to 5000 Hz is generated in the

SRS (Stanford Research Systems) 335 Signal generator, which drives a full range loud speaker through a power amplifier. The heavy wooden rectangular box lined with open cell polyurethane foam in the inside measures 750 mm (L) x 500 mm (W) x 600 mm (D). There are no acoustical leaks in the box, except for a 150 mm x 150 mm square opening in the top for placing the test barrier material. The foam lining ensures that the sound intensity vector is normal to the test material, and cross modes are not excited. The measurement is effective in the frequency bandwidth from 200 Hz to 5000 Hz. The present method of determining the transmission loss of materials is cheap and quick compared to the two-room method. Further this method can also be used to compare the TL of different barrier materials under identical sound field conditions.

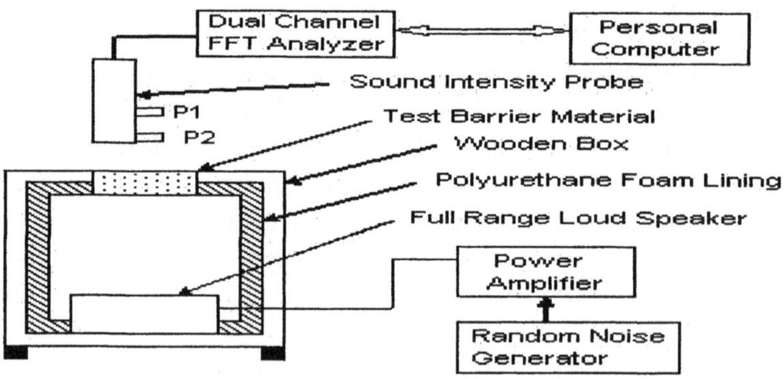

Figure 1. Experimental setup to Determine the Transmission Loss of Sound Barrier Materials

The sound intensity in a direction perpendicular to the barrier material is measured by scanning a two-microphone sound intensity probe over the barrier material.The microphone spacing in the sound intensity probe is 35mm. Brüel & Kjaer Type 4130, $\frac{1}{2}$ condenser microphones are used for the sound intensity measurement. The cross-spectrum between the two microphones is measured in a dual channel FFT analyzer which is connected to a personal computer by GPIB (General Purpose Interface Bus) for transferring the cross-spectrum data. Then the sound intensity computations for the calculation of the transmission loss of the material are done in the computer. The transmission loss is calculated as given in Eq. (4)

$$TL = 10log_{10}\frac{I_i}{I_t} \tag{4}$$

Where Ii is the normal sound intensity without the barrier material and is considered to be equal to the incident sound intensity It is the normal sound intensity with the barrier material present and is equal to the transmitted sound intensity.

4. Results

The theoretical normal specific acoustical impedance of a closed tube of length 'L' is given by the following equation,

$$\frac{Z}{\rho c} = -jcot(kL) \tag{5}$$

Figure. 2 shows the excellent agreement between the normal specific measured impedance of a 50.8mm closed tube and the theoretical value from Eq.(5) at an ambient temperature of 200C.

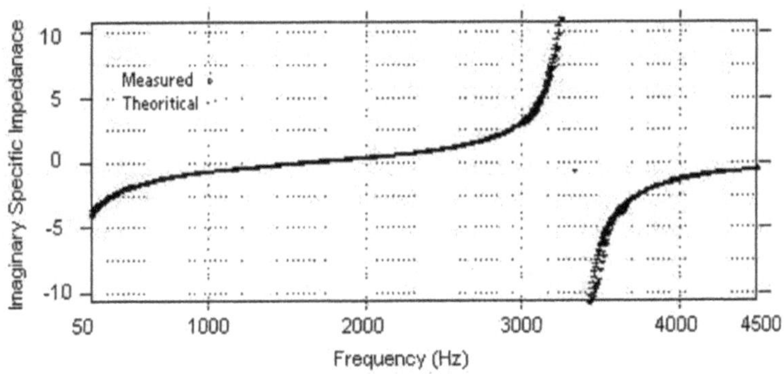

Figure 2. Measured and Theoretical Normal Specific Impedance of a 50.8 mm closed tube at 20°C

In Fig. 3 the normal specific sound absorption co-efficient as measured in the developed test rig is presented for various materials of 25.4 mm thickness with a rigid backing, some of which are used in automobiles for noise control purposes. In general all sound absorbing materials have poor sound absorbing characteristics at low frequencies, and the sound absorbing co-efficient increases with an increase in frequency. The sound absorbing co-efficient also improves by increasing the thickness of the material. Fibre-glass and wool have excellent sound absorbing properties, however in automobiles because of safety issues usually open cell type polyurethane foam (sponge) is preferred.

The values of the complex normal specific impedance of the various sound absorbing materials as a function of frequency is given in Table 1. The materials are all of 25.4 mm thickness and had a rigid backing behind them. Such values help in numerical modeling using the FEM and the BEM techniques [3].

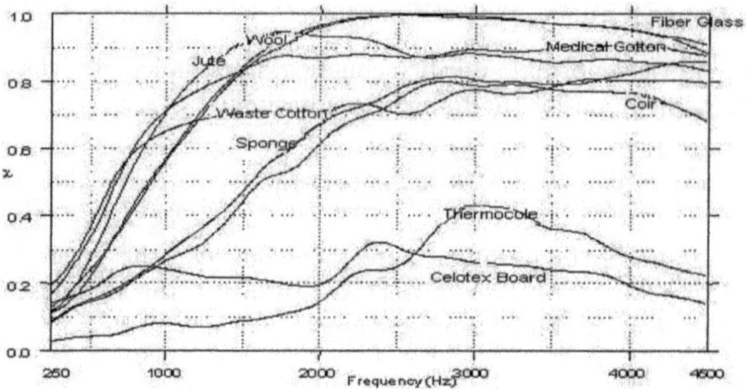

Figure 3. Measured Normal Specific Sound Absorption Coefficient of Different Materials of 25.4 mm thickness with rigid backing.

Table 1. Normal specific Impedance of Various Sound Absorbing Materials of 25.4 mm thickness with a rigid backing

Frequency (Hz)	Jute	Coir	Wool	Waste Cotton	Medical Cotton	Celotex Board	Thermocole	Foam	Fiber Glass
250	1.050–i8.952	1.938–i 8.389	0.057–10.862+	1.958–6.795	0.742–7.254	15.578–8.846	13.347–53.225	–0.889–7.560	0.604–6.187
500	1.209–i3.472	0.675–i4.149	0.067–i0.046	1.998–3.637	1.302_3.334	15.763–9.288	10.441–26.875	0.511–3.976	0.705 3.154
1000	1.221–i1.369	0.370–i 1.923	2.707+i2.319	2.413–1.747	1.450–1.427	13.889+1.321	15.069–20.437	0.245–1.857	0.61–1.366
1500	1.293–0i557	0.287–i1.058	0.489–0.252	2.526–1.370	1.457–1.616–	14.370+1.314	9.765–16.429	0.174–1.0308	0.700–0.705
2000	1.686–i0.232–	0.307–i0.153	1.073–0.371	2.353–1.413	1.616–0.815	17.959–4.311	3.432–9.295	0.176–0.562	0.817–0.323
2500	2.025–i0.050	0.312–i0.153	1.044–0.070	2.557–1.700	1.651–0.781	9.327+3.157	3.976–7.376	0.184–0.177	1.085–0.0755
3000	2.068–i0.060	0.375+i0.273	0.956–0.192	1.909–1.157	1.576–0.617	12.879+3.398	5.866–2.970	0.193+0.185	1.156+0.087
3500	2.207–i0.238	0.512+i0.636	0.769+0.241	1.928–1.127	1.533–0.624	10.866+8.016	9.199+0.035	0.252+0.504	1.390+0.164
4000	1.999–i0.584	0.873+i1.1050	1.468–0.416	1.617–1.095	1.401–0.686	13.604–9.150	10.379–5.181	0.352+0.859	1.555–0.065
4500	1.739–i0.958	2.823+i1.378	0.494+0.049	1.363–1.168	1.207–0.818	9.230–9.933	4.251–6.676	0.880+1.582	1.572–0.584

In Fig. 4 the measured transmission loss of a barrier material using the developed rig (Loaded rubber sheet of 12.5 mm thickness) is shown. The experimentally measured TL values are shown as circles, and the solid line represents a curve fit line. The TL increases with frequency and since the coincident frequency of the material is beyond the frequency range of the experimental rig, the dip is not seen. And the TL increases

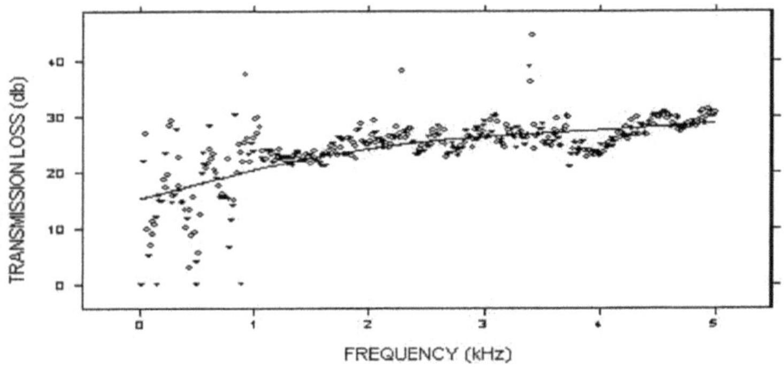

Figure 4. Measured Transmission Loss of a 12.5 mm Loaded Rubber Sheet

by 6 dB per octave. The reasons for the variations in the measured data is due to the fact that no special attention has been made to prevent leakages at the edges of the test material in the present setup.

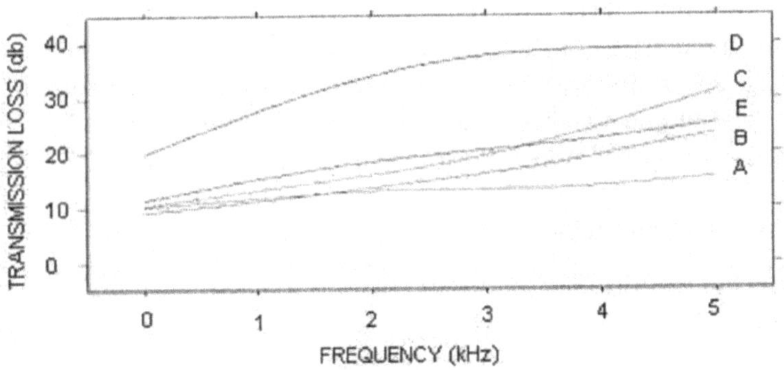

Figure 5. Measured Normal Specific Sound Absorption Coefficient of Different Materials of 25.4 mm thickness with rigid backing

In Figure 5 the measured TL for a family of sound barrier materials is shown. The experimental measured TL values are shown by their respective curve fit lines in Fig. 5. The samples of the material were made available from a barrier material manufacturer. Table 2 gives a description of the sound barrier materials.

Table 1. Description of the Sound Barrier Material

Material No.	Sound Barrier Material Description
A	100 mm thick open cell foam with metalized mylar film
B	PVC Barrier sandwiched between two layers of open cell foam.
C	PVC Barrier attached to open cell foam on one side
D	PVC Barrier with **closed** cell foam on one side and ribbed wear surface on other
E	PVC Barrier with open cell on one side and woven carpet on the other side.

It is seen from Fig. 5, that sound barrier material D has the highest transmission loss. This is because of the fact that the PVC material is backed by a closed cell foam. Material A is actually best for sound absorbing properties and has a relatively low TL value. The specific gravity of the PVC material used in the sound barriers described in Table 2 has a specific gravity of 2.5.

5. Conclusions

The details of experimental rigs for the determination of the sound absorbing co-efficient and the transmission loss of acoustical materials in NVH reduction in an automobile have been provided. The sound absorbing and the sound barrier materials for which their acoustical properties have been measured can be used in various places in an automobile interior for NVH reduction. The developed test and measurement facility will also enable one to select and use different acoustical materials available commercially for noise control purposes in general.

Acknowledgments

The author would like to thank the R & D Department of Mahindra & Mahindra Limited, Automotive Division, Nashik, India for partially funding the project in order to develop the experimental rigs at the Indian Institute of Technology, Kharagpur.

Nomenclature

c	Speed of sound, [m/s]
H_{12}	Transfer functions between microphones at locations 1 and 2
k	Wave number, $[m^{-1}]$
I_i	Incident sound intencity, $[W/m^2]$
I_t	Transmitted sound intensity, $[W/m^2]$
l	Distance between the specimen and the nearest microphone, [m]
L	Length of the tube, [m]
R	Complex reflexion coefficient
s	Microphone spacing, [m]
TL	Transmission Loss, [dB]
Z	Complex acoustic impedance, $[kg/m^2 - s]$
α	normal specific absorption co-efficient
ρ	Density of air, $[kg/m^3]$

References

ASTM-E-1050, (1990) "Standard Test Method for Impedance And Absorption of Acoustical materials using a tube, two microphones and a digital frequency analyzer".

ASTM-E-384, (1987) "Standard Test Method for Impedance and Absorption of Acoustical Materials by the Impedance Tube Method".

Mohanty, A.R., St.Pierre, B.D., and Surulii-Narayansami, P. (2000) "Structure-borne noise reduction in a Truck Cab Interior using Numerical Techniques", Applied Acoustics, Vol.59,No.1, pp.1-17.

Kinsler, L.E. and Frey, A.R. (1982) "Fundamentals of Acoustics," 3e, John Wiley & Sons.

ASTM E90, (1990) "Standard Method for Laboratory Measurement of Air-borne Sound Transmission Loss of Building Partitions."

SAE J1400, (1990) "SAE Recommended Practice - Laboratory Measurement of Airborne Noise Reduction of Acoustical Materials."

ASTM E756, (1980) "Standard Method for Measuring Vibration Damping Properties of Materials."

Chung, J.Y. and Blaser, D.A. (1980) "Transfer Function Method for Measuring Induct Acoustical Properties, Part I-theory, II-Experiments", Journal of the Acoustical Society of America, 68(3), pp. 907-913,914-921.

Seybert, A.F. and Ross, D.F. (1977) " Experimental Determination of Acoustic Properties using a Two-Microphone Random Excitation Technique", Journal of the Acoustical Society of America, 61(5), pp. 1362-1370.

Mohanty, A.R., Seybert, A.F., and Strong, W.F. (1991) "Acoustical Property determination by a personal Computer System", Proceedings of NOISE-CON 91, Tarry Town, New-York, pp. 597-602.

AT-DSP 2200 , (1997) User's Manual, National Instruments Inc., USA.

ACOUSTIC CHARACTERIZATION OF ONE-PORT SOURCES AND ITS APPLICATION TO DUCT NOISE CONTROL SYSTEM DESIGN

M. G. PRASAD
Noise and Vibration Control Laboratory
Department of Mechanical Engineering
Stevens Institute of Technology
Hoboken, New Jersey 07030 USA

1. Introduction

In the design of a duct system acoustic aspect is very important for its efficient and quiet performance. Electro-acoustic speakers, engines, fans, blowers, pumps and fluid machines are some of the commonly used sources (figure 1) in duct systems. A literature survey indicates that there have been continuous efforts in theoretical and experimental modeling of duct noise control systems such as muffler systems. Although there are several acoustic studies on duct noise control systems, still the acoustic characterization of sources and their interaction with loads has been a challenging task.

This is because of the interactive behavior of source and its corresponding load. It is for this reason that the performance of a given muffler cannot be specified by a user independent of the engine source on which the muffler is used. This paper presents an overview of the linear system approach in which acoustic characteristics of an one-port acoustic source is measured and then used as an input to the muffler system design and performance evaluation.

2. System Model

A duct system can be modeled as a source-path-termination model or a source-load model as shown in figure 2. These models are interrelated when the path-termination is treated as a load. The source-path-termination model is commonly used as it can be directly related to engine-exhaust pipe-muffler-tailpipe-radiation system.

M.L. Munjal (ed.), IUTAM Symposium on Designing of Quietness, 33–46.

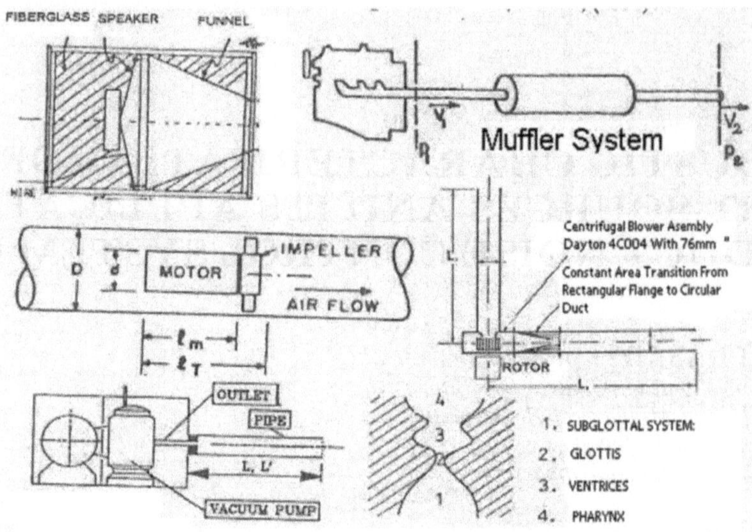

Figure 1. Various Acoustic Sources

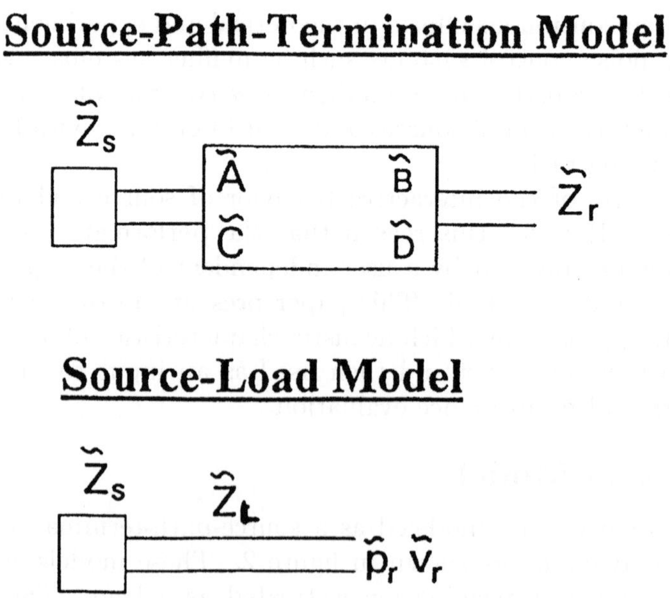

Figure 2. Duct system models

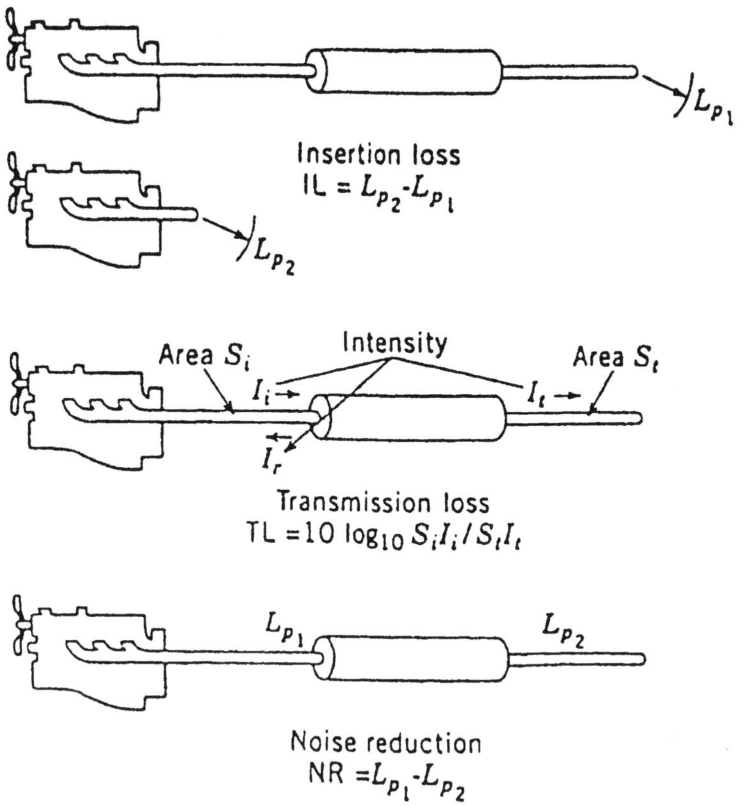

Figure 3. Muffler system performance descriptors

The three commonly used descriptors (figure 3), namely Insertion Loss (IL), Transmission Loss (TL) and Noise Reduction (NR) are given by equations 1, 2 and 3.

$$IL = 20\log_{10}\left|\frac{\tilde{A}\tilde{Z}_r + \tilde{B} + \tilde{C}\tilde{Z}_S\tilde{Z}_r + \tilde{D}\tilde{Z}_S}{\tilde{A}'\tilde{Z}_r + \tilde{B}' + \tilde{C}'\tilde{Z}_S\tilde{Z}_r + \tilde{D}'\tilde{Z}_S}\right| dB \tag{1}$$

$$TL = 20\log_{10}\left|\frac{1}{2}\left(\tilde{A} + \frac{\tilde{B}S}{\rho c} + \frac{\tilde{C}\rho c}{S} + \tilde{D}\right)\right| dB \tag{2}$$

$$NR = 20\log_{10}\left|\left(\tilde{A} + \frac{\tilde{B}}{\tilde{Z}_r}\right)\right| dB \tag{3}$$

Where $\tilde{Z}_S$ and $\tilde{Z}_r$ are the source and radiation impedances respectively and $\tilde{A}$, $\tilde{B}$,$\tilde{C}$ and $\tilde{D}$ are the four-pole parameters of the muffler including its upstream and downstream ducts. It is needless to say that the IL is the most useful descriptors among these three descriptors. It is well known that the knowledge of acoustic characteristics of the source is essential for the evaluation of the IL. Radiated Sound Pressure Level (Lp) is also very useful as it gives the system output which can be used to determine IL.The other two descriptors namely Transmission Loss (TL) and Noise Reduction (NR) do not need the knowledge of acoustic characteristics of the source. By definition, both TL and NR are independent of the influence of the source-load interaction. Although, IL is easier to measure but very difficult to predict because of its dependence on the source impedance, $\tilde{Z}_s$ [1-4].

Linear transfer matrix formulation using four-pole parameters has proved to be quite efficient and useful. Particularly due to its modular basis and the ease of use in the industrial environment, the approach has been widely used in literature. However, the validity of the linear system model and analysis to various real-life systems has been investigated. Except for certain discrete frequencies and operating conditions the linear black-box approach is found to be valid. Also, the overall system performance in terms of insertion loss (IL) could be insensitive to the non-linearity of the source especially when viewed from the engineering considerations. Certainly the non-linear effects need to be investigated on a system basis with the purpose of its inclusion in the black-box approach. The complexity of the real-life system has two aspects namely the dynamics of the source itself and the medium characteristics namely flow and temperature effects [2,5-7].

3. Source Characteristics

The electrical analog of a one-port ducted system with pressure and volume velocity source is shown in figure 4.

Various source characterization methods can be classified based on two approaches namely direct and indirect methods [8]. There have been consistent developments in both direct and indirect methods particularly in experimental aspects and the corresponding error analysis [9-19]. A brief overview of principles and developments in both direct and indirect methods is presented (figure 5).

4. Direct Methods

The direct methods are based on the application of the well-known techniques of measuring passive acoustic impedance of sound absorbing

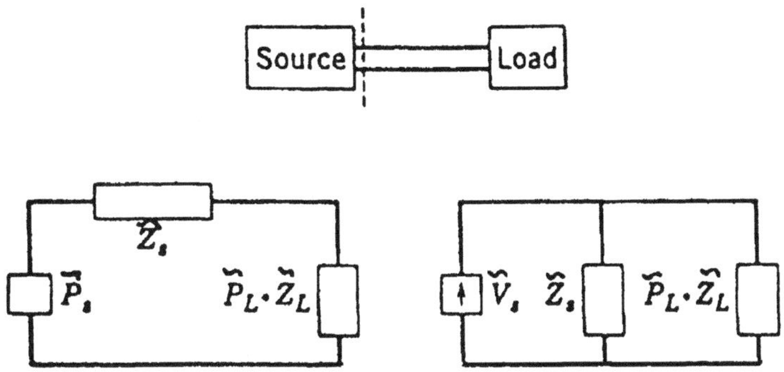

Figure 4. Electrical analog model with pressure and volume velocity sources

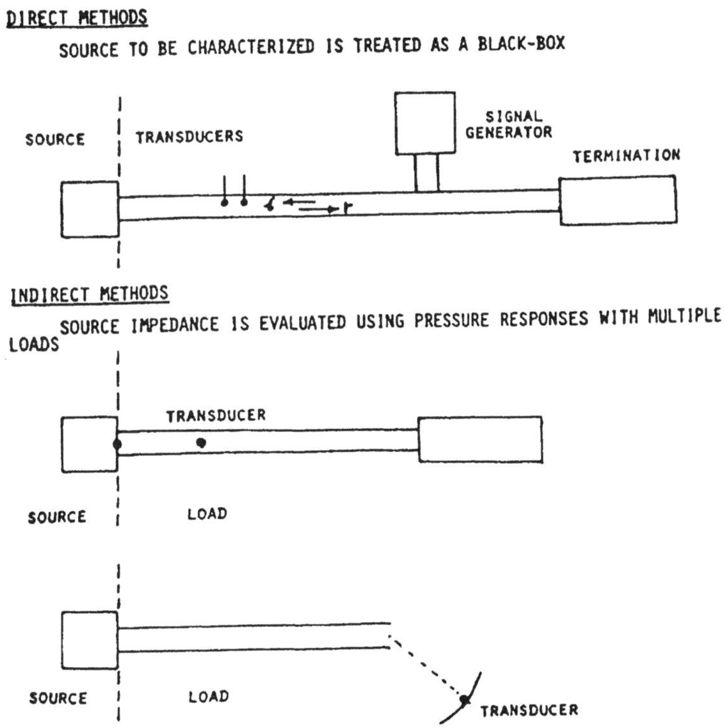

Figure 5. Block diagram representation of direct and indirect methods

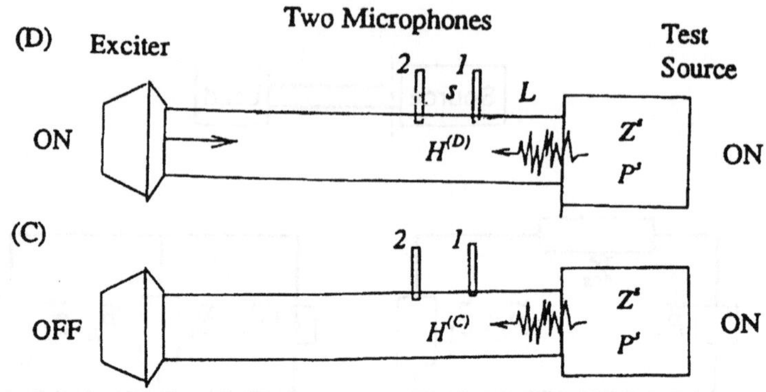

Figure 6. Additional calibration measurement for modified transfer function [12].

materials. The commonly used two techniques are standing wave ratio and two-microphone transfer function methods. the transfer function method is found advantageous due to the random excitation approach which makes it a faster method. The general equations for transfer function and reflection coefficient are given by equations (4) and (5).

$$\tilde{R} = e^{j(k_i+k_r)l}\frac{\tilde{H}_{12} - e^{-jk_is}}{e^{-jk_rs} - \tilde{H}_{12}} \tag{4}$$

$$\tilde{Z}_n = \frac{1+\tilde{R}}{1-\tilde{R}} \tag{5}$$

Both these methods have been applied for measuring acoustic impedance of dynamically operating sources. The features of both these methods are that they need a secondary acoustic source, a high signal-to-noise ratio and in-duct acoustic measurements [10]. Reliable results have been obtained on simple and complex sources [9,10]. The transfer function method has been modified with an additional calibration measurement that effectively deals with the difficulties due to low signal-to-noise ratios (figure 6) [11,12]. The reduction in the required high signal-to-noise ratio helps in systems with flow.

5. Indirect Methods

The main advantages of indirect methods are that they do not need secondary acoustic sources and acoustic measurements can be done using microphone outside the duct. The indirect method depends on number of acoustic loads used on the test source. The general equation for indirect methods based on the electrical analog (figure 4) are given by equations (6) and (7).

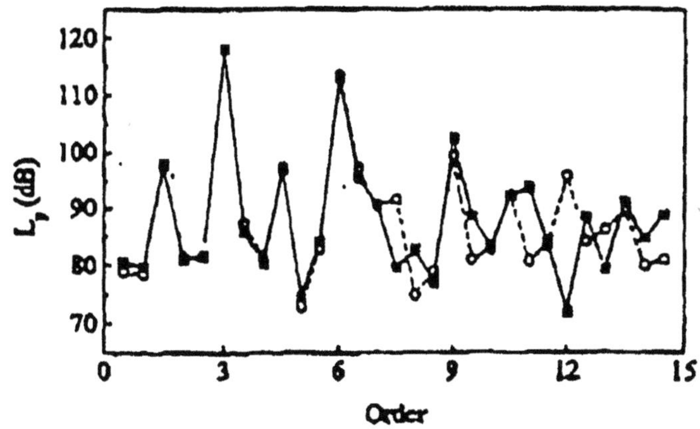

Figure 7. Sound levels at 0.5m from exhaust opening, predicted -•-, measured -o-[20]

$$\tilde{P}_L = \frac{\tilde{P}_S \tilde{Z}_L}{\tilde{Z}_S + \tilde{Z}_L} \tag{6}$$

$$\tilde{V}_L = \frac{\tilde{V}_S \tilde{Z}_S}{\tilde{Z}_S + \tilde{Z}_L} \tag{7}$$

The commonly used acoustic loads are straight pipes. The use of two and three acoustic loads require measurements of complex acoustic pressures in relation to a reference source signal. The use of four loads needs only the measurements of magnitudes of sound pressures. This can be obtained by the auto-spectra in terms of sound pressure levels [13-15]. Although the four-load method is quite simple in its use but it has instability problems. Hence the four-load method is highly sensitive to errors due to measurements and computations. In addition to error analysis of the four-load method, several developments with four or more loads (multi-loads) have been presented. They are weighted four-load method, least square method, multi-load method and refined multi-load method [12, 16-19]. These newer developments have yielded reliable measurements of source impedance which have been effectively used for system performance evaluation (figures 7 and 8).

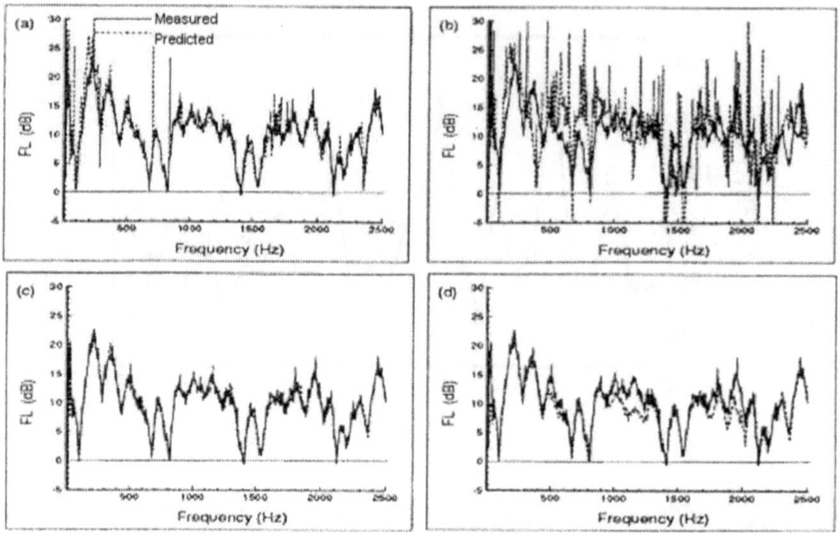

Figure 8. Comparison of filtering loss of a muffler using source impedance from a) weighted four-load method, b) four-load method, c) modified transfer function, and d) transfer function method. All from in-duct measurements. [12].

6. System Performance

The most useful performance descriptor of a filter or a muffler is Insertion Loss (IL). It is calculated using equation (1) using the measured or assumed source impedance, four-pole parameters of the muffler with associated ducts and exit radiation impedance. Figure 9 shows the comparison of IL for an electro-acoustic system for various assumed and measured source impedances. It is clearly seen that among the comparisons, the IL predicted using measured source impedance provides the best comparison. Figure 10 shows a similar result for the case of an eight-cylinder engine. Figure 8 shows that the weighted four-load method [12] based on the modifications made to conventional four-load method [15] yields better results. Figure 11 shows that the least square method [17] gives better agreement than the conventional four-load method [15]. Figure 12 shows the results of the application of the system approach presented on a vacuum pump. The results indicate that the linear system approach is useful in acoustic modeling of systems with various types of sources [20-25].

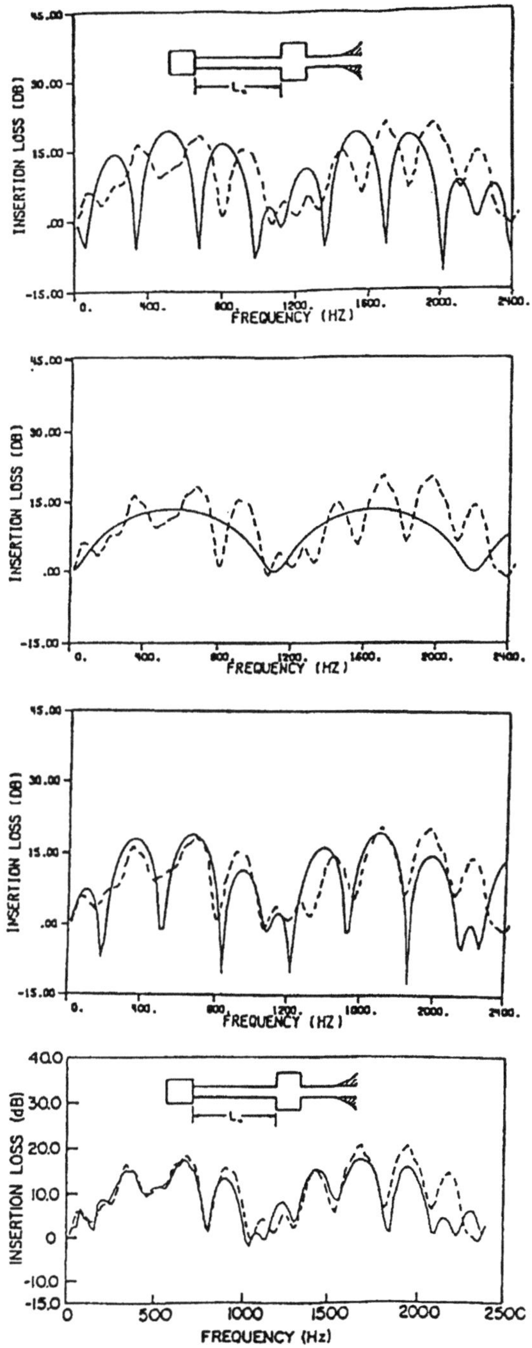

Figure 9. Insertion loss predictions with various assumed and measured source impedances for an electro-acoustic speaker-expansion chamber system [10]. From the top, Source impedance assumed as Zero, Characteristic, Infinite and Measured. Upstream length is 0.502m and downstream is anechoic. - - - Expt, — Predicted

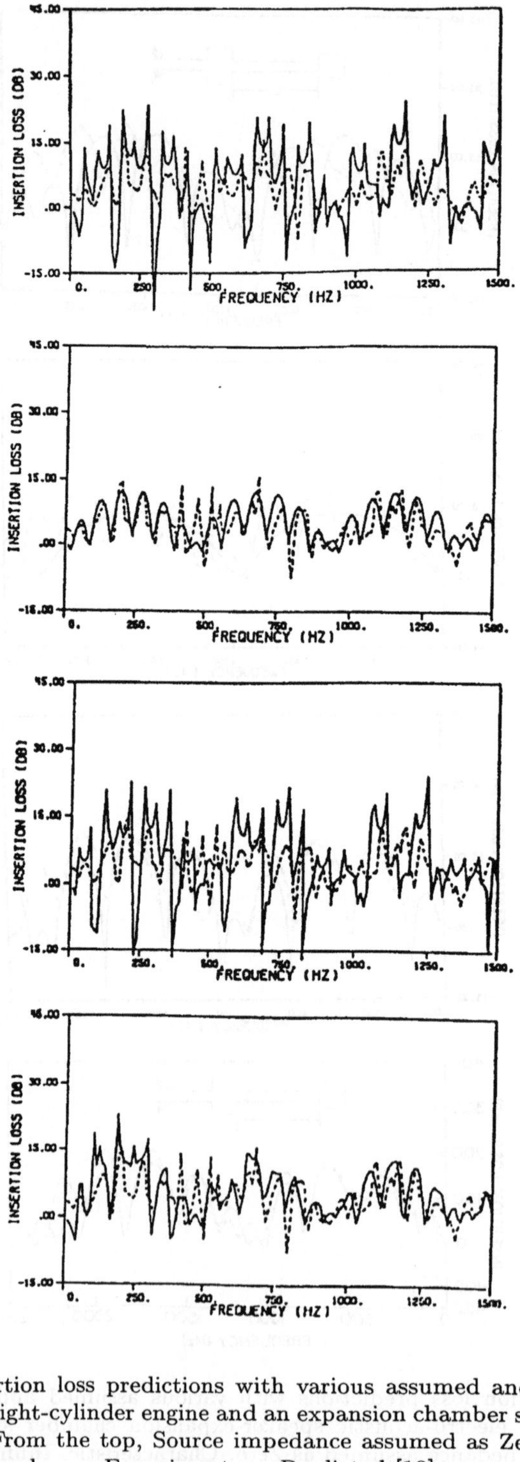

Figure 10. Insertion loss predictions with various assumed and measured source impedance for an eight-cylinder engine and an expansion chamber system at 2000 rpm and 10 inch Hg. From the top, Source impedance assumed as Zero, Characteristic, Infinite and Measured. - - - Experiment, — Predicted [10].

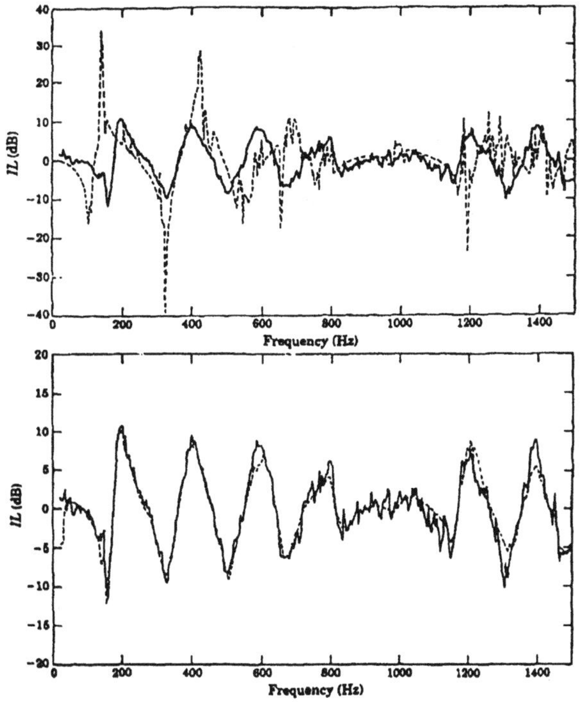

Figure 11. Insertion loss by Four-load and Least squares method - - - Expt, — Predicted [18]

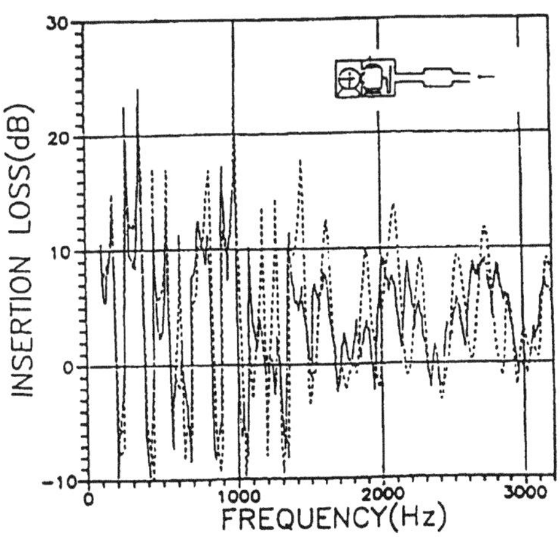

Figure 12. Insertion loss for a pump-expansion chamber, - - - Expt, —- Predicted [22]

7. Concluding Remarks

Thus the studies reviewed and presented show that the linear system approach is effective. The approach uses black-box model for source impedance evaluation. Although wave propagation details are not accounted in this approach, the model provides insight into the source-load system and the interactions. Studies on various systems such as fans as two-port sources, pumps as fluid machinery sources, vacuum pump and glottis as source in speech analysis have been effective and useful in system design [19-29]. We can see in figure 10 that even using measured source impedance, the agreement at low frequencies is not very good. This is due to the fact that at the firing frequencies the waves becomes non-linear due to very large amplitudes and high temperature varying along the exhaust pipe. At these discrete low frequencies, one could try supplemental approaches like standing wave ratio method. Although an average higher temperature is used in linear modeling of four-pole parameters, the temperature variation along the exhaust pipe constrain the linear models particularly at low frequencies. The linear modeling approaches presented here are based on black-box approach. This black-box approach provides an effective modeling and an overall system design methodology for duct systems. Further studies are needed for understanding the non-linearity in sources, wave-propagation at discontinuities, effects of temperature and flow, transient sources etc. The industrial use of such modeling and analysis of duct systems used in engineering application has to be noted [29].

References

1

Davis, D. D. Jr. (1957) Acoustical filters and mufflers, Chapter 21, Handbook of Noise Control (ed: C.M.Harris), McGraw-Hill Book Co., New York.

Munjal, M.L. (1987) Acoustics of ducts and mufflers, John Wiley and sons., New York.

Prasad, M.G. and Crocker, M.J. (1998) Acoustic modeling (ducted source systems), Handbook of Acoustics (ed: M.J. Crocker), John Wiley and sons, New York

Sridhara, B.S., and Crocker, M.J. (1994) Review of theoretical and experimental aspects of acoustical modeling of engine exhaust systems, J. Acoust. Soc. Am.., **95(5)** , 2363-2370.

Jones, A.D., Van Moorhem, W.K., and Voland, R.T. (1986) Is a full nonlinear method necessary for the prediction of radiated engine noise?, Noise Control Engineering Journal, **26(2)**, 74-80.

Boden , H., and Albertson, F., (2000) Linearity tests for in-duct acoustic one-port sources, J. Sound Vib., **237(1)**, 45-65.

Davies, P.O.A.L. (1996) Piston engine intake and exhaust system design, J. Sound Vib.,**190 (4)** 677-712.

Prasad, M.G. (1991) Characterization of acoustical sources in duct systems-progress and future trends, Proceedings of Noise-Con 91, Tarrytown, New York, Institute of Noise Control Engineering, New Jersey.

Boden, H., and Abom, M. (1995) Modeling of fluid machines as sources of sound in duct and pipe systems, Acta Acustica, **3**, 549-560.

Prasad, M.G., and Crocker, M.J. (1983) Acoustical source characterization studies on a multi-cylinder engine exhaust system and Studies of acoustical performance of a multi-cylinder engine exhaust muffler system, J. Sound Vib., **90(4)** 479-50.

Kim, W.S., and Prasad, M.G., (1993) A modified transfer function method for acoustic source characterization in duct systems, paper no. 93-WA/NCA-5, ASME Winter Annual Meeting.

Kim, W.S., (1994) Modified transfer function and weighted four-load method for source characterization with applications to system modeling in duct acoustics, Ph.D dissertation, Stevens Institute of Technology, Hoboken, New Jersey.

Kathuriya, M. L., and Munjal, M. L. (1979) Experimental evaluation of the aeroacoustic characteristics of a source of pulasating flow, J. Acoust. Soc.Am., **65** ,240-248.

Gupta, V.H., and Munjal, M.L. (1992) On numerical prediction of the acoustic source characteristics of an engine exhaust system, J. Acoust. Soc. Am., **92(5)** 2716-2725.

Doige, A.G., and Alves, H.S. (1986) Experimental characterization of noise sources for duct acoustics, Paper No. 86-WA/NCA-15, ASME Winter Annual Meeting.

Prasad, M.G., (1987) A four load method for evaluation of acoustical source impedance in a duct, J. Sound Vib., **114(2)**, 347-356,1987.

Sridhara, B.S., and Crocker, M.J., (1992) Error analysis for the four-load method used to measure the source impedance in ducts, J. Acoust. Soc. Am., **92(5)** , 2924-2931,

Desmonds, L., and Hardy, J. (1994) A least squares method for the evaluation of characteristics of acoustical sources, J Sound Vib., **175(3)** , 365-376.

Boden, H. (1995) On multi-load methods for determination of the source data of acoustic one-port sources, J Sound Vib., **180(5)** , 725-743.

Jang, S.H., and Ih, J.G. (2000) Refined multi-load method for measuring acoustical source characteristics of an intake or exhaust system, J Acoust. Soc. Am., **107(6)** , 3217-3225.

Whitson, R.J., and Benson, I.M. (1993) Investigation of the fluidborne noise source flow and source impedance characteristics of a single stage centrifugal pump using a secondary noise source, Proceedings of the First International CEETIM symposium on Pump Noise and Vibration, Clamart, France.

Sridhara, B.S., and Crocker, M.J. (1996) Prediction of sound pressure radiated from the open end of a pipe and muffler insertion loss using a single, efficient scheme: Applications to a vacuum pump, J. Acoust. Soc. Am., **99(3)** , 1333-1338.

Lavrentjev, J., Abom, M., and Boden, H. (1995) A measurement method for determining the source data of acoustic two-port sources, J sound Vib., **183(3)** , 517-531.

Terao, M., and Sekine, H. (1992) Fan acoustical characteristics required for reliable HVAC duct sound prediction, CEETIM Fan noise symposium, France.

Ananthapadmanabha, T.V., and Fant, G. (1982) Calculation of the true glottal flow and its components, Speech Communication, **1**, 167-184.

Prasad, M.G., and Ananthapadmanabha, T.V. (1989) Acoustic noise synthesis and its application to the design of a duct system, Proceedings of Inter-Noise 89.

Munjal, M.L., and Prasad, M.G. (1986) Plane wave propgationin a uniform pipe in the presence of a mean flow and and a temperature gradient, J. Acoust. Soc. Am., **80(5)** , 1501-1506.

Prasad, M. G., and Munjal, M. L. (1986) Evaluationb ofb mechanical source impedance with Application to vibration isolation system design, PVP Vol **106**,ASME Conference on Pressure Vessels and Piping.

Minutes and Summary of an International workshop on fluid- and structure borne sound source characterization, (1996) University of Liverpool, England, August.

THEORETICAL AND EXPERIMENTAL STUDIES OF THE ACOUSTICAL DESIGN OF VEHICLE CABS–A REVIEW OF TRUCK NOISE SOURCES AND CAB DESIGN USING STATISTICAL ENERGY ANALYSIS

M. J. CROCKER, A. R. PATIL and
J. P. ARENAS
Mechanical Engineering Department,
Auburn University,
Auburn, AL 36849, U.S.A.
mcrocker@eng.auburn.edu

Abstract In this paper the main noise sources on heavy truck vehicles are discussed. Methods of identifying these sources and their dependence on engine and vehicle speed are described. Some analytical studies and experimental results for the sound transmission into an idealized cab enclosure from a homogeneous sound field are presented. Statistical energy analysis (SEA) is utilized to obtain the time- and space-averaged information needed in the design stage. The sound transmission into an enclosure including the effects of apertures is analyzed using a theory for the diffraction of sound, which gives improved results in the mid-frequency range. The effects of system parameters on the frequency-averaged attenuation are considered. The constrained optimization method is used for this purpose. In addition, some design guidelines are given for minimizing the sound power transmitted into the cab enclosure.

Keywords: Truck noise sources, vehicle cab design, statistical energy analysis, sound transmission

1. Introduction

Noise inside heavy vehicle cabs is a problem because of its speech interference effects. It is also a cause of annoyance and if it is sufficiently intense it can be a hazard to the hearing of the vehicle operator and cab occupants. In the USA the regulation of the in-cab (interior) noise of heavy commercial vehicles (trucks) is based on limits of noise set in the

M.L. Munjal (ed.), IUTAM Symposium on Designing of Quietness, 47–65.

workplace by OSHA. Noise is limited to 90 dB(A) for eight hours with a halving of allowable exposure time for every 5 dB(A) increase in noise. Noise inside vehicle cabs also masks warning signals from other vehicles and from railroad locomotives at crossings. The masking effect of noise has been shown to be the cause of numerous accidents between vehicles and at railroad intersections, etc. It is important to reduce noise inside cabs for the reasons stated.

The transmission of sound from outside noise sources into enclosure interiors is of interest not only in vehicle cabs, but also in the design of automobile compartments, aircraft and ship cabins and factory personnel enclosures. In many practical situations, the possibility of sound transmission through apertures or slits (such as direct noise leaks from the engine compartment and partially open windows in vehicles) into the interior of the enclosure exists. The noise sources in heavy vehicles are numerous and depend upon engine speed and load and upon road speed and other parameters. The noise inside the vehicle cab will depend upon geometrical parameters and the structural materials used. The complete analysis of the noise level inside a particular vehicle thus is complicated. In order to obtain some insights into the important parameters in cab design we have undertaken some analytical studies of sound transmission into an idealized cab enclosure from a homogeneous sound field. And instead of seeking detailed frequency and spatial information about the interior sound field, which would be inaccurate and unnecessary in the preliminary design stages, statistical energy analysis (SEA) is utilized. The SEA provides frequency and time- and space-averaged information about the sound field, which is often all that is needed in the design stage.

The sound transmission into an enclosure including the effects of apertures has been analyzed previously using SEA by Cole and Crocker [1,2,3]. They used the theories of Ingerslev [4] and Mullholland [5] for calculating the transmission coefficient of an aperture. These theories do not accurately predict the aperture transmission coefficient at mid- and high-frequencies. In this paper, the theory for the diffraction of sound by a circular aperture developed by Wilson [6] is used. This gives somewhat better results in the mid-frequency range. The total transmission of sound into the enclosure can be divided into resonant and non-resonant transmission. The relative contributions of each are different in the low-, medium- and high-frequency ranges. The resonant and non-resonant contributions depend on system parameters such as aperture radius, panel thickness, area of the interior absorbing material, and the enclosure panel density. In this paper, these effects on the frequency average transmission loss are considered simultaneously. The constrained optimization method is used for this purpose. Some design guidelines are also given for minimizing the sound power transmitted into the cab enclosure.

2. Noise and vibration sources in heavy vehicles

The main noise sources in trucks, which affect both the vehicle cab interiors and the community, are shown schematically in Fig. 1. The strength of these noise sources depends primarily on engine speed (RPM) and/or vehicle speed. The mechanical noise sources are mainly associated with the engine (power source) and are related to engine RPM, while the tire noise and aerodynamic noise are related to vehicle speed. Mechanical noise dominates at low vehicle speed, while at high speed over about 50 mph (80 km/h), tire and aerodynamic noise sources dominate. These noise and vibration sources can be classified into six main categories: a) Engine, b) Exhaust System, c) Induction System (air intake), d) Other Vehicle Components (gear box, transmission, cooling fan, brakes), e) Tires, and f) Aerodynamic.

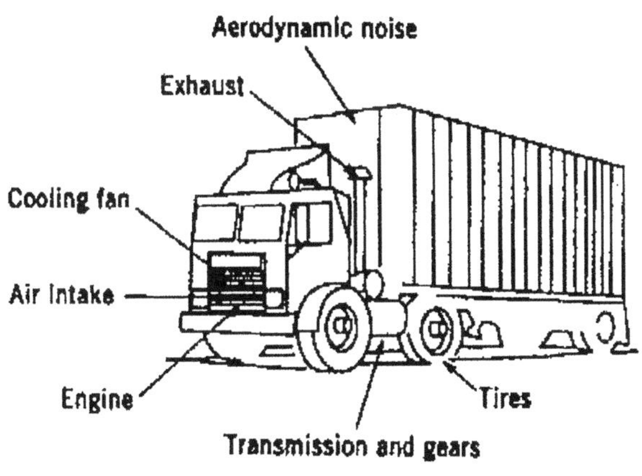

Figure 1. The main noise sources in a heavy truck.

Table 1 shows the dependence of some of these noise sources on engine and vehicle speed.

Various experimental studies have been conducted to determine the noise sources on heavy vehicles. Figure 2 shows an example of the vehicle used in one such study [7]. In order to determine the magnitude of the main sources on the truck, they were individually "silenced" by 1) the use of four high volume exhaust system mufflers which were also wrapped with fiberglass and leaded vinyl, 2) use of oversize intake si-

Table 1. Dependence of Major Truck Noise Sources on Engine and Vehicle Speed.

	Noise sources	*Source dependence*
Mechanical Noise	Engine	Related to engine power and speed
	Transmission and gears	Dominates at low speed
Tire and Aerodynamic noise	Tire noise	Related to vehicle speed, depends on roadway surface, tire tread design and other factors
	Aerodynamic noise	Turbulent boundary layer fluctuations; proportional to speed to fourth power

lencers which were also wrapped, 3) complete wrapping of the engine with 10 cm thick fiberglass and leaded vinyl, 4) removal of the cooling fan, 5) complete wrapping of the drive-line, differential housing, and 6) damped plywood covers used over the wheels. The fully silenced truck produced a maximum noise of 71 dB(A) at 15m during an acceleration drive by test.

Figure 2. IH COF-4070A truck with artificially silenced exhaust and intake systems for noise source identification work

To determine the contributions of each noise source on the original truck, each noise-producing source was unwrapped and individually returned to its original state. Figure 3 shows the results for the maximum sound levels recorded at 15m for each source during an acceleration test. The total sound level of 88 dB(A) was made up principally of fan noise and exhaust noise.

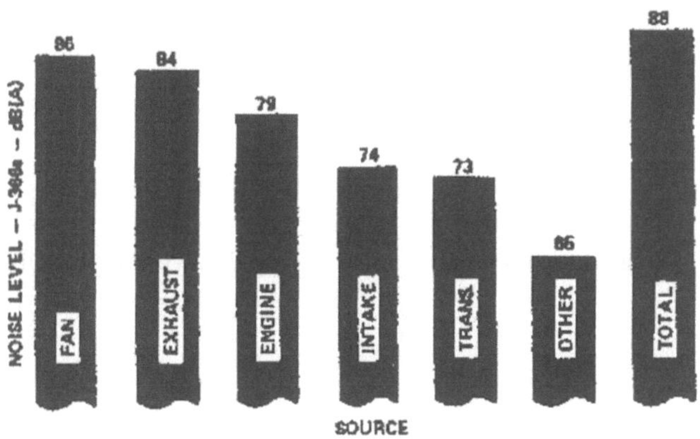

Figure 3. A-weighted sound levels of "baseline" truck. Fan noise source is dominant.

The next figure, Fig. 4 shows how the three major noise sources varied during the drive-by acceleration test. The distances are given in feet. At the beginning of the test (for 50 feet before the microphone) the cooling fan, engine and exhaust noises are seen to be about equal. But soon the exhaust and engine noise exceed the cooling fan noise and eventually the cooling fan noise becomes predominant when it reaches its maximum speed at the end of the test run.

Further understanding of the total truck noise can be gained by examining Fig. 5 which shows the spectra of the same three major noise sources when the total truck has reached maximum near the end of its test run. It is seen that at low frequency from 250 to 630 Hz, the fan noise dominates, in the mid frequency range from 800 to 3000 Hz the three major sources are about equal, while at high frequency above 4000 Hz the exhaust noise dominates.

Figure 6 shows the noise levels of the main sources of the same truck after engineering changes had been made to reduce the drive-by noise to the goal of 80 dB(A) at 15m during the same acceleration drive-by J-366a test.

The next table, Table 2, shows cruise-by noise levels of different size vehicles traveling at about 50 mph (80 km/h). It is observed that although there was some reduction of noise levels in medium and heavy

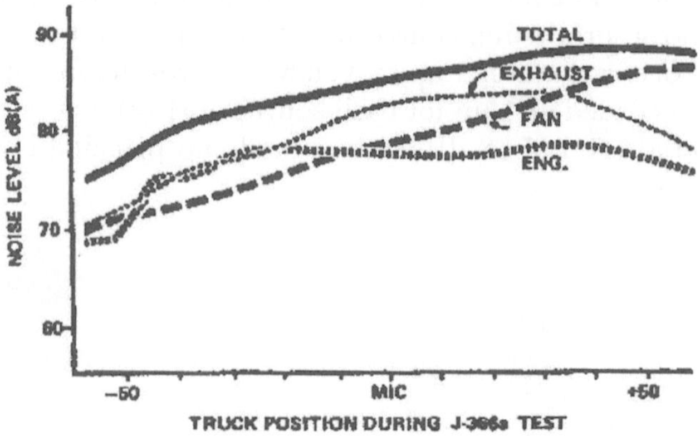

Figure 4. Fan contributes most noise during drive-by of "baseline" truck.

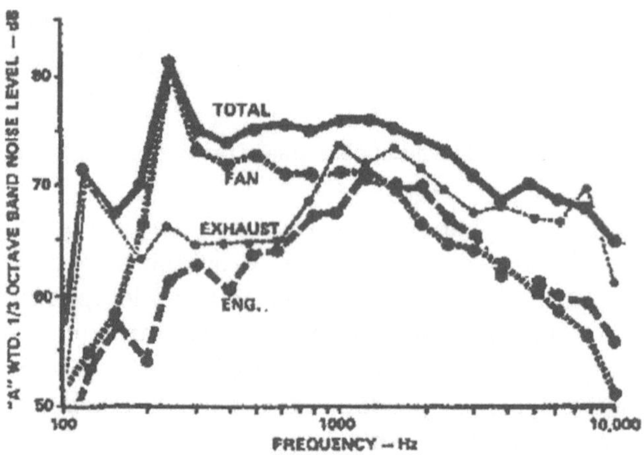

Figure 5. One-third octave, frequency analysis of truck noise sources during drive-by of "baseline" truck.

trucks between 1975 and about 1985, there was no reduction or even a slight increase in noise levels for the ten years after from 1985 to 1994.

The EU has been steadily reducing the noise emission levels allowed for trucks and manufacturers have been using some of the methods to reduce noise shown in Table 3 and Fig. 7. Table 3 lists some of the noise and vibration control measures used by manufacturers to control sources in four major noise categories. Tire noise is not shown in this table but is a very important source whose strength depends not only upon vehicle speed but also on roadway surface, tire tread design and

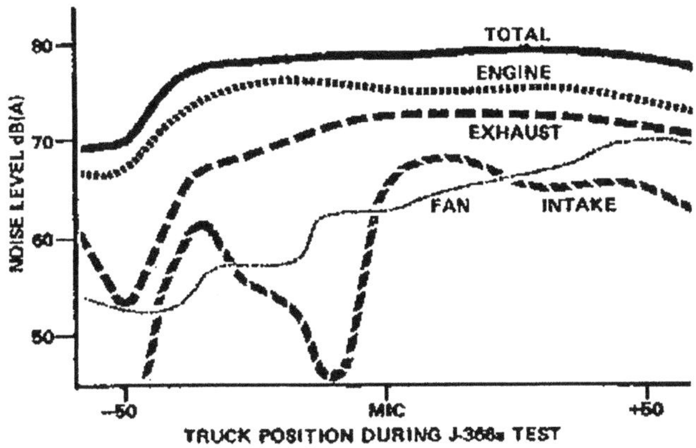

Figure 6. Drive-by sound levels of final quieted truck. Compare these with Fig. 4.

Table 2. Cruise-by noise levels measured at various sites in the USA.

Mean A-weighted sound levels (dB)				
Year	*Numbers of vehicles*	*Automobiles*	*Medium trucks*	*Heavy trucks*
1975	3500	71.7	82.4	86.3
1982-85	2500	72.8	79.9	83.8
1994	6000	73.8	79.9	84.0

other factors. Figure 7 illustrates the time during which some of these noise control measures have been applied.

3. SEA and its use in automotive heavy vehicle cab design

The discussion about vehicle noise and vibration sources and their variation with time during acceleration tests shows that the prediction of the noise inside a vehicle cab is a very complicated problem. There are several major noise sources and their individual strengths depend upon engine speed and road speed in a complicated way. In addition these sources feed energy into the cab through various paths that may be mainly airborne or mainly structure-borne or a combination of both. Often precise predictions of the sound pressure variation with location and detailed frequency information is unnecessary and in fact beyond the capabilities of existing engineering methods. In the low frequency region, engineers currently use numerical methods such as the finite element method (FEM) and the boundary element method (BEM). These

Table 3. Examples of Noise Reduction Methods Used by Truck Manufacturers.

Engine in general	Switch-over to turbo-charged engines Optimization of the engine combustion process, e.g. by using electronics or by improving the shape of the combustion chamber Encapsulation or shielding of entire engines or specially noisy parts of them Use of hood blankets or laminated covers Sound absorptive material in the engine compartment Optimization of the stiffness of the cylinder block Use of structure-borne noise reduction material
Exhaust system	Minimization of outlet and mantlet emission of exhaust silencer, e.g. by increased volume Introduction of more than one silencer Optimization of pipes to/from the silencer, e.g. by equal length pipes or air gap pipes Dual mode mufflers Use of absorptive materials Active noise control
Induction system	1/4-wave tuners or other resonators Thicker duct walls, and/or lined ducts Increased volume of air cleaners Intake covers or shields Active noise control
Other vehicle components	Improvement of gear boxes, damped propeller shafts Improved rear axle transmission Shielding of transmission components Regulation of the fan by thermostat Decreased speed of fan by using larger fan, or optimization of fan shape Silencers for air compression outlet noise Improvement of brakes for reduction of brake squeal Improved aerodynamics

deterministic predictions however can never be completely reliable since many of the structural and acoustical boundary conditions and material properties can never be known precisely. In the high frequency region above a few hundred hertz, the numerical FEM and BEM methods become unwieldy because of the immense amount of computing time required. In the high frequency region statistical methods such as the statistical energy method (SEA) become attractive. A study on an idealized truck cab enclosure follows in the next sections of the paper.

3.1 SEA model

The idealized enclosure system is divided into eight subsystems where the first subsystem is the outside room, the eighth subsystem is the enclosure interior and other subsystems are panels as shown in Fig. 8.

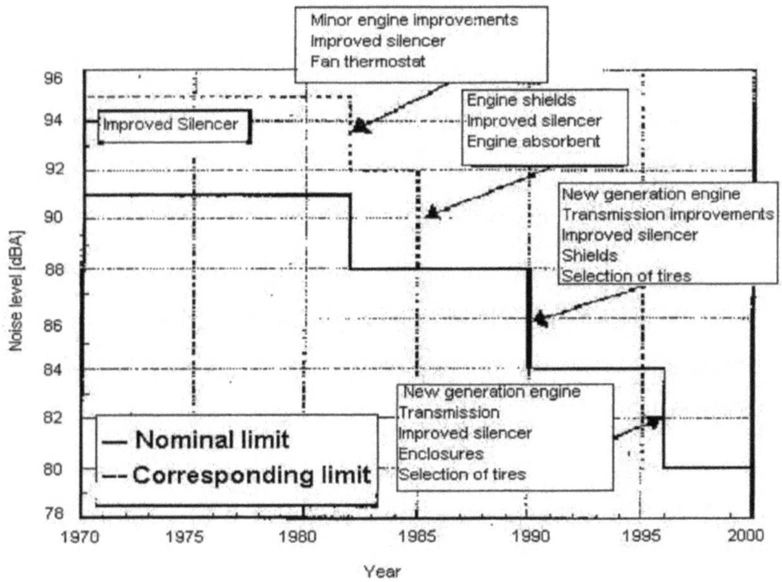

Figure 7. Noise reduction measures being employed in order to meet the tightened noise emission limits in the EU.

The system was divided into eight subsystems because they represent separate wavefields. The power input Π_{in_1} is given only to the first subsystem as acoustic excitation. Power flow between different subsystems is shown as Π_{ij} to be the power flow from the ith subsystem to the jth subsystem. Power dissipated for the ith system is shown as Π_{diss_1} . The power balance equations are written for each subsystem where E_i^m is the modal energy, n_i is the modal density, Π_{in_i} is the power input to the ith system and η_{ij} is the coupling loss factor between the ith and the jth systems. These systems of equations can be rewritten using the relation [8],

$$\eta_{ij} n_i = \eta_{ji} n_j, \tag{1}$$

where n_i is the modal density of the ith system. The effect of an aperture is included in the coupling loss factor η_{18} . The transmission coefficient needed for calculating this coupling loss factor is calculated using Wilson's theory of sound diffraction through an aperture [6].

Power flow equations are written in matrix form as follows,

$$\begin{bmatrix} E_1^m \\ E_2^m \\ E_3^m \\ E_4^m \\ E_5^m \\ E_6^m \\ E_7^m \\ E_8^m \end{bmatrix} = \frac{-1}{\omega \Delta f} \begin{bmatrix} -\eta_{t_1} n_1 & \eta_{21} n_2 & \eta_{31} n_3 & \cdots & \eta_{61} n_6 & \eta_{71} n_7 & \eta_{81} n_8 \\ \eta_{12} n_1 & -\eta_{t_2} n_2 & \eta_{32} n_3 & \cdots & \eta_{62} n_6 & \eta_{72} n_7 & \eta_{82} n_8 \\ \eta_{13} n_1 & \eta_{23} n_2 & -\eta_{t_3} n_3 & \cdots & \eta_{63} n_6 & \eta_{73} n_7 & \eta_{83} n_8 \\ \eta_{14} n_1 & \eta_{24} n_2 & \eta_{34} n_3 & \cdots & \eta_{64} n_6 & \eta_{74} n_7 & \eta_{84} n_8 \\ \eta_{15} n_1 & \eta_{25} n_2 & \eta_{35} n_3 & \cdots & \eta_{65} n_6 & \eta_{75} n_7 & \eta_{85} n_8 \\ \eta_{16} n_1 & \eta_{26} n_2 & \eta_{36} n_3 & \cdots & -\eta_{t_6} n_6 & \eta_{76} n_7 & \eta_{86} n_8 \\ \eta_{17} n_1 & \eta_{27} n_2 & \eta_{37} n_3 & \cdots & \eta_{67} n_6 & -\eta_{t_7} n_7 & \eta_{87} n_8 \\ \eta_{18} n_1 & \eta_{28} n_2 & \eta_{38} n_3 & \cdots & \eta_{68} n_6 & \eta_{78} n_7 & -\eta_{t_8} n_8 \end{bmatrix}^{-1} \begin{bmatrix} \Pi_{in_1} \\ 0 \\ 0 \\ 0 \\ 0 \\ 0 \\ 0 \\ 0 \end{bmatrix} \tag{2}$$

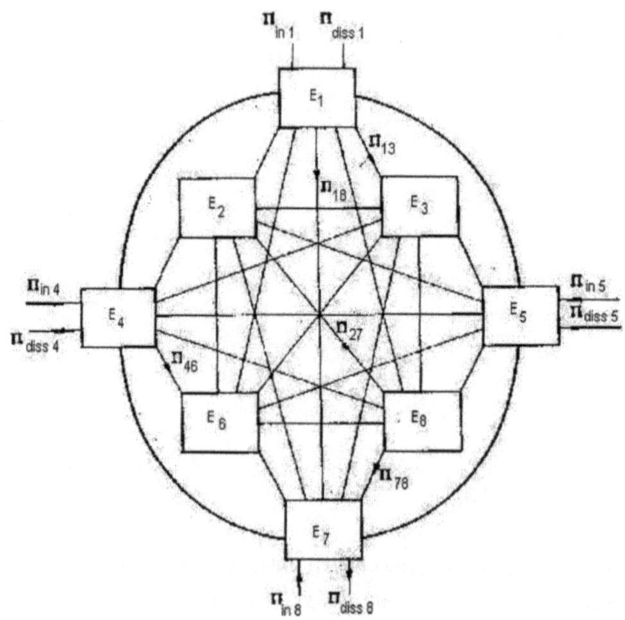

Figure 8. Power flow diagram.

The frequency range of 100 to 20,000 Hz is divided into a number of bands each with a bandwidth of 100 Hz. The system of matrices can be solved for each band to give the modal energy densities of each subsystem. These values are further analyzed to give the sound pressure level in the room and the enclosure. The attenuation for each frequency band is calculated as

$$\text{Attenuation} = 10 \log_{10} \frac{E_1^m n_1 V_8}{E_8^m n_8 V_1}, \tag{3}$$

where V_i is the volume of the ith system. Fig. 9 shows the attenuation of the system used by Cole and Crocker ? with aperture radius of 35mm.

The analysis in which Wilson's theory is used to calculate η_{18} gives better results than other theories in the low- and mid-frequency ranges. The theoretical results do not agree well with experiment in the high frequency region because Wilson's theory does not predict the transmission coefficient of an aperture with sufficient accuracy in the high frequency range.

3.2 Resonant and non-resonant transmission

The sound power is transmitted into the enclosure by two different ways i.e. resonant and non-resonant sound transmission. Non-resonant transmission of sound power is attributed to two ways of transmission i.e. transmission through the panels and transmission through an aperture. Non-resonant transmission through the panels is caused by the non-resonant response of modes that are resonant outside the frequency band of interest.

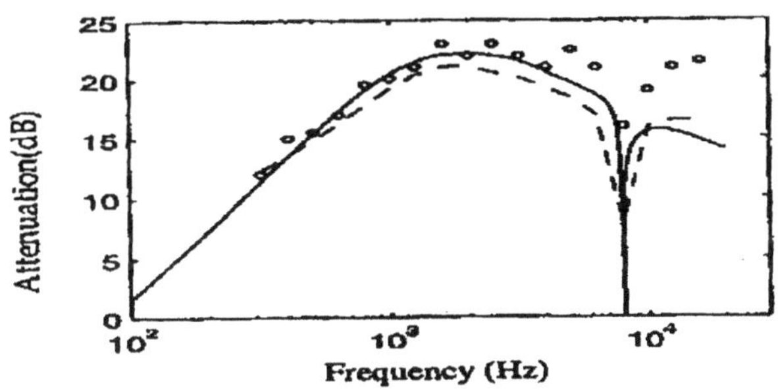

Figure 9. Attenuation of the enclosure. — theory; -- theory[1]; ∘ experiment [1]

The transmission coefficient of each panel (without an aperture) is calculated by assuming that the panel responds as a limp mass with negligible stiffness and damping. The transmission coefficient of the aperture alone is calculated assuming the panels to be rigid. These two transmission coefficients are used to calculate the coupling loss factor (η_{18}) between the room and the enclosure interior. The resonant transmission of sound power is attributed to the response of the modes of the panels that are resonant in the frequency band of interest. These modes are excited by the incident sound energy. The vibration caused by these panel modes excites the acoustic modes in the enclosure interior. The energy balance equation for the enclosure interior can be rewritten as follows by assuming the sound power input Π_{in_8} to be zero,

$$\Pi_{total} = \Pi_{diss_8} = \Pi_{nonres_8} + \Pi_{res_8}, \tag{4}$$

where

$$\Pi_{diss8} = \omega\eta_8 n_8 E_8^m,$$

$$\Pi_{nonres8} = \omega\eta_{18} n_1 (E_1^m - E_8^m),$$

$$\Pi_{res8} = \sum_{i=2}^{7} \omega\eta_{i8} n_i (E_i^m - E_8^m),$$

where n_i is the modal density of the ith subsystem. The sound pressure level inside the enclosure is dependent on Π_{diss8} which is the total sound power input to the system. The total power input to the enclosure is a summation of the non-resonant sound power transmitted $\Pi_{nonres8}$ and the resonant sound power transmitted Π_{res8}. The sound power levels $\Pi_{nonres8}$, Π_{res8} and Π_{diss8} are plotted for a typical configuration in Fig. 10. The whole frequency range can be divided into three frequency ranges as follows

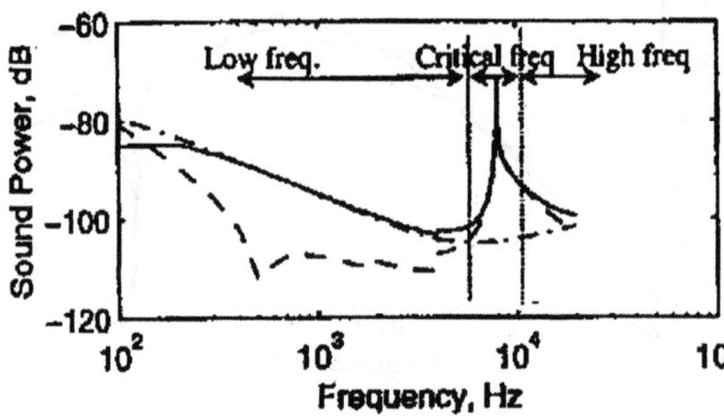

Figure 10. Contribution to the total sound power transmission. Panel thickness $h = 2$ mm. Aperture radius $a = 0.1$ m. — Π_{total}; — — Π_{res8}; — · — $\Pi_{nonres8}$.

Low frequency region. In the low frequency region, the non-resonant transmission is dominant. Resonant transmission is smaller because the modal density of the panels is low in the low frequency region and radiation of sound occurs only from the corners or edges of the panels, hence resonant transmission is low. The non-resonant transmission is greater due to the transmission through the panels than due to the transmission through the aperture.

Critical frequency region. The resonant transmission is dominant because of the "coincidence effect". There is a perfect matching of the bending wavelengths on the panel and the trace wavelengths of the sound radiated. The radiation efficiency of the panels is very high in this region.

High frequency region. The non-resonant transmission is considerable in this frequency region because the sound power transmitted through the aperture is very high. In the absence of an aperture, the resonant transmission would have been dominant in this region.

3.3 Effect of system parameters

The sound transmission into the enclosure depends on system parameters such as aperture radius, panel thickness, area of absorptive material, density of the panel, etc. These parameters are used to calculate the parameters needed for the SEA calculations. Resonant and non-resonant sound power transmission is dependent on these physical parameters. These parameters have different effects in the different frequency regions. These parameters were varied one at a time to study their effects on the transmitted sound power.

Effect of the aperture radius. The sound transmission through an aperture is essentially non-resonant. The sound power transmitted by an aperture increases with an increase in its radius (Fig. 11). It dominates the non-resonant transmission at higher frequencies because the transmission coefficient of an aperture increases as the frequency increases.

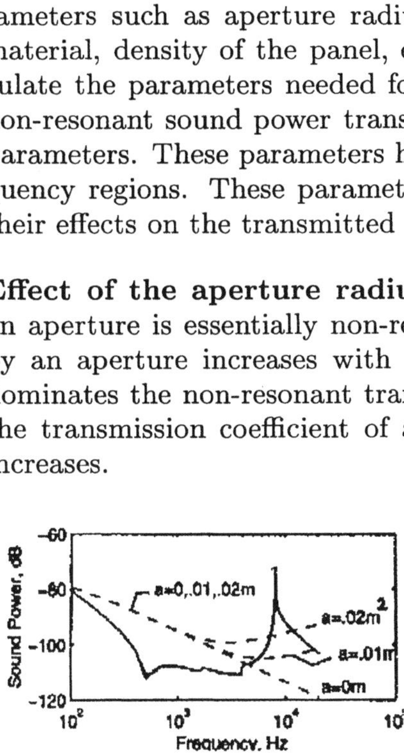

Figure 11. (1)Effect of an aperture radius. —Π_{res}; – – Π_{nonres}.

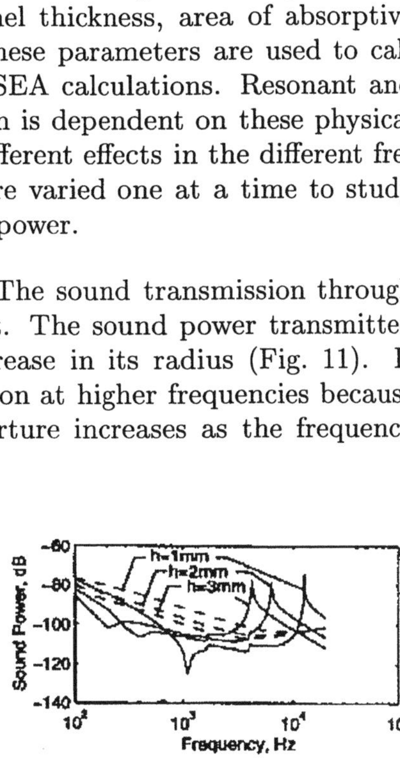

Figure 12. Effect of panel thickness. —Π_{res}; – – Π_{nonres}.

Effect of panel thickness. The coincidence frequency depends strongly on the panel thickness (Fig. 12). So the frequency region in which the resonant and non-resonant transmission is dominant depends on the panel thickness. The panel thickness also has a strong effect on the frequency-averaged transmission loss over the whole frequency range. In the low frequency region, non-resonant sound power transmission reduces with an increase in thickness because essentially it is proportional to the mass per unit area of the panel.

Effect of absorption area. The absorption area is seen to have a limited effect on the sound power transmission (see Fig. 13).

Effect of panel material density. A reduction in the panel density causes a slight reduction in the resonant sound power transmission. It also slightly changes the value of the critical coincidence frequency (Fig. 14).

4. Total attenuation of an enclosure

The discussion so far has been concerned with non-resonant and resonant flows of sound energy into an idealized cabin enclosure. The next section of this paper involves the analysis of the total enclosure attenuation, which includes both resonant and non-resonant sound transmission into the enclosure.

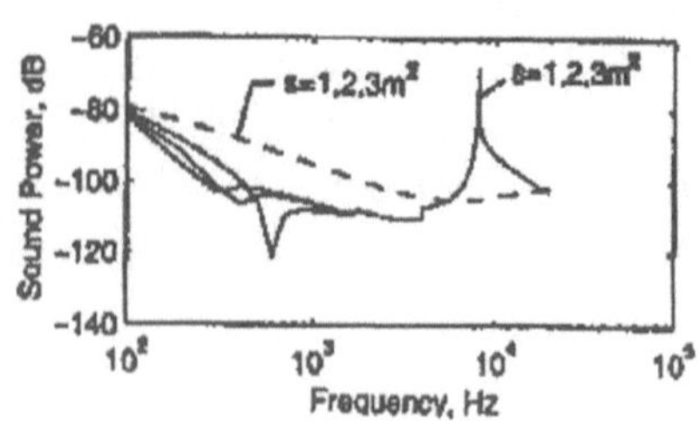

Figure 13. Effect of absorption area. — Π_{res}; – – Π_{nonres}.

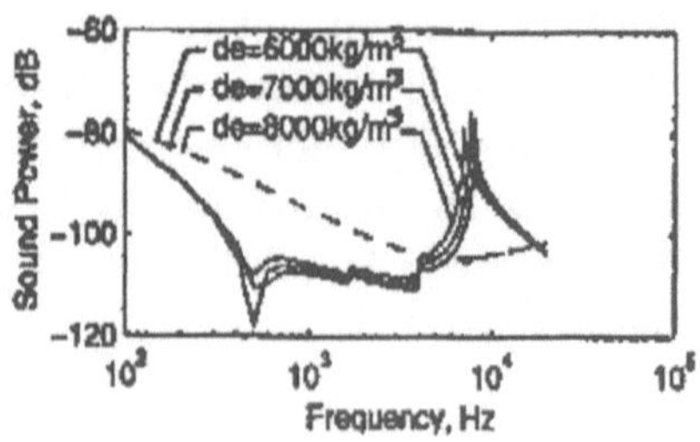

Figure 14. Effect of panel material density. —Π_{res}; – – Π_{nonres}.

A number of parameters of the enclosure are varied while the others are kept constant. Figures 15 through 18 show the effects of varying panel damping, thickness, internal sound absorption area, and aperture radius in the idealized truck cab enclosure while each of these other variables are kept constant.

Finally Fig. 19 shows the difference in experimental results between the attenuation of an idealized cab enclosure with the leaks between panels unsealed and sealed with clay. The attenuation is defined as the difference in space-averaged sound pressure levels outside and inside the enclosure.

5. Optimization of enclosure attenuation

As seen in the earlier analysis, the sound power transmission into the enclosure has a complicated relationship with the different parameters of the system. These effects vary widely with frequency. If the system

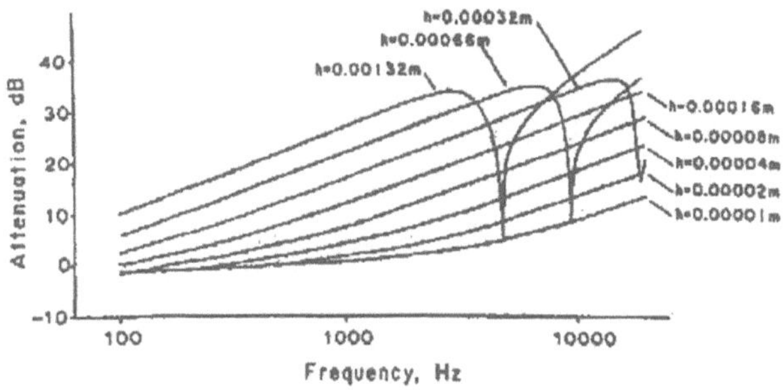

Figure 15. Effect of changing panel thickness h ($\bar{S}$=1.19 m^2, ρ_s=7700 kg/m^3).

is subjected to an external diffuse sound field, it is convenient to take frequency averages of the attenuation over the whole frequency range.

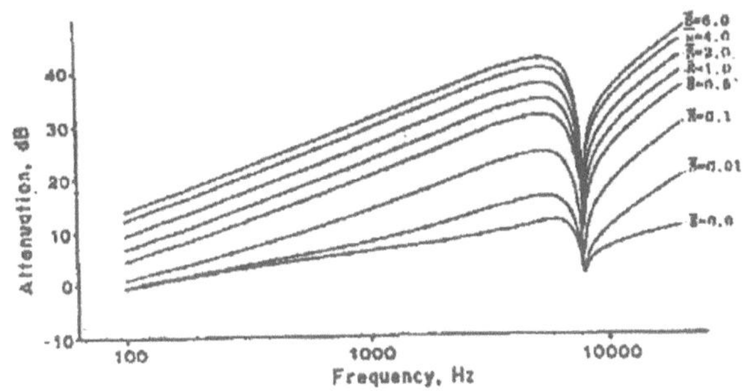

Figure 16. Effect of changing panel damping coefficient (ρ_s=7700 kg/m^3).

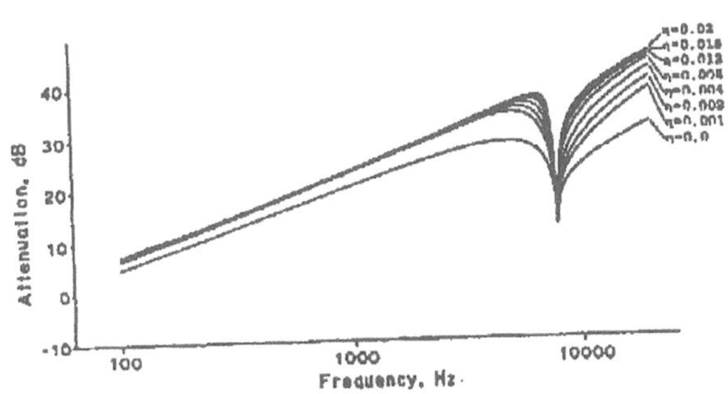

Figure 17. Effect of changing circular aperture radius, a (in m)

A frequency average is first calculated for each one-third octave frequency band by considering the transmission coefficient $\tau$for each subband within the one-third octave band under consideration,

$$\tau = \frac{\sum_{i=1}^{NSB} \Delta f_i \tau_i}{\sum_{i=1}^{NSB} \Delta f_i}, \tag{5}$$

where *NSB* is the number of sub-bands within a one-third octave band. The *overall attenuation* over the entire frequency band is termed the *attenuation level* (AL) and is calculated from the *attenuation* Att_i of each one-third octave band,

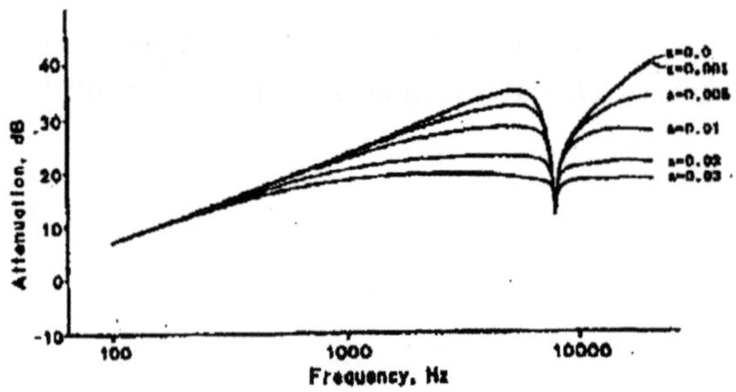

Figure 18. Effect of changing circular aperture radius, *a* (in m)

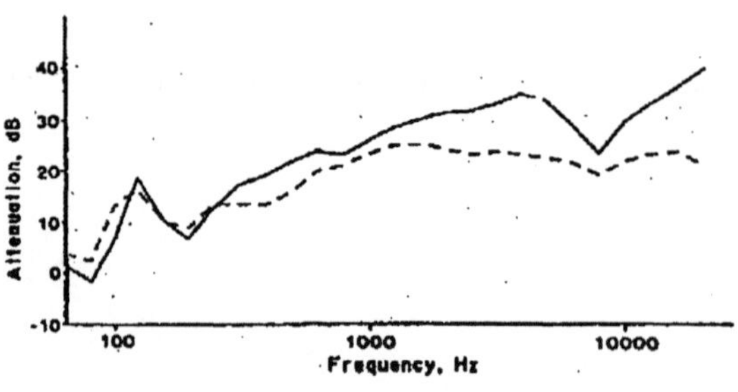

Figure 19. Experimentally measured attenuation of the sealed model enclosure. — joints sealed; –– joints unsealed

$$\text{AL} = 10 \log_{10} \sum_{i=1}^{NB} 10^{(Att_i/10)}, \tag{6}$$

where *NB* is the number of one-third octave bands. For an effective design of the system to minimize the AL, all parameters must be considered simultaneously by performing the optimization. The constrained optimization method was used to consider the bounds on the parameters. The design variables for the optimization problem were set with x_1 as the aperture radius in m, x_2 as the panel thickness in m, x_3 as the absorption area in m^2 and x_4 as the panel density in kg/m^3. The optimization problem can be described as

Find a vector of design variables $\mathbf{x} = (x_1, x_2, x_3, x_4)$
which minimizes the cost function $f(\mathbf{x}) = \text{AL}$,
subject to the upper bound on $\mathbf{x}$ as $\mathbf{x}_u = (.03, .002, 3, 8000)$ and a lower bound on $\mathbf{x}$ as $\mathbf{x}_l = (.005, .001, 0.1, 6000)$,
with an initial estimate of $\mathbf{x}$ as $\mathbf{x}_0 = (.01, .00159, 0.5, 7700)$.

The Constrained Optimization Module of MATLAB [9] was used for the optimization. A typical configuration $\mathbf{x}_0$ as described above was used as an initial estimate for the optimization. This configuration gives an AL of 35 dB. The system parameters after the optimization are: an aperture radius of 0.005 m, a panel thickness of 0.0011 m, an absorption area of 3 m^2, and a density of 8000 kg/m^3. The optimized design gives an AL of 44 dB. The effect of the optimized design on the attenuation is shown in Fig. 20 [10]. There is a substantial increase in attenuation over the entire frequency range. It shows the importance of considering all the parameters simultaneously for optimization.

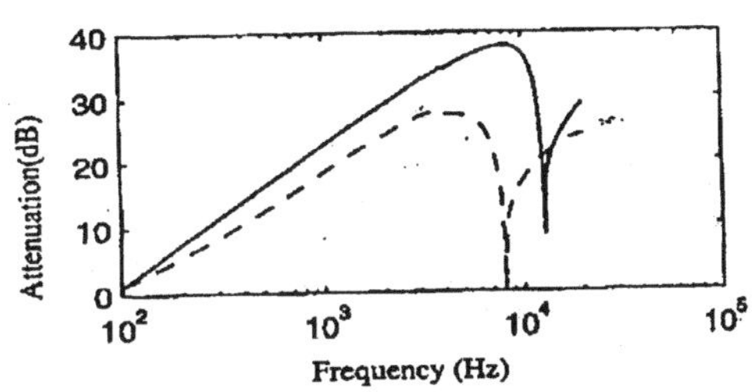

Figure 20. Effect of optimization on attenuation.— — *beforeoptimization*; — after optimization.

6. Conclusions

In this paper, the statistical energy analysis was used to simulate the sound transmission into an idealized enclosure from a noise source outside. In addition, sound diffraction theory was used to predict the coupling loss factor between the two acoustic spaces. The model produced predicts the attenuation of the idealized cab enclosure more accurately than most previous analytical models. The aperture radius has a strong effect on the non-resonant transmission at high frequencies. The absorption area affects the resonant transmission. The panel thickness determines the critical frequency and it also affects the resonant and the non-resonant sound transmission. The panel material density slightly affects resonant as well as non-resonant sound transmission. It is useful to consider all of these parameters simultaneously to obtain a good design which maximizes the attenuation. The Constrained Optimization Method has proved successful for this purpose. The results obtained in the above analysis can be used as guidelines for the design of automotive cabins and factory enclosures and optimization methods can be used to obtain the best design among numerous possibilities.

References

Cole, V., Crocker, M.J. and Raju, P.K. (1983). Theoretical and experimental studies of the noise reduction of an idealized cabin enclosure, *Noise Control Engineering Journal*, 20:122–132.

Cole, V., Crocker, M.J. and Cheng, M.K. (1984). The use of statistical energy analysis to predict the noise reduction of idealized cabin enclosures, *SECTAM XII Proceedings*, Auburn University, 63-68.

Crocker, M.J. (1994). A systems approach to the transmission of sound and vibration through structures, *Noise-Con '94 Proceedings*, 525-540.

Ingerslev, F. and Nielsen, A. (1944). On the transmission of sound through small apertures, *Ingeniorvidenska-beliga Skrifter*, 5.

Mulholland, K.A. and Parbrook, H.D. (1967). Transmission of sound through apertures of negligible thickness, *Journal of Sound and Vibration*, 5:499-508.

Wilson, C. and W. Soroka, W. (1965). Approximation to the diffraction of sound by a circular aperture in a rigid wall of finite thickness, *Journal of Acoustical Society of America*, 37:286-297.

Staadt, R.L. (1974). Truck noise control, *Society of Automotive Engineers*, Paper SP-386.

Lyon, R.H. and DeJong, R.G. (1995). *Statistical Energy Analysis*, Butterworth - Heinemann, 2nd edition, New York.

MATLAB, Version 5.2, Users Manual, Mathworks Inc.

Patil, A.R. and Crocker, M.J. (1999). Sound transmission into an enclosure with an aperture using statistical energy analysis, *Inter-noise 99 Proceedings*, 529-534.

DISTURBANCE DAMPING OF AN UNFOLDED SPACE MIRROR

V. I. BUJAKAS
Astro Space Center of P. N. Lebedev Physical Institute
Russian Academy of Science
Profsoiuznaya st.84/32, Moscow, 117810, Russia.
bujakas@asc.rssi.ru

Abstract The problem of disturbance damping in the unfolded petal-type space antenna is considered. The mathematical model for the dynamics of the design is built and investigated.

Keywords: Large space structures, deployable antenna, mathematical models, quieting of disturbances.

Introduction

Important problems for various large space structures are vibration damping and quieting of disturbances. These questions are especially important for large space antennas needed for radio astronomy. Two radio astronomical space projects, Radioastron [1] and Millimetron [2], are now under consideration in the Astro Space Center of P.N.Lebedev Physical Institute (ASC) (http://www.asc.rssi.ru). In both cases petal-type deployable space antennas are planned to use. The problem of disturbances quieting of unfolded large petal type space structure proposed as a version for Millimetron project, is studied in the paper.

1. Deployable petal-type space antennas

Large space antennas are required for implementing a number of advanced scientific and technical projects. They cannot be placed into orbit in their working configuration due to the limited capacities of transport facilities. That is why folded antenna structures, which may exist in two states, transport and working, are developed. Overall dimensions of an antenna in the transport state allows placing inside a launch rocket. After the antenna is placed into orbit, a deployment system transforms

M.L. Munjal (ed.), IUTAM Symposium on Designing of Quietness, 67–77.

the mirror into the working state. For the Radioastron [1], Millimetron [2], other projects[3], [4] a transformed mirror is a solid body petal-type structure which consists of a central mirror and a number of petals. The Radioastron petal type antenna mirror in the transport state is shown in Fig.1 and in the working state, in Fig.2.

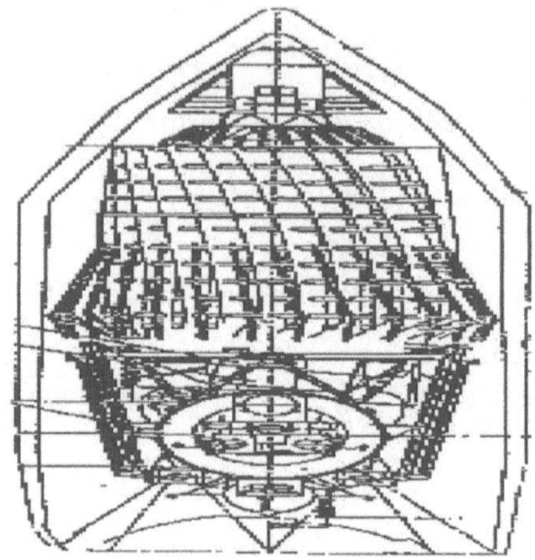

Figure 1. Petal-type antenna in a folded state.

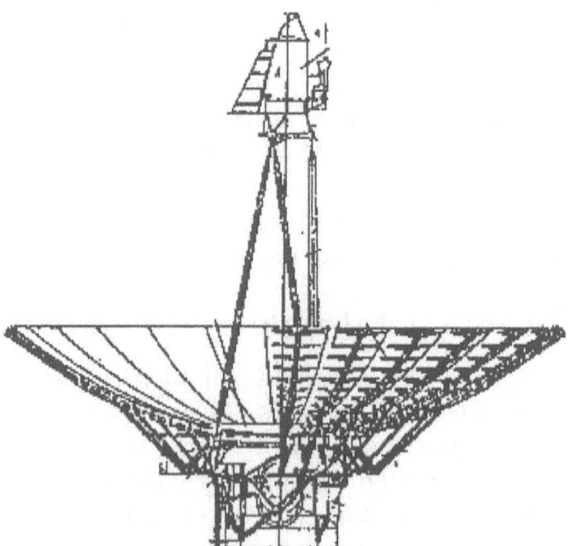

Figure 2. Fully deployed Radioastron mirror.

Computer simulation of the antenna opening is shown in Fig.3. The deployment of three petals, from fully folded to fully deployed, is shown in the first pictures. The completely deployed antenna is shown in the last picture.

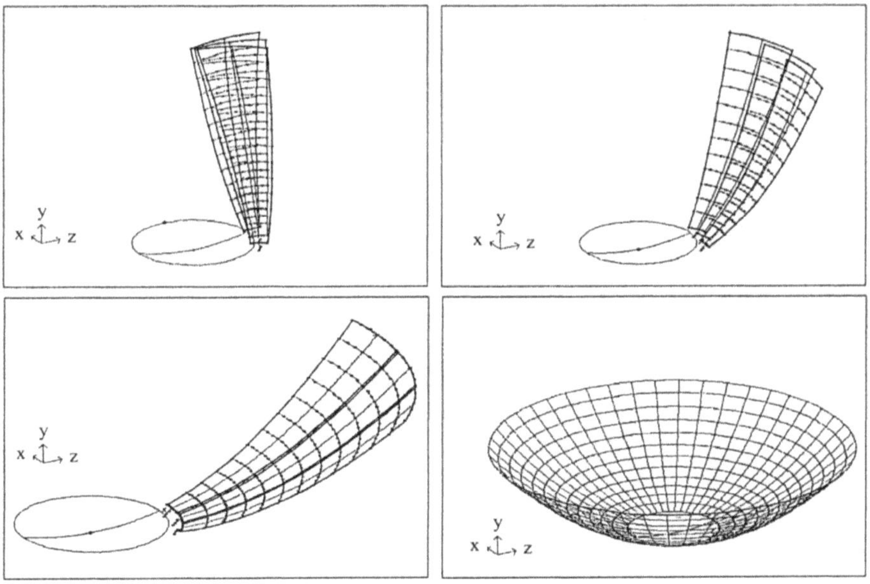

Figure 3. Simulation of antenna opening.

We consider the structures, which after deployment in the equilibrium state form statically determined designs. Such structures used together with self-adjusting locks, may ensure high accuracy of antenna's reflecting surface, high repeatability of antenna's deployment [5,6], and permit to avoid the stresses in the petals due to the errors of deployment and thermal deformations of the links.

2. Kinematic scheme of the unfolded design.

The kinematic scheme of a deployed multimirror antenna design dealt with in the paper is shown in Fig.4.

Each petal of the design is connected with the central mirror via two cylindrical hinges. The axes of rotation of the hinges $\omega_1(i), \omega_2(i)$ are not in the same plane. These hinges are used for deploying of the petals. The total number of kinematics constraints introduced by these hinges are:

$$S_1 = 4n,$$

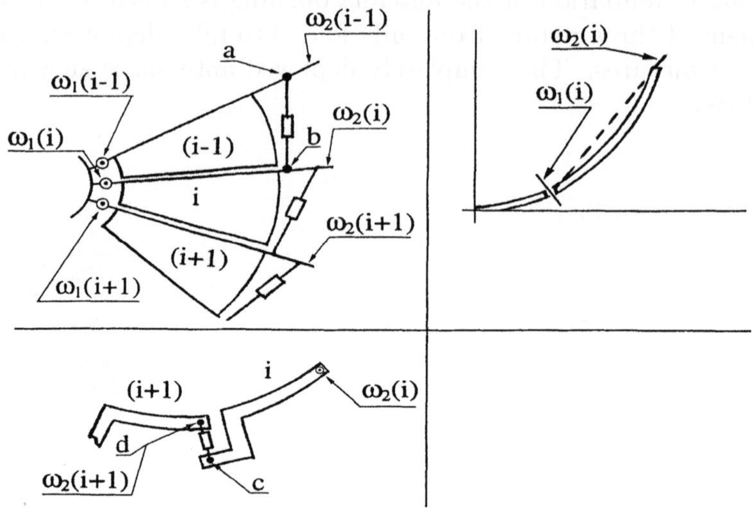

Figure 4. Kinematic scheme of unfolded structure.

where n - is the number of petals. The paper assumes that petals are absolutely rigid and hinges are perfect.

At the final stage of antenna deployment two additional links between each pair of neighbor petals are introduced. The links are placed along the periphery of an unfolded mirror. The number of peripheral links is:

$$S_2 = 2n$$

The total number of kinematics constraints in the open structure is:

$$S = 6n$$

and Maxwell's rule [7] is fullfiled.

The links between petals are elastic and are provided with dampers. Disturbances in the structure, that accompany changes in the radiotelescope orientation, should be damped by peripherial links. The peripheral links are linear elements, their ends are fixed on the neighbor petals by the spherical hinges.

The locations of fixing points and directions of axes $\omega_1(i), \omega_2(i)$ are chosen so as to ensure:

-geometrical invariability of the deployed stucture in the equilibrium state,

-disturbed motion of the petals around axes $\omega_1(i), \omega_2(i)$ to be independent (in the linear approximation).

We now describe the directions of axes of rotation and the location of the points of fixation:

a.The axis $\omega_1(i)$ is orthogon to the reflecting surface of the mirror.

b.The axis $\omega_2(i)$ passes along the left edge of the i-th petal through its upper and lower edges (Fig.4a,b)

Two types of links between petals are used.

c. Links of type 1 connect points a(i) and b(i+1) of neighbor petals (Fig.4a). These points are placed on the continuation of axes $\omega_2(i)$ and $\omega_2(i+1)$ at different distances to the foot of the petal. Hence links of tipe 1 do not produce moments of forces with respect to $\omega_2(i)$ axes.

d. Links of type 2 connect points c(i) and d(i+1) of neighbor petals (Fig.4c). The directions of these links are parallel to the axes $\omega_1(i)$, and point d(i+1) is placed on $\omega_2(i+1)$ axis. Hence links of type 2 do not produce moments of forces with respect to $\omega_1(i)$ axes.

Thus, in the linear approximation, motions around axes ω_1 and ω_2 are independent.

3. Mathematical model of petal-type structure dynamics

The dynamics of the disturbed motion of the unfolded petal-type mirror is described by a system of differential equations:

$$I_1 \frac{d^2\alpha_1(i)}{dt^2} = M_1(i) \tag{1}$$

$$I_2 \frac{d^2\alpha_2(i)}{dt^2} = M_2(i) \tag{2}$$

here n is the number of petals. Each petal is regarded as a solid body, the rotation angles $\alpha_1(i), \alpha_2(i)$ of the petal with respect to the axes $\omega_1(i), \omega_2(i)$ are read from the unfolded structure in the equilibrium state. I_1, I_2 are moments of inertia of the petal with respect to ω_1, ω_2 axes. Elastic damping peripheral links develop moments of forces M_1, M_2 when the mirror is changing its orientation.

The directions of the forces developed by the links are as in Fig.4, and their values - in the linear approximation - are:

$$f_1 = k_{11}[l_1\alpha_1(i+1) - l_2\alpha_1(i)] + k_{12}[l_1\dot{\alpha}_1(i+1) - l_2\dot{\alpha}_1(i)] \tag{3}$$

$$f_2 = k_{21}\alpha_2(i) + k_{22}\dot{\alpha}_2(i) \tag{4}$$

here k_{11}, k_{12} are elasticity coefficients of the links, k_{12}, k_{22} are their damping coefficients.

The moments of the forces acting upon the i-th petal are:

$$M_1 = k_{11}[l_1 l_2 \alpha_1(i-1) - (l_1^2 + l_2^2)\alpha_1(i) + l_1 l_2 \alpha_1(i+1)]$$
$$+k_{12}[l_1 l_2 \dot\alpha_1(i-1) - (l_1^2 + l_2^2)\dot\alpha_1(i) + l_1 l_2 \dot\alpha_1(i+1)] \quad (5)$$

$$M_2 = k_{21} l_3 \alpha_2(i) + k_{22} l_3 \dot\alpha_2(i) \quad (6)$$

Thus the equations of motion (1), (2) are independent, and the system (2) disintegrates into n independent subsystems. We will call the system (1),(5) the equation of mirror's frame dynamics and the system (2), (6) the equation of reflecting surface dynamics.

Later on the following notations are used.

a. The motion around the axes $\omega_1(i)$, $i = 1, ..., n$:

$\alpha_1(i) = x(i)$ - rotation angle of the i-th petal with respect to the $\omega_1(i)$ axes,

$\dot\alpha_1(i) = y(i)$ - angular velocity of the i-th petal with respect to the $\omega_1(i)$ axes,

$Z = (z(1), .., z(n))$, $z(i) = (x(i), y(i))$.

B. Motion around the axes $\omega_2(i), i = 1, ..., n$

$\alpha_2(i) = u(i)$ - rotation angle of the i-th petal with respect to the $\omega_2(i)$ axes,

$\dot\alpha_2(i) = v(i)$ - angular velocity of the i-th petal with respect to the $\omega_2(i)$ axes,

$W = (w(1), ..., w(n))$, $w(i) = (u(i), v(i))$.

4. Disturbed motion of the frame

Let us substitute (5) into (1) and rewrite the equation in the new notations:

$$\dot x(i) = y(i)$$

$$\dot y(i) = k_{11}[l_1 l_2 x(i-1) - (l_1^2 + l_2^2)x(i) + l_1 l_2 x(i+1)] \quad (7)$$

$$+k_{12}[l_1 l_2 y(i-1) - (l_1^2 + l_2^2)y(i) + l_1 l_2 y(i+)],$$
$$i = 1, ..., n;$$

$$x(n+1) = x(1); \quad y(n+1) = y(1); \quad x(0) = x(n); \quad y(0) = y(n).$$

We will show that at $l_1 \neq l_2$ system (7) has only a trivial solution

$$x(i) = 0; \qquad i = 1, .., n,$$

therefore, in the equilibrium state the frame is geometrically invariable. The equilibrium conditions lead to the system:

$$l_1 l_2 x(i-1) - (l_1^2 + l_2^2)x(i) + l_1 l_2 x(i+1) = 0 \quad (8)$$

$$i = 1, .., n; \quad x(0) = x(n); \quad x(n+1) = x(1).$$

Let us confirm that at $l_1 \neq l_2$ system (8) has only a trivial solution $x(i) = 0$. Written in a matrix form the system (7) looks as

$$Lx = 0$$

$$x = (x(1), ..., x(n))$$

$$L = \begin{vmatrix} -(l_1^2 + l_2^2) & l_1 l_2 & 0 & \dots & 0 & l_1 l_2 \\ l_1 l_2 & -(l_1^2 + l_2^2) & l_1 l_2 & 0 & \dots & 0 \\ 0 & \ddots & \ddots & \ddots & \vdots & \vdots \\ \vdots & \vdots & \ddots & \ddots & \ddots & \vdots \\ \vdots & \vdots & \vdots & l_1 l_2 & -(l_1^2 + l_2^2) & l_1 l_2 \\ l_1 l_2 & 0 & \dots & 0 & l_1 l_2 & -(l_1^2 + l_2^2) \end{vmatrix}$$

It is sufficient to show that the determinant of the cyclic matrix L differs from zero. We have:

$$DetL = \lambda_1 * ... * \lambda_n$$

Here λ_i - are the eigenvalues of matrix L. The eigenvalues of cyclic matrix may be determined in explict form [8]:

$$\lambda_s = -(l_1^2 + l_2^2) + 2Cos(s\gamma) * l_1 * l_2$$

$$\gamma = 2\pi/n$$

The following estimate is valid for $l_1 \neq l_2$:

$$\lambda_s \leq -(l_1^2 + l_2^2) + 2l_1 l_2 = -(l_1 - l_2)^2 < 0$$

Thus, all eigenvalues of the cyclic matrix L at $l_1 \neq l_2$ are not zero, system (7) has only a trivial solution and the design in the equilibrium state is geometrically invariable.

5. The damping of frame disturbances

Let us rewrite the system (7) in matrix form:

$$\dot{z} = Az \qquad (9)$$

The $(2n * 2n)$ matrix A has a block-cyclic form

$$L = \begin{vmatrix} A_0 & A_{n-1} & 0 & \dots & 0 & A_1 \\ A_1 & A_0 & A_{n-1} & 0 & \dots & 0 \\ 0 & \ddots & \ddots & \ddots & \dots & \dots \\ \vdots & \vdots & \ddots & \ddots & \ddots & \vdots \\ \vdots & \vdots & \vdots & A_1 & A_0 & A_{n-1} \\ A_{n-1} & 0 & \dots & 0 & A_1 & A_0) \end{vmatrix}. \qquad (10)$$

The $(2 * 2)$ blocks A_0, A_1, A_{n-1} are:

$$A_0 = \begin{vmatrix} 0 & 1 \\ -k_{11}(l_1^2 + l_2^2) & k_{12}(l_1^2 + l_2^2) \end{vmatrix}; \qquad A_1 = A_{n-1} = \begin{vmatrix} 0 & 0 \\ k_{11}l_1l_2 & k_{12}l_1l_2 \end{vmatrix}.$$

It is known [9], that by nonsingular linear tansformation the block-cyclic matrix A may be reduced to a block-diagonal form:

$$B = \begin{vmatrix} B_1 & 0 & \dots & 0 \\ 0 & \ddots & \dots & 0 \\ \dots & \dots & \ddots & \dots \\ 0 & \dots & 0 & B_n \end{vmatrix},$$

$$B_j = A_0 + \epsilon^j A_1 + \epsilon^{j(n-1)} A_{n-1}, \qquad j = 1, .., n,$$

where $\epsilon = e^{i2\pi/n}$ and matrices A and B have the same eigenvalues.

The numbers ϵ^j $\epsilon^{j(n-1)}$ are complex conjugate and $A_1 = A_{n-1}$, therefore the matrices $\epsilon^j A_1 + \epsilon^{j(n-1)} A_{n-1}$ and B_j are real.

$$B_j = \begin{vmatrix} 0 & 1 \\ k_{11}p & k_{12}p \end{vmatrix},$$

$$p = -(l_1^2 + l_2^2) + 2Cos(j\gamma) * l_1 * l_2, \qquad \gamma = 2\pi/n.$$

The eigenvalues of matrix B_j are the roots of characteristic equations:

$$det|B_j - \lambda_j E| = \lambda_j^2 - k_{12}p\lambda_j - k_{11}p = 0$$

and determine the rate of decay of j-th harmonics of disturbed motion of the frame. The eigenvalues of the system may be found in explicit form:

$$\lambda_j^{1,2} = (k_{12}p)/2 \pm \sqrt{(k_{12}p)^2/4 + k_{11}p}$$

The last formula permits to calculate the rates of disturbance decay when the parameters of the frame $(l_1, l_1, k_{11}, k_{12})$ are chosen.

6. Dynamics of the reflecting surface

By substitution (6) into (2) and rewriting the system in the new notations we get the equations of reflecting surface dynamics:

$$\dot{W}(i) = CW \tag{11}$$

$$C = \begin{vmatrix} C_1 & 0 & \dots & 0 \\ 0 & \ddots & \dots & 0 \\ \dots & \dots & \ddots & \dots \\ 0 & \dots & 0 & C_n \end{vmatrix},$$

where

$$C_j = \begin{vmatrix} 0 & 1 \\ k_{21}l_3 & k_{22}l_3 \end{vmatrix}, \qquad j = 1, 2, .., n$$

The rate of decay of initial disturbance of reflecting surface is determined by eigenvalues μ_j of marices C_j.

$$det|C_j - \mu E| = \mu^2 - k_{22}l_3\mu - k_{21}l_3 = 0$$

We have

$$\mu_j^{1,2} = (k_{22}l_3)/2 \pm \sqrt{(k_{22}l_3)^2/4 + k_{21}l_3}$$

The desirable velocity of surface damping may be achieved by appropriate choice of the parameters l_3, k_{21}, k_{22}.

7. Conclusion

The results obtained in the paper are given below:

1. The problem of disturbance quieting in petal-type structure is considered.

2. The new scheme of reflecting elements connection within the design is proposed.

3. The mathematical model of the dynamics of a petal-type structure is built.

4. The eigenvalues of the matrices which define the quickness of disturbances decay, are found in explicit form.

References

Kardashev,N.S. and Slysh,V.I(1986) *The Radioastron Project, in The Impact of VLBI on astronomy and geophysics*, Kluwer Academic Publishers, 1988, pp. 433 - 440.

Kardashev,N.S. et. al. (1995) Orbiting very long base interferometer (OVLBI) for radio and optical astronomy.*Acta astronautica,***37**, pp.271 - 278.

Helwig G. (1986), Analysis of high precision composite sandwich antennas. *Proc. 2nd ESA Workshop on Mechanical technology for antennas*, Noordwijk: Europ.Space Agency,.p.51-55 .

Tsutsumi N.S.and Kasahara A.,1986 Study on solid surface, deployable antenna reflector,*Proc. of the 2nd ESA Workshop on Mechanical technology for antennas*, Noordwijk: Europ. Space Agency, p.41-45.

Bujakas V.I (1998)., Shape Control and Kinematic Waves in Large Statically Determinate Space Structures, *Internationa Journal of Space Structures***13**, n 1, 13-21.

Bujakas, V.I. and Rybakova, A.G. (2000) *High precision deployment and shape correction of multimirror space designs*, Proc. IUTAM/IASS Deployable Structures Symposium, Cambridge, (Ed. S. Pellegrino and S.D.Guest,), Kluwer acad. publish.,2000, p 55-62.

Parkes,e.W.(1974) *Braced Frameworks*, Pergamon Press, Oxford.

Marcus,M. and Minc H. (1964) *A Survey of Matrix Theory and Matrix Iequalities*,Allyn and Dacon Inc., Boston.

Davis P.J. (1979) *Circulante Matrices*, J. Wiley, N.Y.

IN PURSUIT OF QUIETER SHIP DESIGN

V. BHUJANGA RAO
Associate Director
Naval Science and Technological Laboratory
Visakhapatnam 530 027, India

1. Introduction

The design and construction of a ship - whether a merchant ship, passenger ship, exploratory vessel, warship, or any other vessel - takes many years, encompassing many stages of design and construction, before entering into its chartered service. The importance of building acoustically silent ships needs no emphasis. It could be airborne noise reduction for better habitability, or reduction of underwater radiated noise for stealthy features. With the development of various acoustical materials, active noise and vibration cancellation techniques, various acoustic mitigation measures (Table 1), one can today design and construct much more quieter ships. This is effectively possible only if all acoustical aspects form an integral part of the vessel design from conceptual stage to construction stage. Preceding noise mitigation measures, one needs to predict the radiated noise, self-noise, and airborne noise of the ship under consideration at the design stage itself.

A number of computational tools using techniques such as Finite Element Method, Boundary Element Method, Statistical Energy Analysis, Regression models, etc. have been well developed in the past two decades. In figure 1, the typical frequency ranges are indicated which correspond with various motives for noise control in ships. Efforts for noise reduction onboard ships have the following backgrounds:

- High noise levels on bridge wings caused by the exhaust systems, wind forces, etc. may lead to difficult situations when navigation sound signals become masked.

- Within the ship, the noise may interfere with comfort and communication, whereas in engine rooms, it also presents a hazard with respect to hearing damage.

M.L. Munjal (ed.), IUTAM Symposium on Designing of Quietness, 79–97.

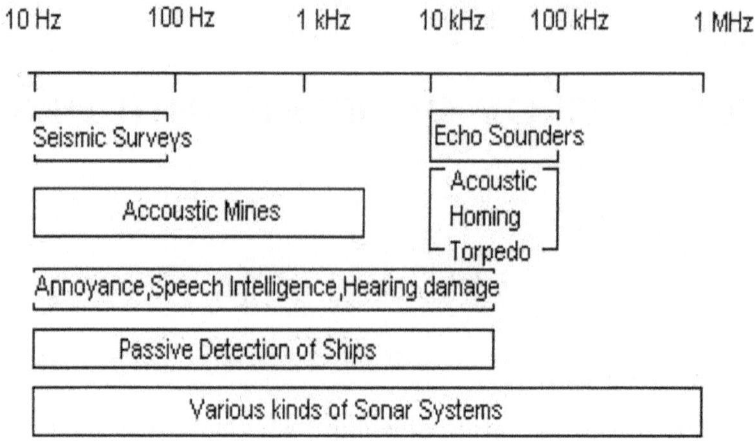

Figure 1. Typical acoustic frequency ranges of marine interest

Table 1 (Ref. 6)

SHIP QUIETING TECHNIQUES

EXTERNAL TO HULL	
• NON CAVITATING SKEW PROPELLERS	• ASTERN SHAPE FOR UNIFORM WAKE
• BUBBLE SCREENS/MASKER SYSTEMS	• HULL LININGS
• COMPLIANT COATINGS	• DECOUPLERS
• ACTIVE NOISE CANCELLATION	• HULL FORMS
• WATERJET PROPULSION	• MULTILAYER DOMES WITH BAFFLES
INTERNAL TO HULL	
• MACHINERY ENCLOSURES OR HOODS	• CLADDINGS/ANECHOIC LININGS
• CONSTRAINED LAYER DAMPING	• ACTIVE VIBRATION CANCELLATION
• COMPOSITE SHAFTS	• RAFT/FLEXIBLE MOUNTS
• ISOLATED COMPARTMENTS	• LOW NOISE AC DUCTS
• EXHAUST SILENCERS/ MUFFLERS	• DETUNERS
• IMPEDANCE MISMATCHES	• SILENT FLOORS
• MASTIC COATINGS	• FLOATING FLOORS
• DOUBLE WALLS	• ABSORBING COATINGS

- For naval vessels, fishery vessels, and seismic research vessels, low underwater sound levels are of importance. Underwater sound radiation degrades the sonar performance of a ship, and moreover in naval vessels, it increases the vulnerability. Acoustic mines may be actuated, and more generally, the probability of detection and being targeted by enemy's passive sonar systems increases.

This paper outlines various issues involved in quieter ship design with reduced acoustic signatures at different stages of a ship-building pro-

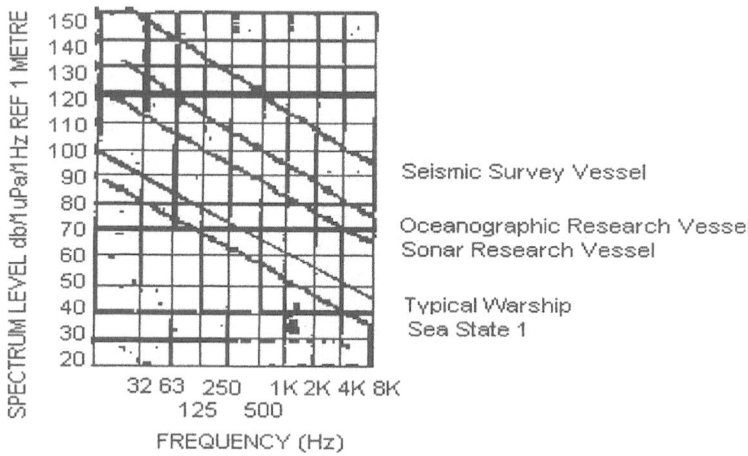

Figure 2. Underwater Noise limiting curves of different types of vessels compared to ambient noise at Sea State 1 (Ref.5)

gramme. The role of theoretical acoustics in this process is to help provide the means to carry out required optimisation at each stage of design.

2. Acoustical Design Considerations

Acoustical design constitutes the following major aspects of any marine vessel design.

2.1 Noise limiting curves

It is necessary to lay down 'limiting noise levels' or 'target levels' of radiated, self-noise and airborne noise of vessel under design. These levels form the basis for acoustical design. One way of setting these noise levels is based on historical data of different classes of ships and experience. In case of warships, one can stipulate these noise levels based on the premise: 'Enemy cannot detect it before it can detect the enemy'. Figure 2 refers to one such limiting curve in the design of different ocean going vessels. Importance is also to be given to achieving airborne noise levels shown in Table II from habitability considerations.

2.2 Noise level prediction models

With preventive shipboard noise control becoming an important technology field, a number of prediction techniques which can rightly be brought under one heading called 'Computational Acoustics' are avail-

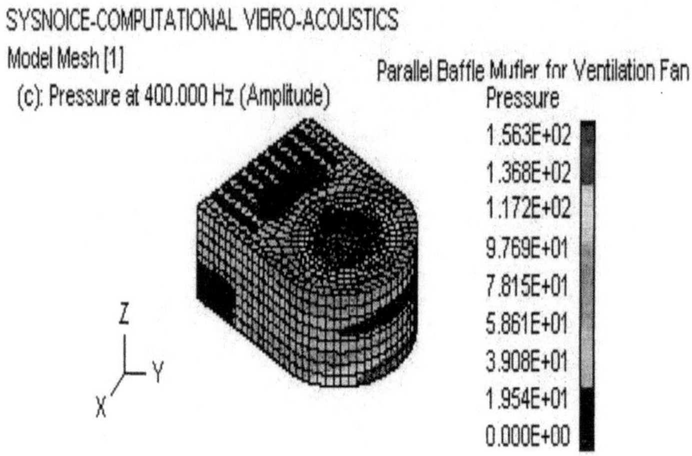

Figure 3. Prediction of silencer performance using FEA (Ref. 7)

Table 2 (Ref. 1)

	SWEDISH REG.	DANISH	WEST GERMANY	ISRAEL	USSR	DSRK EAST GERMANY	PRS POLISH REG. OF SHPG.	NORSK MARITIME DIRECTORATE	MARAD	USN	NKF RECS.	HS-7 RECS.
ACCOMMODATIONS												
CABINS, OFFICES	55	60	60	55	50	60	60	60	56	60	60	60
GALLEY, PANTRY, HOBBY ROOM	65	70						70	-			65
MESSES, DAYROOM	65	65	65	65	55 (DR)	60 (DR)	60 (DR)	70	56 (DR)			65
PASSAGEWAYS									62	75	80	NO REC.
BRIDGE												
BRIDGE WINGS	70	70	65					70	-			NO REC.
WHEEL HOUSE	65	65	60	60	55	60	65	65	-	55	60	60-65*
RADIO ROOM	65	65	60	60	50	60	60	65	-	60		60*
MACHINERY SPACES												
MANNED W/O CONTROL RM.	85	90	90	85	80	-	90	90	85	85	90	85
CONTROL ROOM	70	75		70	65	80	75	75	70-75		75	75
WORKSHOP & STORES	75	85		75	65	90	90	85	-			85

NOTE: All Noise Levels shown in dbA.
*65 w/Doors Open
60 w/Doors Closed

able today. They include Finite Element Analysis, Boundary Element Analysis, Statistical Energy Analysis,

analytical methods dealing with flow noise, design of mufflers, low noise ducts, regression models on structure-borne sound, empirical models on overall noise levels, etc. This list is by no means exhaustive. Except a few of them, most of these computational and analytical techniques are scattered in literature and not available as a single software package dedicated to the application of ship design. This is one aspect in which acoustics as an applied technology is lagging behind other computational areas such as CFD/CAD/CAM application softwares etc. Figure 3 is an example in predicting the performance of an intake silencer using FEM. Similarly, in predicting radiated noise levels of an engine room, SEA has been employed. (Figure 4). Figure 4 gives computed data for a typical case.

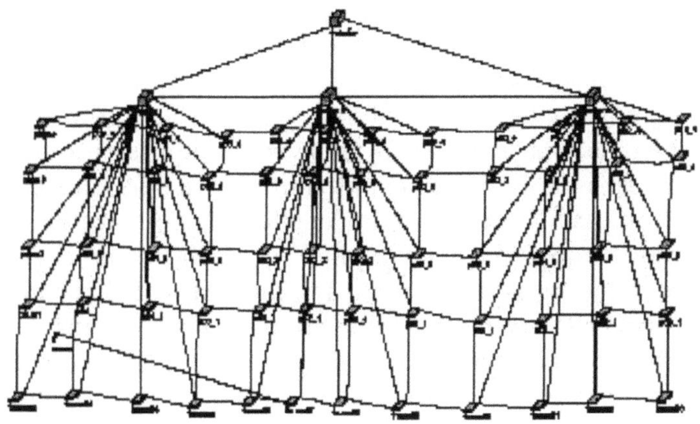

Figure 4. Engine Room of a ship - SEA Model (Ref. 7)

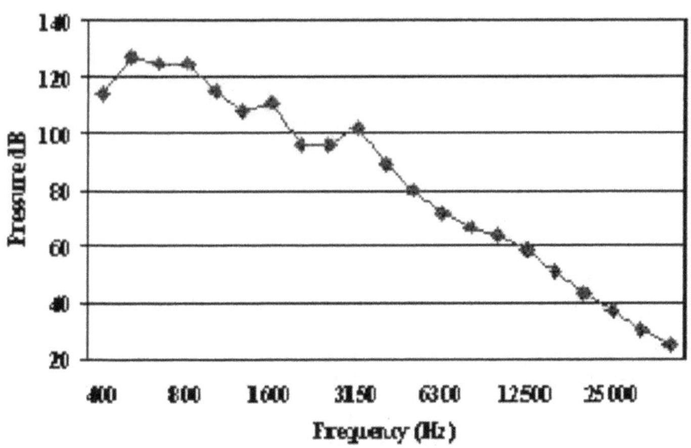

Figure 5. Radiated Noise Estimation using SEA Acoustic Pressure @ 50 m from Engine Room (Ref. 7)

Statistical Energy Analysis (SEA) and Finite Element Analysis (FEA) are to be used complementarily to cover the complete frequency range of noise prediction. SEA can go down to a low frequency of 250 Hz, and further below, one has to use FEA. SEA also can be used below 250 Hz, but for studying the trends and possible effects of damping on noise control.

2.3 Acoustic quieting techniques

Table 1, though not an exhaustive one, gives a list of techniques by which a ship's underwater noise can be controlled. All these techniques, some of them complementary to each other, need to be optimised, based on strong theoretical estimation, if one has to achieve the 'limiting values'. There are many aspects revolving around these quieting technologies. For example, one method of noise reduction is through damping layers. There is some indication that damping helps, and there is also some indication that it is not worth the effort. There are places in ships where structural damping can be applied successfully and there are other places where damping is just a dead weight. All these techniques need good insight into the noise generating sources, propagating mechanisms and radiations, etc. Otherwise they may lead to counter-productive results. Our understanding of these aspects is still not complete, because, there are still many questions such as low frequency behaviour, nonlinearities, instabilities, radiation problems, flow noise effects, etc which need a lot of future research. Notwithstanding these aspects, the cheapest way of reducing noise is to use all possible means that require no acoustical treatments. This will require careful thought at the earliest stage in the design of the ship. For instance, the quietest machinery should be used, mounted in the most satisfactory way. At the same time, it should be realised that the designer may not have the freedom of action he would like.

2.4 Model experiments

In the design process, model tests are needed in certain areas of acoustics to check design calculations and to predict phenomena for no reliable computational tool exists. Model experiments are reliable methods to predict the various aspects of acoustical performance of a ship before it is built. In the preliminary design phase, use can be made of rules of thumb or empirical methods based on results of previous models. In view of the cost and time involved, the amount of model testing in the early stages of design is usually restricted, unless important innovations are involved in the design. A physical scale model, a numerical model, and a regression model are models of full-scale physical reality. All models represent only a limited part of the real system and are based on a conceptual model. Propeller noise, flow noise, structure-borne noise attenuation, characterisation of acoustical materials, silencers, study of efficacy of active noise/vibration cancellation, etc can all be studied using these models. In addition to purely physical and purely numerical models, there are many kinds of hybrid models, which are numerical models based on a mathematical description of physical phenomenon supplemented with measured data.

2.5 Involvement of shipbuilding yards

The most important aspect is that 'noise-consciousness' must permeate the whole shipbuilding process, not only throughout design but also into construction shipyard. It is not sufficient for the design office, which includes the acoustical consultant to specify noise reduction measures, if they are not carried out properly into shipyards. The noise reduction measures should be properly supervised and checked by experienced and qualified shipyard staff. The fact that noise reduction at the design stage is cost-effective but will be expensive or even impossible in an already built ship should be appreciated. Shipbuilding being a matter of experience and skill, development of a new ship type is a gradual and extremely slow process. So is the case with acoustical design.

2.6 Full scale noise measurements

Feedback from full-scale noise measurements will provide the ultimate validation of both computational and scaled down physical models. Full-scale data encompassing all aspects of noise are scarce because these measurements of real ships are expensive, difficult and time consuming. With modern hardware, it becomes feasible to put dedicated equipment onboard for continuous monitoring of all noise performance characteristics. This will provide the ship not only the validation of the models and design tools, but also provide better insight into the relative importance of the different noise sources. By periodically monitoring the noise levels, the 'acoustic hygiene' of the vessel can be assessed at any instant in any regime of operation. However, sensors and the equipment are to be carefully selected and installed at appropriate locations onboard ship during the ship's construction. This needs careful planning during detailed design stage of the ship.

To give an example, hydrophones can be fitted flushed to the hull structure at the stern to measure the cavitation noise, and microphones in the engine room can measure the machinery noise. On an average,

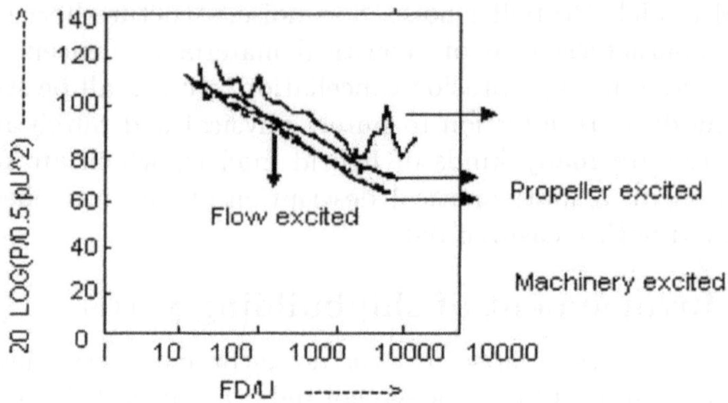

Figure 6. Full-scale noise measured for a typical ship (Ref. 7)

nearly 100 to 150 sensors are required to compute the acoustic fields of the ship. Figure 6 shows a plot of full-scale noise measured in a specially built noise range for a typical ship.

2.7 Structural design to overcome transmission paths

Once noise energy is introduced into a structure as complex as that of a ship, there is very little that may be done to attenuate it effectively. It is also a difficult measurement proposition to determine how much energy is flowing via any one particular transmission path. All transmission paths cannot be severed and isolation of noisy sections achieved without sacrificing structural integrity, and so, the only measure that may be utilised is the application of structural damping.

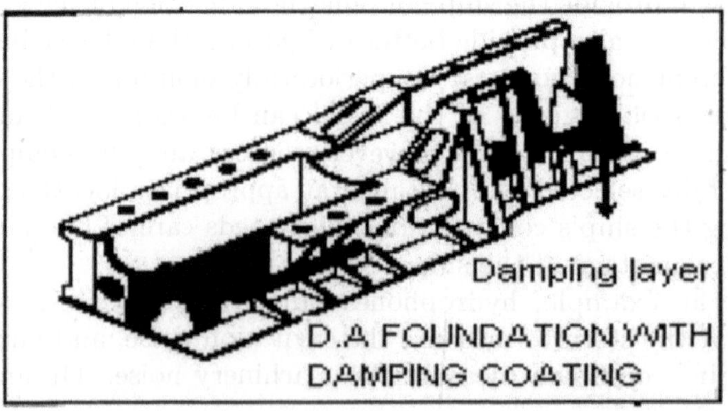

Figure 7. Diesel Alternator Foundation with damping coating (Ref. 7)

The best use of structural design with regard to noise control is in the design of machinery foundations and seatings such that they are as stiff as possible, particularly beneath machinery connection points. This will generally minimise the amount of noise entering the structure, particularly if resilient mounts are utilised. (Fig.7). Because of the multitude of sound transfer paths in ships, noise isolation is most effective in regions close to the sound source as well as the receiving space.

2.8 Vibration Isolation

For the purpose of noise and vibration control, it is essential to minimise the transmission of vibrational energy from machinery to ship structure as far as can be accomplished, within the practical limits of machinery primary functional requirements. Because of the contribution of structureborne noise, it would be inadequate to treat only the walls of the machinery spaces. The empirical formulae which has been found useful in estimating power transmission loss between deck and engine is

$$\begin{aligned} IL &= 19 + 5logf & \quad 22 \langle f \langle 707 \\ &= 22 + 5logf & \quad 708 \langle f \langle 2828 \\ &= 25 + 5logf & \quad 2829 \langle f \langle 11313 \end{aligned}$$

where f is frequency in Hertz.

2.8.1 Double Stage Mounting System with Acoustic Enclosure. If very low levels of fluctuating forces are required to be transmitted to the hull via a mounting system over a wide frequency range, a two-stage mounting system of type shown in Figure 8 is to be considered. The expected isolation one can achieve from theory is shown in Figure 9.

2.9 Propeller noise

As far as propeller noise design goes, it is reasonable to suggest that the problem is one for hydrodynamicists rather than acousticians, in that, once cavitation is present in any of its numerous forms, any typical water-borne noise specification is likely to be exceeded. The first problem in designing ship propulsion characteristics to meet water-borne noise specifications should be that of ensuring that cavitation is not present. Data-based prediction models are available to estimate the noise from both cavitating and non-cavitating propellers. But, the crucial design problem is that of avoiding cavitation, which usually results in a very sudden increase in noise levels of 20 to 30 dB across a wide frequency range from 100 Hz to 10 kHz. As an alternative solution to noisy pro-

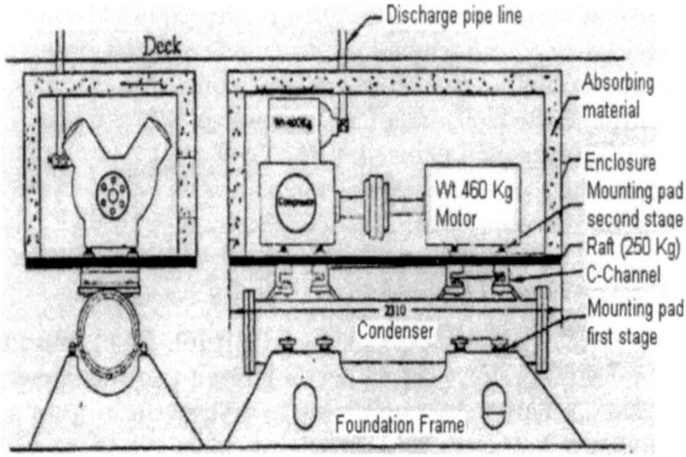

Figure 8. Two-stage Mounting system for a shipboard machine (Ref. 7)

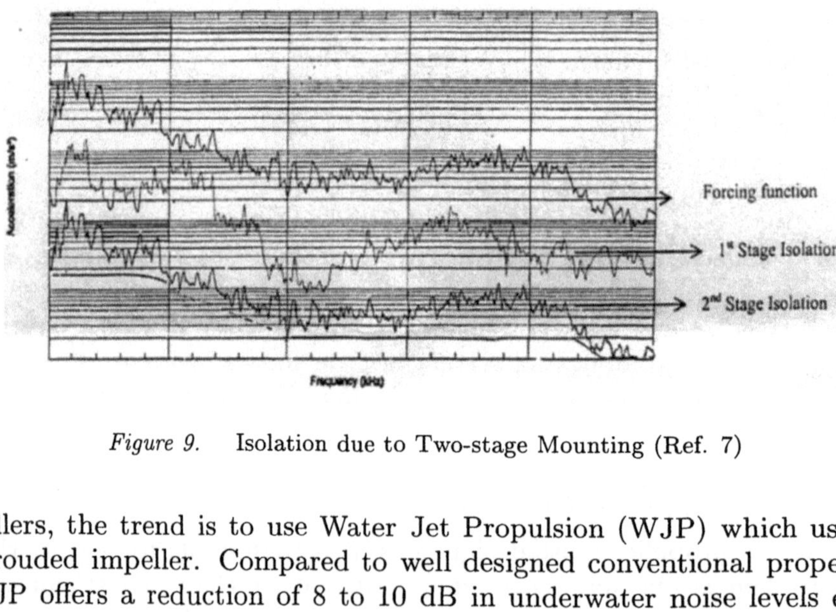

Figure 9. Isolation due to Two-stage Mounting (Ref. 7)

pellers, the trend is to use Water Jet Propulsion (WJP) which uses a shrouded impeller. Compared to well designed conventional propeller, WJP offers a reduction of 8 to 10 dB in underwater noise levels even at high speeds, where conventional propellers normally cavitate. From airborne noise point, the noise due to water jet propeller is from casing vibration. The empirical formula used to estimate airborne noise level is

$$L_p = C_p + A log H_p$$

Where C_p is a function of frequency octave band (given in table below) and H_p is the horsepower of the pump. The large differences at high horsepower level are due to the different values of the coefficient A, which

Freq. (Hz)	32	63	125	250	500	1000	2000	4000	8000
C_p	61	61	62	62	63	65	62	59	54

varied between 7 and 10. Because of the effort expended to develop a cavitation-free, efficient pump design, a numerical average of 12 was used as a conservative estimate for the horsepower dependency of the pump noise.

2.10 Radiation from hull structure

If a point force is exciting the hull as sketched in Figure 10, acoustic radiation takes place at ultra-low frequencies emanating from a large area of the hull. This is to say, the total hull participates in vibration at ultra-low frequencies.

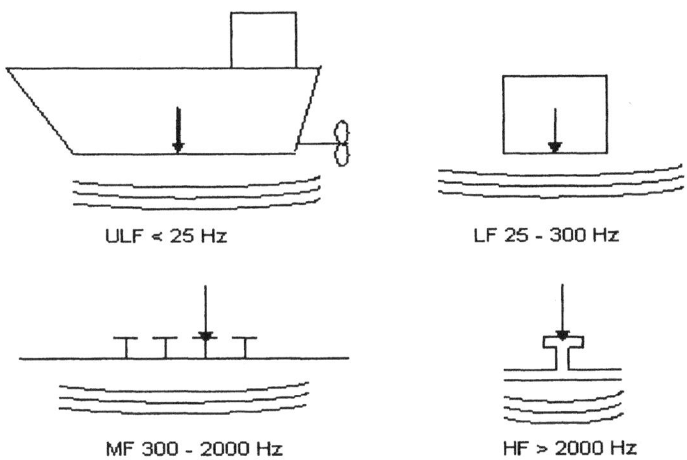

Figure 10. Radiation due to point excitation of hull

At low frequencies (LF), the radiation is largely confined to a compartment and shell flexural vibrations are dominant, the baseline level being nearly twice that of the previous value. At medium frequencies (MF), the radiation is modified by the stiffening frame structure, and plate type flexural vibrations of the shell and frames are dominant. At high frequencies (HF), coincidence lobe radiation from supersonic wave motion on the plate surface is dominant and the radiation is from a small region of the hull. At low frequencies, very little of the energy that gets into the shell structure is radiated, at high frequencies most of the energy is radiated.

At ultra low (ULF) and low frequencies, where the level of noise is dictated by the force exciting the hull, the use of isolation in the transmission path is a vital element in reducing the low frequency noise. As one goes up in frequency progressively, localisation of the source of sound radiation allows damping materials and decoupling coatings to be effective in reducing the radiated noise.

2.11 Structureborne sound due to acoustic excitation and transmission to adjacent compartments

Knowing airborne noise power levels, structure-borne noise as well as spectral density of machinery random vibration can be predicted.

The ratio, R, of aero-noise power to intrinsic structure-borne noise power is derived from the following semi-empirical formula:

$$R \approx \tfrac{1.5}{\rho c}\langle \rho_s \rangle \langle t \rangle \langle v + 300 \rangle$$

where
ρ = density of air
c= velocity of sound in air
ρ_s = average density of the machine casing material
$\langle t \rangle$ = average thickness of machine casing
ν = frequency, Hz
Any consistent set of units may be used in the above equation, except that c must be in distance/seconds.

The structureborne noise analysis is much more complicated and less well understood than its airborne counterpart. The ability to predict structure noise power based on the airborne noise power of a machine is a major step in the understanding of this important area. In addition to the intrinsic structure-noise power, three additional calculations must be completed to relate machinery noise to sound power radiated into a receiving space. These are
a)Power insertion lossof the machinery support structure
b)Acoustical power flow through the ship structure
c)Radiation efficiency of the vibrating panels

The radiated power P_{rad} is obtained from the expression

$$\frac{P_{rad}}{P_{SB}} = \frac{2\rho c \sigma_{rad}}{\rho_s 2\pi f \eta_t + 2\rho c \sigma_{rad}}$$

where
ρ_s = surface density of plate
P_{SB} = structureborne noise power

σ_{rad} = radiation ratio
ρ = density of surrounding medium
c = speed of sound in surrounding medium
f = frequency

The above equation is dimensionless and hence, any consistent set of units may be used.

Noise intensities from both the inside and outside paths are influenced by bulkhead transmission loss. The outside airborne path is additionally influenced by the source proximity, mufflers, and shell plating transmission loss. Transmission Loss (TL) of a simple panel is primarily limited to the surface density, Ms , and is approximately

$$TL = 23 + 15 log M_s$$

for the frequency bands of 125 through 2000 Hz.

2.12 Noise transmitted between compartments

All noise sources onboard ship are isolated from parts of the ship structure using anti-vibration mounts. Although primarily used to reduce vibrations being transmitted through the structure to the hull where they would be radiated as overboard noise, these mounts effectively decouple the machines from the structure at audio frequencies reducing noise transmission to adjacent areas. The level of noise transmitted between two compartments is given by (Ref. 4)

$$L_{p2} = L_{p1} - 6 + 10 log S_B - R - 10 log A_2$$

where

L_{p1} = Sound Power Level of source area
L_{p2} = Sound Power Level of receiver area
A_2 = Acoustic absorption of receiver area
S_B = Area of compartment
R= Sound Reduction Index

2.13 Ventilation Noise

Noise entering a ventilation supply or exhaust lead can be transmitted to many compartments throughout a ship. To calculate the level of sound transmitted between compartments via ventilation ducts, one must take the attenuation of noise along ducts, the generation of flow noise and the effects of break-in and breakout noise into account. The noise generated by flow of air through ducts is given by (Ref. 4)

$$L_{W1} = K + 10 log S_X + 55 log \nu - 45$$

Reducing the flow rate of air and smoothing constrictions within the ducts will reduce flow noise. Noise that breaks-in is calculated by (Ref. 4)

$$L_{w2} = L_{P1} - R_D + 10log\left[\frac{1}{4} + \frac{S_W}{A_D}\right] + 10logS_X$$

Breakout Noise is given by (Ref. 4)

$$L_{w2} = Lw1 - R_D + 10log\left[\frac{S_W}{S_X}\right]$$

where
L_{W1} = Sound Power Level, Source Area
L_{W2} = Sound Power Level, Receiver Area
S_W = Duct Cross Section S_X = Duct Surface Area
R_D = Sound Reduction Index, Duct Wall
A_D = Acoustic Absorption within Duct
K = Duct factor (dB)
ν = Air flow velocity (m/s)

2.14 Acoustic Enclosures

Of late, acoustic enclosures for shipboard machinery are being used to reduce not only airborne noise levels, but also underwater-radiated noise levels. An acoustic enclosure designed jointly by NSTL and M/s. Noise and Heat Control Systems Pvt. Ltd., Bangalore, is shown in Figure 11.

Figure 11. Acoustic Enclosure for shipboard machinery (Ref. 7)

The enclosure is made of sliding doors to facilitate easy access to the machine during emergency repairs.

2.15 Flow Noise

Flow noise is an important parameter that increases the self-noise of the ship's sonar, especially at higher speeds of the ship. The turbulent flow on the ship's sonar dome excites its structure, leading to acoustic radiation into the interior of the sonar dome. There are well-established numerical methods, using which one can estimate the flow noise of the sonar dome. The predicted value along with measured value of flow noise, for a typical ship at frequency of operation of sonar is given in figure 12.

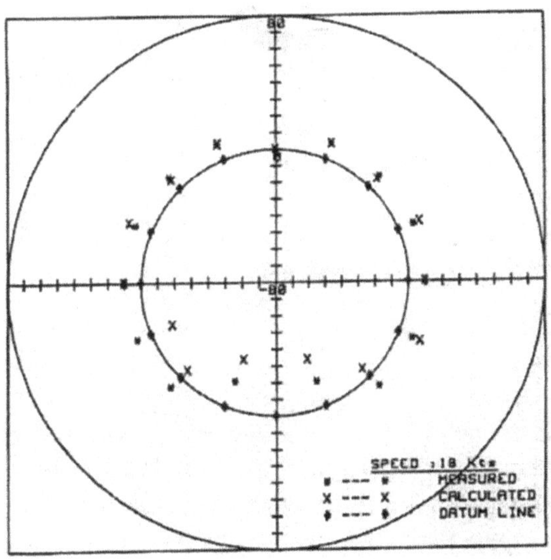

Figure 12. Flow Noise of ship's sonar dome - Computed vs. Measured values (Ref.6)

2.16 Acoustical design scheme of a ship

The arduous task of the ship designer in designing more and more silent ships is a highly demanding one. Ship silencing measures have certain cost. However, these costs can be kept to minimum if acoustical aspects are considered in the early design stages of the ship as integral part of the total design. Figure 13 shows the design spiral involved in acoustical design with attendant interactive processes.

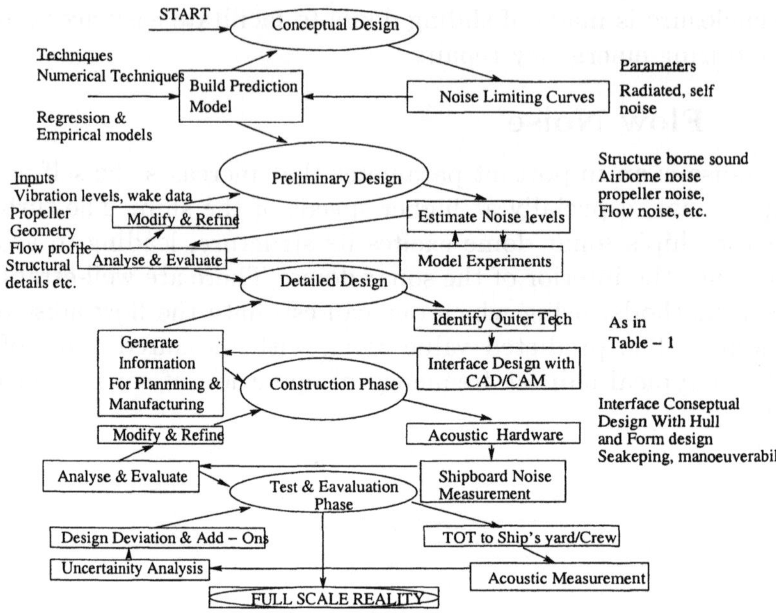

Figure 13. Acoustical design - An International Process (Ref. 6) values (Ref. 6)

3. Concluding Remarks

This paper has highlighted various issues involved in the acoustical design of silent ships. A number of noise prediction schemes have been briefly outlined. Apart from underwater noise, environmental and operational aspects also need to be considered. It is needless to emphasize that even high airborne noise levels can result in habitability problems, thereby reducing the efficiency of workforce. Continued research in ship acoustics and numerical techniques over the last decade has resulted in increasingly powerful and accurate computational codes for the analysis of noise aspects of ships. Modern CFD/CAD/CAE etc. softwares allow the joint consideration of hull design, powering performance, sea keeping and maneuvering in early stages of ship design. Integration of Computational Acoustics softwares will enhance the possibilities of optimisation of ship design from an acoustic point of view.

Acknowledgements

The author would like to thank Rear Admiral S. Mohapatra, Director, NSTL, for giving permission to present this paper. Thanks are also due to Shri. KVVSS Murty, Shri. PVS Ganesh Kumar, and others of Vibration Group, NSTL, for the technical support. The author is also thankful to Mr. YVN Chandra Sekhar, for his help in preparing this manuscript.

References

Noonan, EF., "State of the Art for Shipboard Vibration and Noise Control", SNAME, *Ship Vibration Symposium*, **Oct 1978**

Gary, L. Fox. "Noise and Vibration Control Techniques for the US Navy 3000LT Surface Effect Ship", *Marine technology* **Vol 17, No. 1, Jan 1980.**

Naidu, NS. et. al., "Shallow Water Ambient Noise Measurements", *National Symposium of Acoustics*, **1997.**

Clampton, SJ., "Identification of Airborne Noise Problems onboard Submarines", *Undersea Defence Technology*, **1994**,p . 172-175.

Morrow, R.T., "Noise Design Methods for Ships", *Meeting of the Royal Institution of Naval Architects*, **April 1988**

Bhujanga Rao, V., "Underwater Acoustical Planning as an Integral Part of Design of Naval Vessels - Some Strategies", *UDT Pacific98*,**1998**,Sydney, Australia

Technical Reports of Naval Science and Technological Laboratory, Visakhapatnam,**1980-2000**

ACTIVE CONTROL OF STRUCTURE-BORNE NOISE IN HELICOPTER CABIN TRANSMITTED THROUGH GEARBOX SUPPORT STRUT

S. GOPALAKRISHNAN and D. ROY MAHAPATRA
Department of Aerospace Engineering, Indian Institute of Science, Bangalore 560012, India
krishnan@aero.iisc.ernet.in

Abstract A model based study on active control of multiple wave transmission in helicopter gearbox support struts is presented. The problems associated with transmission of structure-borne noise into helicopter cabin from its rotating components such as main rotor hub, gearbox, tail rotor etc. pose severe constraints in the service life of the structural components and on-board instrumentation. There is also a low frequency disturbance due to transient aerodynamics surrounding the rotor plane in addition to the steady state rotor harmonics. In this paper the analysis of active gearbox support strut mechanics is carried out and related complexities in modeling the closed-loop structure-control interaction is addressed. Unlike the previously available studies, finite cylindrical strut with distributed actuator dynamics and near-field effects on the sensors are considered. The computational strategy is based on a recently proposed model called Active Spectral Element Model (ASEM). This model, which is a displacement based, finite element type discretized model in the Fourier domain, can be used for developing "realistic" models of integrated active strut systems on which the gearbox is mounted. The development of a multi-input-multi-output feedback controller based on ASEM is also presented along with a discussion of the numerical implementation. Different cases illustrating control of axial wave, flexural wave, and axial-flexural coupled wave transmission through gearbox support struts are presented. The importance of considering actuator dynamics, coupling between axial, flexural and torsional waves, and actuator-sensor configurations are brought forth in this work. The efficiency of a closed-loop scheme based on velocity feedback is also demonstrated. This study shows that such accurately modeled mechanics integrated with control system parameters appears to be far better behaved when compared with the control system purely based on estimation of error signals.

M.L. Munjal (ed.), IUTAM Symposium on Designing of Quietness, 99–128.

Keywords: Structure-borne noise, Gearbox strut, Wave transmission, Path treatment, Active control, Spectral element, Sensor-actuator configuration, Open-loop, Closed-loop.

1. Introduction

Reduction of structure-borne noise in helicopter cabin has received considerable attention over the last two decades. With several strategies to interior noise reduction, passenger comfort as well as operational ease have been focused along with enhancement of structural fatigue life and performance of on-board precision instruments. In many large helicopters, it is found that the main rotor and the gearbox mechanisms serve as the main sources of persistent acoustic disturbances while quick changes in gear teeth movement and aero-elastic events associated with maneuver transitions can result in transient acoustic disturbances. In many of the helicopters, the entire gearbox assembly is mounted on a set of vertical and inclined struts (e.g., Staple and Wells [1], Vignati [2]. The transient and persistent disturbances generated by the gearbox mechanisms as well as the main rotor are transmitted through the struts into the fuselage; this transmission is one of the main sources of structure-borne noise in a helicopter cabin. The frequency components associated with this noise lie in the bandwidth extending up to 6 kHz (e.g., Balachandran and Sampath [3], Pelinescu and Balachandran [4].

Among many approaches pursued to realize cabin noise reduction, control of waves transmitted through the noise path such as struts has been found to be viable. An early initiative in this regard is due to Westland Helicopters Ltd. During the late 1980s, they demonstrated that considerable reduction of the average vibration level can be achieved by using an Active Control of Structural Response (ACSR) system at the blade passing frequency of 17.5 Hz (Reference [1,5]. An electro-hydraulic actuator is used in this low frequency application. Recently, application of active wave control in the gearbox support strut at high frequencies was carried out in a major group effort as a part of the Reduction of Helicopter Interior Noise (RHINO) project. This project involved Westland Helicopters Ltd., Agusta, and others (Sutton *et al.* [6]. As reported in this work, an actual helicopter strut was tested in a test rig under realistic loading conditions. A primary source in the form of a shaker was used to excite the system at the upstream end-plate of the strut, and a group of three secondary sources in the form of magnetostrictive actuators were mounted in parallel to the strut. The choice of the driving voltage inputs (i.e., amplitudes and phases) to the actuators was based on minimizing the sum of the squared responses measured by six error sensors at the downstream end-plate. The location of the compliant ring for the secondary actuator group housing was chosen based on earlier studies on structural wave control in ducts and beams by Nelson and Elliott [7], Elliott and Billet [8], and Brennan *et al.* [9]. As reported in the study [6], an adaptive multiple actuator feedforward control scheme produced energy attenuation of up to 40 dB at the downstream interface of strut over different discrete frequencies in the range extending up to 1.25 kHz. Development of an Active Noise Control (ANC) system for S-76 helicopter by Sikorsky Aircraft Corporation has also been reported (Yoerkie *et al.* [10] and Millot *et al.* [11].

Although active materials integrated finite-length struts have been used extensively in experimental studies of full-scale systems, related analytical efforts have primarily focused on infinite-length struts. Recently, Pelinescu and Balachandran [4] developed integrated mechanics models for finite-length struts and used them for open-loop studies of longitudinal and flexural wave transmission. In their models, the strut is treated as a homogeneous cylindrical waveguide, and the actuators are also modeled as wave guides. In the related study of Ortel and Balachandran [12], these models are used to develop closed-loop schemes for control of flexural wave transmission in finite-length struts. Although these studies represent progress towards analytical development of an integrated active system, in order to use them for realistic support strut system, the following features need to be taken into account in the modeling: (a) effects of structural boundaries, joints, and other local discontinuities that can be important in the high-frequency range, (b) static axial and transverse loading, and (c) coupling between longitudinal waves and flexural waves. In addition, the developed models need to be amenable for establishing schemes for controlling wave transmission over a broad frequency range. However, models of realistic systems inclusive of features such as that mentioned above are not amenable to analytical solutions. In this particular context, recently, the authors have proposed a frequency domain based finite element methodology integrated with multi-input-multi-output (MIMO) generalized PID feedback algorithm and is called Active Spectral Element Model (ASEM) [13]. The efforts of the current work have been directed towards developing models of active integrated systems inclusive of realistic features and computational means to find numerical solutions.

Here, in developing the computational models, a set of partial differential equations (PDEs) is first used to describe the spatial continuum of interest and the spectral methodology is used to reduce the set of PDEs to a set of ordinary differential equations (ODEs). The reduction is carried by using Fourier transformation of the dynamic variables. Assuming wave-type solution for the set of ODEs, the wave number corresponding to each wave component at each discrete frequency is determined. Many applications of this method in structural wave propagation problems can be found in reference [14]. Analysis of wave propagation in finite rods and beams by using this spectral method has been carried out by Doyle [15], Doyle and Farris [16], and Gopalakrishnan *et al.* [17]. Also, similar methods based on Fourier domain representation and the transfer function concept were developed by Miller and Flotow [18] and Flotow [19]. These transfer function based methods were later extended to adaptive controlof structural waves for minimizing the transmitted power flow through one dimensional elastic members [20,21]. For distributed structural networks, the concept of low authority control of traveling waves has also been developed and studied by using a feedback compensator [22]. However, the aspects of distributed sensing and actuation and issues such as actuator dynamics have not been addressed in the above studies based on the transfer function concept.

From the abovementioned studies, it is noted that as complexity in structural geometry and applying the transfer function analysis increases, the control analysis involves extensive computations. On the other hand, the spectral methodology can be efficient from a computational viewpoint due to its generalized assembly technique and easy post-processing. These attributes have been important for choosing spectral element models for the active finite-length strut problem in this study. However, one has needs to choose appropriate mechanics model without loosing any important information of the actual behavior of the strut system. Since, here we bring the issues related to high frequency behavior, the simple beam model may not be sufficient to gain complete physical insight, since these struts are generally monolithically constructed hollow thick cylinders with winged cross section at their bearing ends. Such configuration may cause considerable ovaling and warping of the cross section at high frequency regime. Another, futuristic aspect is that to substitute the conventional aluminum struts by composite struts. The major advantage here is the structural stiffness-to-weight ratio. But the immediate challenge becomes the rationalization between complicated analysis and rapid input of crucial information for reliable control system performance.

In the next section, first the mathematical model is developed by appropriate judgement based on three dimensional configuration of a thick cylindrical strut. In Section 4, the formulation of ASEM is presented along with discussion of computational complexities involved. Next, in

Section 5, the case studies on active broad band feedback control of waves transmission at the strut-fuselage interface are carried out. Finally, the paper is concluded with excerps of the special features of the overall strategy, its limitations and further improvement.

1.1 Active Local Control for Structural Wave Guides

Tremendous progress in the field of structural dynamics and its control related technologies have changed the perspective of looking at the problems of controlling waves. In recent time, the reported literature shows a successful materialization of the concept of wave control in many fields of science and engineering. Although earlier studies on structural wave propagation in infinite or semi-infinite rods, beams and plates form the basis for understanding the overall system behavior, they are not adequate for modeling of the performance of the integrated structure-control system. This is because, structural boundaries and material interfaces may produce distortion and scattering of the original traveling waves and additional reflected waves whose magnitudes may be of the order of the source wave magnitude. Consideration of the trapped energy in a structural member will be an important factor for a control system, designed for spatially local attenuation of kinetic energy components at different frequencies. In the studies [20,21] on finite beams, it is predicted that a single wave (axial type or flexural type or torsional type) can be controlled completely by minimizing the power flow corresponding to the control force with the associated degree of freedom (DOF). But in many situations, considerations such as actuator dynamics and coupling between different types of waves may require one to use configurations of multiple actuators to simultaneously control transmission of multiple waves.

As a special interest in the present study, simultaneous control of axial and flexural waves in a finite-length strut by using a single actuator is addressed and the optimal performance that can be realized is discussed. As suggested by Gardonio and Elliott [21], coupling between axial and flexural waves can be modeled by an equivalent asymmetric scattering termination for one dimensional structural members. An experimental study [21] supporting this model reveals that control of one wave type (axial or flexural) results in standing waves in both types, if scattered waves from terminations are not effectively controlled. Since, exact coupling between various wave types was not studied in available literature, it was concluded in the study [6] that effective control of all the waves would require at least one actuator per wave type. An experimental investigation supporting this conclusion can be found in reference [23]. The experimental study of Pelinescu and Balachandran [24] also indicates that coupling between axial and flexural waves will be important to deal with finite-length struts with dynamic axial and transverse loading. In light of the above studies, it is noted that actual implementation of control of multiple waves (axial and flexural waves) in a helicopter gearbox strut is a challenging task. Here, an attempt is made to address some of the basic modeling complexities in carrying out this task.

2. Modeling the Mechanics of Finite-Length Cylindrical Strut

From observations in practice, helicopter gearbox support struts can be idealized as monolithic cylindrical shells with appropriate end sections to support bearings. In the study of Sutton *et al.* [6], it is reported that the ring frequency of the EH101 strut is typically 22.5 kHz. A high ring frequency is indicative of the dominance of torsional and warping modes in the overall response. Based on the disturbance frequency range, it was concluded in this study [6] that the torsional motions are unlikely to be excited to a great extent in practice and that the resonances associated with warping motions are too high to affect the control performance significantly. Hence, for simplicity, as in the study of Pelinescu and Balachandran [4], only axial and flexural modes of the strut are considered in the model development based on elementary beam with equivalent solid cross-sectional properties. The effect of coupling between these two types of waves has been studied in [13].

Here, the cylindrical strut is modeled as a finite length thin beam with a stiffness equivalent solid cross-section. Deviation in the wave propagation characteristics for this model is compared with a general 3D model for wave propagation in thick cylinder (Figure 1) with secondary warping and ovaling of its cross section. In order to allow for dynamic boundary conditions, the strut ends are considered to be in a free-free configuration. As noted in the study [4], the configuration is an idealization made to suit the analysis. This idealization allows to specify the loads transmitted at the gearbox-strut interface and strut-fuselage interface.

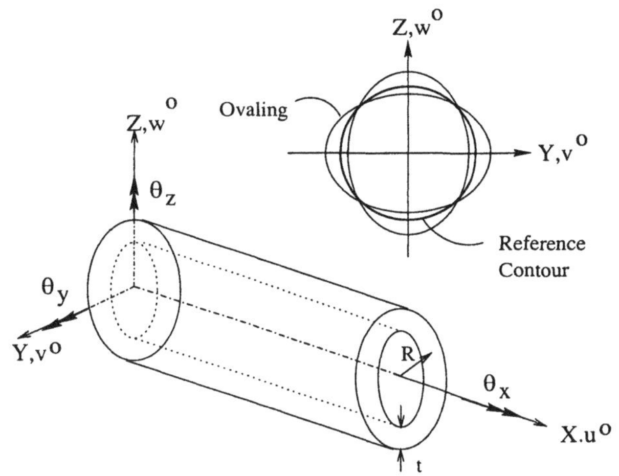

Figure 1. Configuration of a cylindrical strut segment with degrees of freedom

2.1 Active Strut with Magnetostrictive Actuators

Here, following the earlier works [4,13], development of a model for an active strut mounted with magnetostrictive actuators is considered. Instead of the feedforward control schemes considered in the study of Elliott and Billet [8] where the inputs to an array of secondary actuators are guided by the responses of upstream error sensors, here, a feedback scheme based on responses measured by downstream displacement or strain sensors are considered. This model based scheme requires accurate modeling of actuator induced strain in the host structure to capture the near-field effects, etc. With a good model, it is believed that the analysis and controller design effort may be carried out with a lower number of actuators than that required in the feedforward effort.

The linear coupled magneto-elastic equations for the magnetostrictive actuator (Butler [25]) are given by

$$S_1 = s_{11}^H T_1 + d_{11}^m H_1 \, , \qquad B_1 = d_{11}^m T_1 + \mu_{11}^T H_1 \, . \tag{1}$$

The significance of notations and their corresponding units are in accordance with the IEEE Standard [26]. Here, the direction 1 represents the axial direction of the rod shaped magnetostrictive actuator. In Equations (1), S_1 is the axial strain, s_{11}^H is an elastic compliance constant, d_{11}^m is a magnetostrictive constant, and μ_{11}^T is a magnetic permeability component. Furthermore, T_1, H_1 and B_1 are respectively, the axial stress, the magnetic field strength, and the magnetic flux density normal to the rod cross-section. It is assumed that a linear strain-magnetic field strength (S-H) characteristic can be used to describe the actuator. This can be realized by using an appropriate DC bias magnetic field H_0. Further details can be found in reference [25]. The design value of H_0 need to be determined from analysis. In case, the actuator specification does not meet this required value of H_0, then, the use of more number of similar actuators in a group can be an option.

2.2 Kinematics and Equations of Motion for the Cylindrical Strut Segment

Following the comments made in the previous literature e.g. [6], it is practical to neglect the effect of torsional coupling due to its very high natural frequency which is out of the frequency range of interest. However, in the present study, the effect of cross-sectional warping is considered as a possible cause of deviation in the actual response which can be ensured as done in the following derivations. Here, we can consider two displacement field models as
(1) Euler-Bernoulli Beam with equivalent properties:

$$u(x,y,z,t) = u^o(x,t) - zw(x,t)_{,x} \ , \qquad w(x,y,z,t) = w^o(x,t) \ . \tag{2}$$

(2) Cylindrical Shell with First Order Shear Deformation and secondary warping:

$$\begin{aligned} u(x,y,z,t) = u^o(x,t) + z\left[1 + \frac{n}{\sqrt{y^2+z^2}}\right]\theta_y(x,t) \\ +y\left[1 - \frac{n}{\sqrt{y^2+z^2}}\right]\theta_z(x,t) \ , \end{aligned} \tag{3}$$

$$v(x,y,z,t) = v^o(x,t) - z\theta_x(x,t) \ , \tag{4}$$

$$w(x,y,z,t) = w^o(x,t) + y\theta_x(x,t) \ . \tag{5}$$

where, $u, v, w, \theta_x, \theta_y, \theta_z$ are the axial, lateral and transverse displacements and three rotations respectively about the axis system (X, Y, Z). The terms with superscript o represents the primary displacements along the strut axis (shown in Figure 1). n is the normal distance of a material point from the mid-plane contour. Note that primary warping for a circular contour is zero and only secondary warping can occur. In the expression of $u(x, y, z, t)$ in Equation (3), the non-linear terms appear as the combined effect of bending and radial displacements which is termed as ovaling. Also note that in this higher order cylindrical bending model, displacement continuity at the the mid-plane contour is ensured. The bending rotations θ_y and θ_z are independent of the curvature and assumed constant throughout the cross section as in case of imoshenko beam model. Further detail regarding similar higher order cylindrical bending model can be found in [27].

Next, the constitutive relation is derived for the Euler-Bernoulli model which, as a subset, automatically reduces from the cylindrical bending case. The constitutive model in the coordinate system (X, Y, Z) can be written as

$$\left\{ \begin{array}{c} \sigma_{xx} \\ \sigma_{xz} \\ \sigma_{xy} \end{array} \right\} = [\bar{Q}] \left\{ \begin{array}{c} \varepsilon_{xx} \\ \gamma_{xz} \\ \gamma_{xy} \end{array} \right\} \tag{6}$$

where

$$[\bar{Q}] = [\Gamma]^T \left[\begin{array}{ccc} E/(1-\nu^2) & 0 & 0 \\ 0 & GK' & 0 \\ 0 & 0 & GK' \end{array} \right] [\Gamma] \,, \tag{7}$$

$$[\Gamma] = \left[\begin{array}{ccc} 1 & 0 & 0 \\ 0 & cos\phi & sin\phi \\ 0 & -sin\phi & cos\phi \end{array} \right] , \tag{8}$$

ϕ is the polar angle of a material point in the cross-sectional plane YZ. K' is the shear correction factor which can be computed from the the expression proposed by Cowper [28] as

$$K' = \frac{6(1+\nu)(1+t'^2)^2}{(7+6\nu)(1+t'^2)^2 + (20+12\nu)t'^2} \,. \tag{9}$$

Here, t' is the ratio of inner to outer radius. Now, in the process of deriving the governing equations of motion and force boundary equations, we come across many higher order stiffness and mass coefficients, which in compact notation, are expressed respectively as

$$[A_{ij}] = \int_0^{2\pi} \int_{R-t/2}^{R+t/2} \bar{Q}_{ij} \{\Psi\}^T \{\Psi\} \, r \, dr \, d\phi \,, \tag{10}$$

$$[M] = \int_0^{2\pi} \int_{R-t/2}^{R+t/2} \rho \{\Psi\}^T \{\Psi\} \, r \, dr \, d\phi \,, \tag{11}$$

where

$$\{\Psi\} = \{\ 1 \quad y \quad z \quad \bar{y} \quad \bar{z} \quad \bar{y}_{,y} \quad \bar{y}_{,z} \quad \bar{z}_{,y} \quad \bar{z}_{,z}\ \}\ , \tag{12}$$

$$\bar{y} = y\left[R/\sqrt{y^2+z^2}\right]\ , \qquad \bar{z} = z\left[2 - R/\sqrt{y^2+z^2}\right]\ , \tag{13}$$

$$y = r\,cos\phi\ , \qquad z = r\,sin\phi\ . \tag{14}$$

Now, the six wave equations derived from Hamilton's principle pertaining to six primary displacement components can be arranged as follows.

$$\begin{aligned}\delta u^o : 0 = M_{11}\ddot{u}^o + M_{15}\ddot{\theta}_y + M_{14}\ddot{\theta}_z - A_{1111}u^o{}_{,xx} - A_{1611}v^o{}_{,xx}\\ - A_{1511}w^o{}_{,xx} - (A_{1512} - A_{1613})\theta_{x,xx} - A_{1115}\theta_{y,xx} - A_{1114}\theta_{z,xx}\\ - (A_{1519} + A_{1618})\theta_{y,x} - (A_{1517} + A_{1616})\theta_{z,x}\ ,\end{aligned} \tag{15}$$

$$\begin{aligned}\delta v^o : 0 = M_{11}\ddot{v}^o - M_{13}\ddot{\theta}_x - A_{1611}u^o{}_{,xx} - A_{6611}v^o{}_{,xx} - A_{5611}w^o{}_{,xx}\\ - (A_{5612} - A_{6613})\theta_{x,xx} - A_{1615}\theta_{y,xx} - A_{1614}\theta_{z,xx} - (A_{5619} + A_{6618})\theta_{y,x}\\ - (A_{5617} + A_{6616})\theta_{z,x}\ ,\end{aligned} \tag{16}$$

$$\begin{aligned}\delta w^o : 0 = M_{11}\ddot{w}^o + M_{12}\ddot{\theta}_x - A_{1511}u^o{}_{,xx} - A_{5611}v^o{}_{,xx} - A_{5511}w^o{}_{,xx}\\ - (A_{5512} - A_{5613})\theta_{x,xx} - A_{1515}\theta_{y,xx} - A_{1514}\theta_{z,xx} - (A_{5519} + A_{5618})\theta_{y,x}\\ - (A_{5517} + A_{5616})\theta_{z,x}\ ,\end{aligned} \tag{17}$$

$$\begin{aligned}\delta\theta_x : 0 = (M_{22} + M_{33})\ddot{\theta}_x - M_{13}\ddot{v}^o + M_{12}\ddot{w}^o - (A_{1512} - A_{1613})u^o{}_{,xx}\\ - (A_{5612} - A_{6613})v^o{}_{,xx} - (A_{5512} - A_{5613})w^o{}_{,xx} - (A_{5522} - 2A_{5623}\\ + A_{6633})\theta_{x,xx} - (A_{1525} - A_{1635})\theta_{y,xx} - (A_{1524} - A_{1634})\theta_{z,xx} - (A_{5529}\\ + A_{5628} - A_{5639} - A_{6638})\theta_{y,x} - (A_{5527} + A_{5626} - A_{5637} - A_{6636})\theta_{z,x}\ ,\end{aligned} \tag{18}$$

$$\begin{aligned}\delta\theta_y : 0 = M_{55}\ddot{\theta}_y + M_{15}\ddot{u}^o + M_{45}\ddot{\theta}_z - A_{1115}u^o{}_{,xx} - A_{1615}v^o{}_{,xx}\\ - A_{1515}w^o{}_{,xx} - (A_{1525} - A_{1635})\theta_{x,xx} - A_{1155}\theta_{y,xx} - A_{1145}\theta_{z,xx}\\ + (A_{1519} + A_{1618})u^o{}_{,x} + (A_{5619} + A_{6618})v^o{}_{,x} + (A_{5519} + A_{5618})w^o{}_{,x}\\ + (A_{5529} + A_{5628} - A_{5639} - A_{6638})\theta_{x,x} + (A_{1648} + A_{1549} - A_{1557}\\ - A_{1656})\theta_{z,x} + (A_{5599} + 2A_{5689} + A_{6688})\theta_y + (A_{5579} + A_{5669} + A_{5678}\\ + A_{6668})\theta_z\ ,\end{aligned} \tag{19}$$

$$
\begin{aligned}
\delta\theta_z : 0 = {} & M_{44}\ddot{\theta}_z + M_{14}\ddot{u}^o + M_{45}\ddot{\theta}_y - A_{1114}u^o{}_{,xx} - A_{1614}v^o{}_{,xx} \\
& - A_{1514}w^o{}_{,xx} - (A_{1524} - A_{1634})\theta_{x,xx} - A_{1145}\theta_{y,xx} - A_{1144}\theta_{z,xx} \\
& + (A_{1517} + A_{1616})u^o{}_{,x} + (A_{5617} + A_{6616})v^o{}_{,x} + (A_{5517} + A_{5616})w^o{}_{,x} \\
& + (A_{5527} + A_{5626} - A_{5637} - A_{6636})\theta_{x,x} + (A_{1557} + A_{1656} - A_{1549} \\
& - A_{1648})\theta_{y,x} + (A_{5579} + A_{5678} + A_{5669} + A_{6668})\theta_y + (A_{5577} + 2A_{5667} \\
& + A_{6666})\theta_z \ ,
\end{aligned} \tag{20}
$$

Note that in Equations (10)-(11), in case of Euler-Bernoulli beam, the (1,1) components reduce to the axial stiffness and cross-sectional mass and (3,3) components reduce to bending stiffness and rotational inertia. Also note that, for material isotropy as in case of the aluminum strut, $[A_{15}] = [0]$, $[A_{16}] = [0]$, $[A_{56}] = [0]$. Therefore, by eliminating the shear terms associated with $[A_{55}]$, the above equations of motion can be reduced for uni-directional bending $w_{,x}$ in the form of Euler-Bernoulli beam model which is given by

$$\delta u^o : 0 = M_{11}\ddot{u}^o - A_{1111}u^o{}_{,xx} + A_{1113}w^o{}_{,xxx} + A^m_{1111}H_{,x} \ , \tag{21}$$

$$\delta w^o : 0 = M_{11}w_{,xx} - A_{1113}u^o{}_{,xxx} + A_{1133}w^o_{xxxx} + A^m_{1113}H_{,xx} \ . \tag{22}$$

Corresponding dynamic boundary conditions are obtained as

$$A_{1111}u^o{}_{,x} - A_{1113}w^o{}_{,xx} - A^m_{1111}H = N_x \ , \tag{23}$$

$$A_{1113}u^o{}_{,xx} - A_{1133}w_{,xxx} - A^m_{1113}H_{,x} = V_x \ , \tag{24}$$

$$-A_{1113}u^o{}_{,x} + A_{1133}w_{,xx} + A^m_{1113}H = M_x \ , \tag{25}$$

where, N_x, V_x and M_x are respectively the axial force, shear force and bending moment for the unidirectional bending in Euler-Bernoulli beam. In Equations (21) and (22), the derivatives of magnetic field $H = H(x,t)$ appear due to modification of the constitutive model (equation of transduction) to deal with rod type magnetostrictive actuators. Here, for a solid strut actuator, the mechanical actuation is induced through magneto-elastic coupling (Equation (1)) in the (1,1) elastic coefficient as

$$\sigma^m_{xx} = \bar{Q}_{11}(\varepsilon_{xx} - d^m_{11}H) \ , \tag{26}$$

where, the subscript m stands for the active material and Equation (1) can be replaced by appropriate equation for use of other active actuator materials such as piezoelectric stack. Note that in Equations (21) and (22), the stiffness coefficient associated with first area moment has been kept. Obviously, this becomes zero for the cylindrical strut segment made of isotropic material. But, successful experiments have been reported in literature [6,24] using piezoceramic patches and stacks. Therefore, in such cases, one has to consider the asymmetric placement of the actuator patches or stacks on the strut cross-section. Hence, it is convenient to assume nonzero A_{1113} to model generic actuator configuration.

The above derivations with their physical consequence encompass the required information to develop the computational model of the active coupled field problem. Here, we adopt the spectral analysis of wave motion as the basic approach to develop the ASEM.

3. Spectral Analysis and Characteristic Wave motion

The approach is basically to construct a set of ODEs from a coupled PDE which is derived in Section 2.2. Here, It is assumed that the electromagnetic scattering at the material interfaces and boundaries has negligible effect. Therefore, spatial variation of the magnetic field strength H vanishes in the Equations (21)-(22), and (24). Then, it can be said that the characteristics of the elastic and magneto-elastic domains as obtained from the governing differential equations become identical under the above assumption. Next, by using the discrete Fourier transform (DFT), the spectral representation of the primary field variables $U = (u^o, v^o, w^o, \theta_x, \theta_y, \theta_z)$ are obtained as

$$U(x,t) = \sum_{n=1}^{N} \hat{U}(x,\omega_n)\, e^{j\omega_n t} = \sum_{n=1}^{N} \left(\sum_i \tilde{U}_i\, e^{-jk_i x} \right) e^{j\omega_n t} , \qquad (27)$$

where ω_n is the frequency variable, $\hat{(.)}$ represents the spectral amplitude of the field variables, k_i is the wavenumber associated with ith mode of wave propagation, and $\tilde{U}_i$ represents the wave coefficients. These coefficients are to be evaluated from the displacement and force boundary conditions. Substituting the transformation equation of primary variables in the governing wave equations for the Euler-Bernoulli beam model (Equations (21) and (22)), one can obtain the wave number of the axial and bending modes respectively as

$$k_l = \pm\omega_n\sqrt{M_{11}/A_{1111}}\ , \qquad k_b = \pm\sqrt{\omega_n}\left(M_{11}/D_{1133}\right)^{1/4}\ . \tag{28}$$

On the other hand, the structure of wave equations for the higher order cylindrical bending model (Equations (15)-(20)) yields six wave number pairs, each associated with one of the six primary modes plus contribution due to cross coupling with certain other modes. Real part of the these wave number pairs $(k_u, k_v, k_w, k_{\theta_x}, k_{\theta_y}, k_{\theta_z})$ solves the (6×6) determinant of the complex kernel matrix $det[K] = 0$. To compare the nature of wave numbers between the two mechanics model, over the frequency range of interest (order of kHz), we consider the similar modes, i.e., u^o, w^o and θ_y. Then, the characteristic equation from the higher order cylindrical bending model reduces to

$$\begin{bmatrix} -\omega_n^2 M_{11} + A_{1111}k^2 & 0 & -\omega_n^2 M_{15} + A_{1115}k^2 \\ 0 & -\omega_n^2 M_{11} + A_{5511}k^2 & jA_{5519}k \\ -\omega_n^2 M_{15} + A_{1115}k^2 & -jA_{5519}k & -\omega_n^2 M_{55} + A_{1155}k^2 + (A_{5599} + A_{6688}) \end{bmatrix} \begin{Bmatrix} \tilde{u} \\ \tilde{w} \\ \tilde{\theta}_y \end{Bmatrix} = 0\ . \tag{29}$$

After solving the determinant of the kernel matrix in Equation (29), the corresponding roots i.e., wavenumbers are plotted in Figure 2. The solid lines in the figure represents the beam model where the longitudinal wave mode k_u (along the strut axis) propagates non-dispersively. Whereas, the flexural wave modes (two propagating components $\pm k_b$ and two evanescent components $\pm jk_b$) propagate dispersively as a parabolic variation of frequency.

The plot in Figure 2 shows three marked deviation in the propagation characteristics. The first one is the transition of an evanescent mode becoming a propagating mode above a cut-off frequency of $8.8kHz$. Its physical significance that, any excitation spectrum spanning over this cut-off frequency will produce a new dispersive wave whose characteristic is similar to k_b for beam model. However, experimental studies on actual strut shows that the noise spectrum of interest is unlikely to exceed $10kHz$ from active control view point. The second deviation between two result is that the rotational mode k_{θ_y} in the cylindrical strut becomes strongly evanescent with higher dispersiveness at low frequency range. Also it can be observed that at the cut-off frequency, the amount of attenuation it produces is nearly 0.25 times that produced by the evanescent component of flexural wave in the beam model. The third

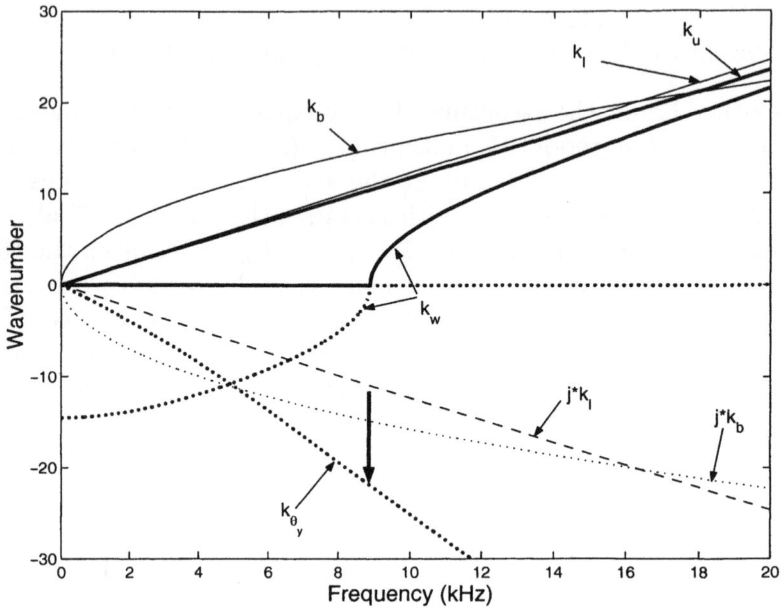

Figure 2. Comparison of the wavenumbers corresponding to axial mode u^o, one transverse mode w^o and associated rotational mode θ_y. Bold lines represents the higher order cylindrical bending model (imaginary components are plotted in the negative Y-axis

deviation between two model is that the axial mode in the beam model is always non-dispersive. Whereas, small dispersiveness in k_u can be observed above the cut-off frequency.

The above results ensure the fact that within the range of nearly $9kHz$, the beam model can be expected to produce reliable result. Only the speed of the flexural waves will be estimated slightly higher. Since in the present study we emphasize on simpler but efficient active control strategy, the beam model is chosen to construct the controller parameters, especially the feedback gain. The basic formulation of the spectral finite element model with generalized feedback algorithm is presented in the next section.

4. Active Spectral Element Model (ASEM)

In this model, the active strut assembly is discretized and classified into three different classes of frequency domain finite elements called spectral elements. These classes are spectral elements for finite beams or struts, distributed or point sensors, and distributed or point actuators. We assume that that controller output for a single actuator can be designed based on a feedback signal constructed from a group of sensors. Furthermore, connectivity between the pth sensor and the qth actuator is made arbitrary, where the sensor response is measured at the local coordinate (X_s^p, Z_s^p) and the actuation force is provided at the local coordinate (X_a^q, Z_a^q). Derivation of the elements and the global assembly procedures are discussed in the following subsections. Detailed derivation of the ASEM can be found in [13].

4.1 Spectral Element for Finite-Length Strut

Under the framework of Euler-Bernoulli beam model, there are six wave coefficients to be evaluated from the element boundary conditions. Therefore, a two noded three degrees of freedom (DOF) per node element is required to capture the exact system dynamics. By using the explicit expression of the displacement field (obtained by using the expression of k_l, k_b and wave coefficient vector $\{\tilde{u}\}$ in Equation (27)), the nodal displacement vector

$$\{\hat{u}\}^e = \{\hat{u}_1^o \ \hat{w}_1 \ \hat{\theta}_1 \ \hat{u}_2^o \ \hat{w}_2 \ \hat{\theta}_2\}^T \tag{30}$$

and the wave coefficient vector

$$\{\tilde{u}\} = \{\tilde{u}_1 \ \tilde{u}_2 \ \tilde{u}_3 \ \tilde{w}_3 \ \tilde{w}_4 \ \tilde{w}_5\}^T \tag{31}$$

can be related as

$$\{\hat{u}\}^e = [\hat{T}]\{\tilde{u}\} \,. \tag{32}$$

Here, $[\hat{T}]$ is a 6×6 non-singular complex unsymmetric matrix representing local wave characteristics of the displacement field. Rotational DOF $\theta_y (= \theta)$ in the element displacement field is brought using the equation

$$\{\hat{u}(x, \omega_n)\} = \{\hat{u}^o(x, \omega_n) \quad \hat{w}(x, \omega_n) \quad \hat{\theta}(x, \omega_n)\}^T = [\hat{J}(x, \omega_n)]\{\tilde{u}\} \,. \tag{33}$$

Making use of Equations (32) and (33), one can obtain

$$\{\hat{u}(x, \omega_n)\} = [\hat{J}][\hat{T}]^{-1}\{\hat{u}\}^e = [\hat{\aleph}]\{\hat{u}\}^e = \begin{bmatrix} \aleph_{11} & \dots & \aleph_{16} \\ \aleph_{21} & \dots & \aleph_{26} \\ \aleph_{31} & \dots & \aleph_{36} \end{bmatrix} \{\hat{u}\}^e \,, \tag{34}$$

where $[\hat{\aleph}]$ is the exact spectral element shape function matrix. Next, by using the expressions for dynamic boundary conditions from Equations (23)-(25), the nodal force vector

$$\{\hat{F}\}^e = \{\hat{N}_1\, \hat{V}_1\, \hat{M}_1\, \hat{N}_2\, \hat{V}_2\, \hat{M}_2\}^T \tag{35}$$

is related to the wave coefficient vector as

$$\{\hat{F}\}^e = [\hat{P}]\{\tilde{u}\}\ , \tag{36}$$

where $[\hat{P}]$ is a 6×6 non-singular complex unsymmetric matrix representing the local wave characteristics of force. After substituting for $\{\tilde{u}\}$ from Equation (32) into Equation (36), the result is

$$\{\hat{F}\}^e = [\hat{P}][\hat{T}]^{-1}\{\hat{u}^e\} = [\hat{K}]^e\{\hat{u}\}^e\ , \tag{37}$$

where $[\hat{K}]^e$ is the exact dynamic stiffness matrix of the spectral element. The complete details of the formulation are provided in reference [13]. Following the conventional procedure for finite elements, the transformation of the above equation from local element coordinates to global system coordinates is carried out as

$$\{\hat{u}\}^e = [t]\{\hat{U}\}\,, \quad \{\hat{F}\}^e = [t]\{\hat{F}\}\,, \quad \{\hat{F}\} = [t]^T[\hat{K}]^e[t]\{\hat{U}\}\,. \tag{38}$$

where $[t]$ is an orthogonal coordinate transformation matrix.

4.2 Sensor Element and Actuator Input

Keeping the modeling objectives in mind that multiple sensors at or near the strut-fuselage interface may be required as well as that a possible sensor array need not be always in the form of conventional strain gauges, we consider the following. Based on the response measured by a displacement sensor (s), which is located at (x_{s_p}, z_{s_p}) in pth sensor element (denoted by subscript s_p), the actuator input spectrum is expressed with the help of Equation (34) as

$$\hat{\eta}_u(x_{s_p}, z_{s_p}, \omega_n) = (j\omega_n)^m \alpha \left[\sum_{i=1}^{6} \left(\aleph_{1i} - z_{s_p}\aleph_{3i}\right) \hat{u}_i^e\right]\ , \tag{39}$$

when the longitudinal displacement is measured, and

$$\hat{\eta}_w(x_{s_p}, z_{s_p}, \omega_n) = (j\omega_n)^m \alpha \left[\sum_{i=1}^{6} \aleph_{2i} \hat{u}_i^e \right] , \tag{40}$$

when the transverse displacement is measured. In Equations (39) and (40), $\alpha(\omega_n)$ is the sensor sensitivity parameter and may have band limited reliability. Power $m = 0, 1$, and 2 stands for displacement spectrum, velocity spectrum, and acceleration spectrum, respectively. Similarly, if a strain sensor is used, one can write the actuator input spectrum as

$$\hat{\eta}_\varepsilon(x_{s_p}, z_{s_p}, \omega_n) = (j\omega_n)^m \alpha \left[\sum_{i=1}^{6} \left(\frac{\partial \aleph_{1i}}{\partial x} - z_{s_p} \frac{\partial \aleph_{3i}}{\partial x} \right) \hat{u}_i^e \right] . \tag{41}$$

4.3 Actuator Element

In the formulation of ASEM, a proportional displacement, velocity, or acceleration feedback control scheme is considered. The controller output in the form of current spectrum ($\hat{I}$) for the qth magnetostrictive actuator, and the resulting magnetic field strength ($\hat{H}$) can be written as

$$\hat{I} = \gamma \hat{\eta} , \quad \hat{H} = \beta \hat{I} , \quad \hat{\eta} = (\hat{\eta}_u, \hat{\eta}_w, \hat{\eta}_\varepsilon) , \tag{42}$$

where the $\hat{\eta}$ are given by Equations (39), (40), and (41), the constant γ is a scalar gain, and the quantity β is an actuator sensitivity parameter introduced to account for the solenoid properties [25]. Next, after substituting for $\hat{H}$ from Equations (23)-(25) into Equation (42) and following the same procedure as that used for discretizing the pure elastic domain (Equations (32)-(37)), the force balance equation for the qth actuator element (denoted by subscript a_q) in local coordinates can be obtained as

$$\{\hat{F}\}_{a_q}^e = [\hat{K}]_{a_q}^e \{\hat{u}\}_{a_q}^e + [A_{1111}^m \quad 0 \quad -A_{1113}^m \quad -A_{1111}^m \quad 0 \quad B_{1113}^m]_{a_q}^T \beta \gamma \hat{\eta} . \tag{43}$$

After substituting $\hat{\eta}$ in terms of the sensor element shape function ($\aleph$) and the corresponding nodal displacement vector ($\hat{u}^e$) from Equations (39)-(41) and with appropriate transformation of sensor local coordinates and actuator local coordinates into the global coordinate system, Equation (43) can be rewritten as

$$\left\{ \begin{array}{c} \hat{F}_{s_p} \\ \vdots \\ \hat{F}_{a_q} \end{array} \right\} = \left[\begin{array}{ccc} [t]^T_{s_p}[\hat{K}]^e_{s_p}[t]_{s_p} & \cdots & 0 \\ \vdots & \ddots & \vdots \\ [t]^T_{a_q}[\hat{K}]^e_{a_q \leftarrow s_p}[t]_{s_p} & \cdots & [t]^T_{a_q}[\hat{K}]^e_{a_q}[t]_{a_q} \end{array} \right] \left\{ \begin{array}{c} \hat{U}_{s_p} \\ \vdots \\ \hat{U}_{a_q} \end{array} \right\}. \tag{44}$$

The notation $[\hat{K}]^e_{a_q \leftarrow s_p}$ is introduced to represent the Sensor-Actuator Stiffness Influence Matrix (SASIM). Hence, the assembled closed-loop active strut system for multiple sensor-actuator configuration in the ASEM is obtained.

5. Case Studies

Following the work of Pelinescu and Balachandran [4], the finite-length strut considered is a 1.0 m long hollow cylindrical aluminum strut with 7.62 cm outer diameter and 6.35 cm inner diameter. Free-free displacement boundary conditions are imposed at the strut ends, and longitudinal and transverse loading with multiple harmonics are specified at each end. Cylindrical magnetostrictive actuators with rod shaped Terfenol-D core are chosen due to their high force capability. It is to be noted that such an actuator has its own housing and reaction mass [24]. These aspects can also be accurately modeled in ASEM, if the relevant stiffness and mass distributions are known. Based on the strut geometry considered here and the loading conditions to be considered, it is known that the longitudinal motion is expected to contribute substantially to the total kinetic energy at the fuselage interface. With this in mind, the first case considered below has to do with control of longitudinal wave transmission in a dynamically loaded strut.

5.1 Control of Axial Wave Transmission

In Figure 3(a), an end-cap actuator located at the strut-fuselage interface is shown along with a point feedback sensor mounted at a distance x_s from the gearbox end of the strut. This system is described by two spectral elements, one of which is an actuator element and the other one which is the strut element (this is also a sensor element as a whole). The results obtained in the uncontrolled cases and the controlled cases are shown in Figure 3(b). The objective of the feedback scheme used is to attenuate the longitudinal displacement at the strut-fuselage interface. With the actuator, the system resonances are shifted to lower values and it appears that the magnitude of frequency shift increases linearly with the mode number.

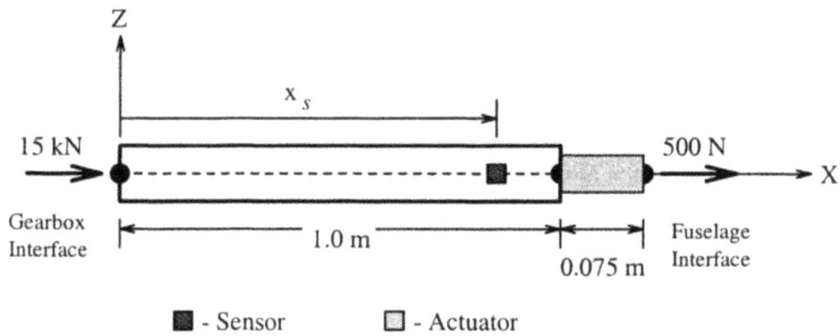

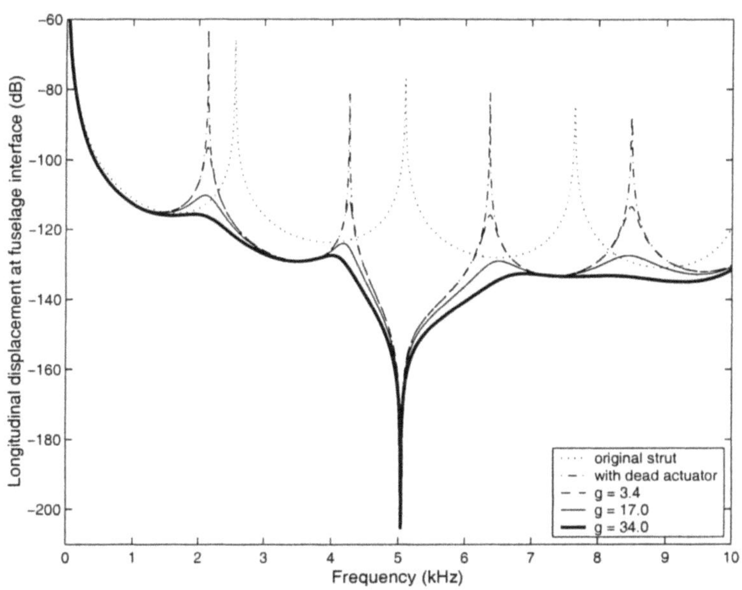

Figure 3. (a) Active strut for control of longitudinal wave transmission. Representative load amplitudes of harmonic loading are shown. (b) Longitudinal displacement at strut-fuselage interface in uncontrolled cases and controlled cases with different gain values. Sensor is collocated with actuator at strut end ($x_s = 1.0\,m$).

To generate the results shown for the controlled cases in Figure 3(b), the non-dimensional scalar velocity feedback gain parameter g is defined as

$$g = (\alpha\beta\gamma)/(n_o I_o/c_o) \tag{45}$$

where γ is the constant controller gain defined earlier, n_o is the number of specific turns in the actuator coil, I_o is a nominal r.m.s. coil current, and c_o is the speed of sound in air. For all of the numerical studies conducted here, the values of different quantities are $\alpha = 1$, $\beta = 1$,

$n_o = 1 \times 10^6\, m^{-1}$, $I_o = 1\, A\, r.m.s.$, and $c_o = 340\, m/s$. The sensor is located at the strut-fuselage end (i.e., $x_s = 1.0\, m$), and the parameter g for velocity feedback is increased in the range from 3.4 to 34. Results for three specific gains are shown in Figure 3(b). It is seen that the assumed collocated configuration can be used to attain considerable displacement response attenuation throughout the frequency bandwidth of interest using velocity feedback. Similar displacement response attenuation at targeted frequencies based on an analog feedforward scheme for the same actuator-sensor configuration was reported in reference [4]. Here, with velocity feedback control, response suppression throughout the frequency bandwidth is addressed. With increase in the gain value, the control force generated by the actuator also increases and this leads to an increased attenuation of the displacement response.

Since it may be difficult to locate the sensor at the actuator base exactly, an alternate configuration is used wherein the sensor is placed away from the actuator base. The results obtained for the configurations $x_s = 0.9\, m$ and $x_s = 0.8\, m$ are shown in Figure 4 along with results previously presented in Figure 3(b) for $x_s = 1.0\, m$. The velocity feedback gain parameter g has the value of 17.0 in all three cases. (This value was chosen as an "optimal" value based on the results shown in Figure 3(b) for the collocated case.) From the results, it is observed that $x_s = 0.9\, m$ may be the most preferable sensor location, since for this choice, the response close to the fourth mode resonance location is completely suppressed. For the non-collocated case $x_s = 0.8\, m$, although there is the largest suppression at the frequency location close to the resonance of the second mode, the suppression at the location close to the resonance of the fourth mode is not as good as that obtained for the case with $x_s = 0.9\, m$. However, in this later case, the suppression at the location near the resonance of the second mode is not as good as that obtained with $x_s = 0.8\, m$. The results shown in Figure 4 are illustrative of the usefulness of the ASEM to determine actuator and sensor locations for control of wave transmission in the gearbox support strut problems.

In some applications it may be necessary to use a group of actuators rather than one actuator to realize the required actuator force. Such actuator groups have been used in earlier work (e.g., Sutton *et al.* [6]). In this study, performance of three secondary magnetostrictive actuators in a group (hosted by a steel end-cap) was studied in an experimental arrangement. The inputs to the actuators was decided based on the measurement from one error sensor located on the upstream side of the actuator. The actuators were driven in phase for controlling longitudinal wave transmission through a strut, and they were not driven in phase for controlling flexural wave transmission. To illustrate the applicability

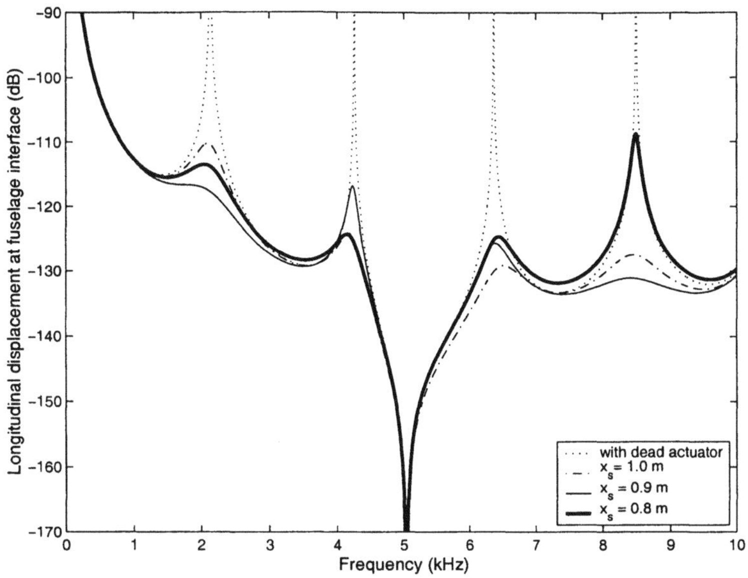

Figure 4. Longitudinal displacement at strut-fuselage interface in uncontrolled case and controlled cases for different sensor locations. Control gain parameter $g = 17.0$ in all cases.

of the ASEM for actuator group configurations, here, the active strut illustrated in Figure 5(a) is considered.

There are two actuators in the actuator group considered here, and these are represented by the spectral elements 3-5 and 4-6 in the modeling, as shown in Figure 5(b). Apart from these two spectral elements, the computational model consists of four other spectral elements, which include one sensor element (2-7) downstream of the actuators. The base ring is modeled as a rigid link (4-2-3). Next, results generated for this configuration with different actuator group locations x_a, different sensor locations x_s, and control gain parameter g values are shown in Figure 5(c). For the results shown here, the point sensor is placed at the strut-fuselage interface ($x_s = 1\,m$). For each actuator, the controller parameter g is set at 8.5 (one half of the "optimal" value of 17.0 obtained previously for the one actuator case) and both of the actuators receive identical inputs. Graphs of longitudinal displacement obtained at the fuselage end are shown as a function of frequency for the uncontrolled case and controlled cases with different values of x_a. In all cases, the attenuation at the location close to the resonance of the first mode is considerable but less than that obtained at other frequency locations. For $x_a = 0.8\,m$, it appears that a zero is introduced in the closed-loop

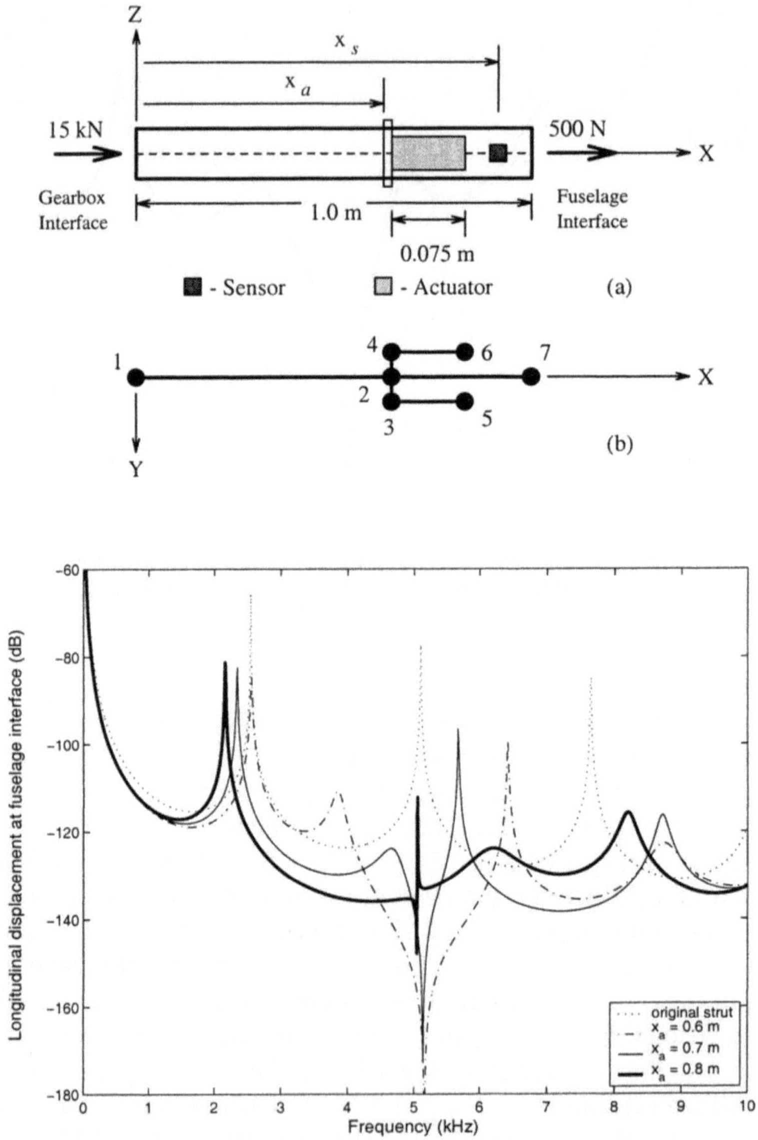

Figure 5. (a) configuration of strut with two intermediate actuators, for control of longitudinal wave transmission and (b) spectral element representation of actuators, sensor, strut, and base ring. (c) Longitudinal displacement at strut-fuselage interface in uncontrolled case and controlled cases for different actuator group locations. Sensor location at $x_s = 1.0\,m$ in all three cases, and control gain parameter $g = 8.5$ for each actuator.

system at a location close to the resonance location of the second mode of the open-loop system.

5.2 Control of Axial-Flexural Coupled Wave Transmission

One of the main objectives of the numerical study was to consider coupled axial-flexural wave transmission which has been reported difficult to model. A two actuator configuration chosen for this purpose is shown along with the strut, the loading amplitudes, and the sensor in Figure 6(a).

One of these actuators is inclined at an angle with respect to the longitudinal axis of the strut, and the other one is located along the longitudinal axis of the strut. The control inputs to both these actuators are based on a single error sensor response. Each actuator and sensor are represented by one spectral element as done earlier in Figure 5. Based on the results presented in the earlier subsections, the sensor location was chosen to be $x_s = 0.9\,m$, as this location was found to be an "optimal" location for the most number of axial and flexural modes. In Figure 6(b), the result obtained for the uncontrolled case with and without the actuators are shown along with the result obtained for a closed-loop case. Consideration of the actuator dynamics produces a considerable change in the pattern of resonances in the open-loop system. The appearance of a number of secondary axial-flexural coupled modes can be seen in this figure. From these figures, it is noted that the additional secondary modes do not contribute significantly to the axial displacement response compared to those due to the primary modes. However, this is not the case for flexural displacement response, where the primary modal amplitudes are influenced considerably by the secondary coupled modes, leading to shifts in the locations of poles and zeros of the closed-loop system.

In constructing the closed-loop system, it has been assumed that the sensor outputs corresponding to both longitudinal and transverse displacements are available from the chosen sensor location. The longitudinal and transverse actuators are driven based on these measured longitudinal and transverse displacement responses, respectively. Among different sets of parametric values considered for velocity feedback gains and sensor locations, the "best" results were achieved for $g_u = 17.0$ and $g_w = 340$. From the results, it is noted that with a constant gain velocity feedback scheme, an increase in effort to control flexural waves leads to less attenuation of axial-flexural coupled modes in the longitudinal displacement response. Localization of additional modes in the

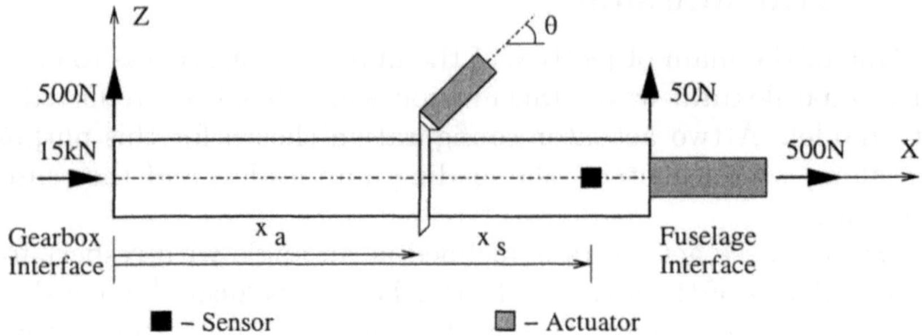

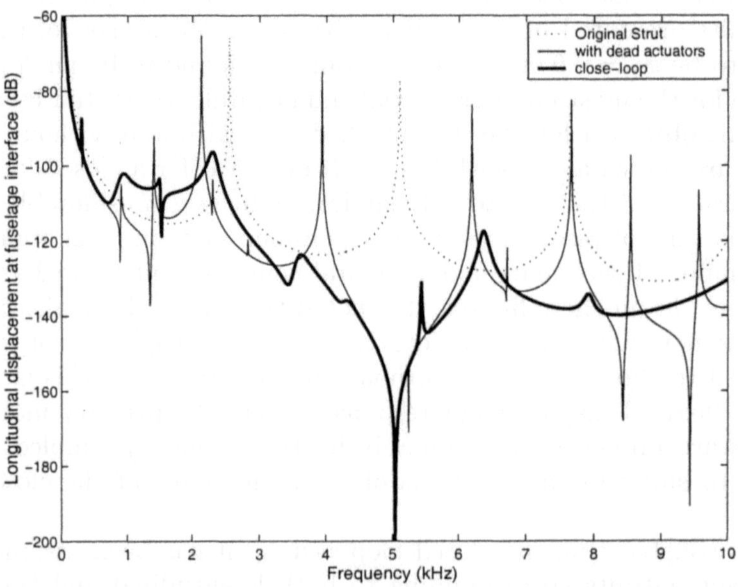

Figure 6. (a) Two actuator configuration for control of coupled axial-flexure wave transmission. (b)Longitudinal displacement at strut-fuselage interface in uncontrolled cases with and without actuator and controlled case. Velocity feedback provided to each actuator based on sensor response measured at $x_s = 0.9\,m$. One actuator is located transverse to the strut ($\theta = 90^o$) at $x_a = 0.6\,m$ and other actuator is located at strut-fuselage interface.

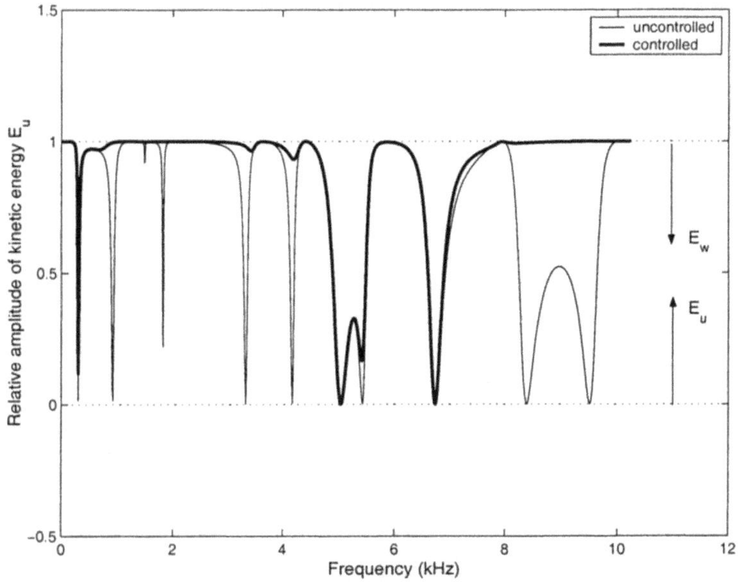

Figure 7. Distribution of kinetic energy between longitudinal and transverse motions at strut-fuselage interface in uncontrolled and controlled cases.

longitudinal displacement response (shown in Figure 6(b)) reflects this aspect.

The presented work may be used as a basis for carrying out path treatment. In such cases, it is of interest to know the level of energy attenuation at the spatial location of interest; here, the strut-fuselage interface. The kinetic energy has contributions from longitudinal and transverse motions. However, the transverse motions do not contribute as much as the longitudinal motions because the primary dynamic loading is along the longitudinal direction. This conclusion is supportive to the numerical results obtained in the earlier work of the author [13]. In order to analyze the redistribution of kinetic energy in the controlled case, Figure 7 is presented.

Plots of the normalized relative amplitudes of kinetic energy spectrum E_u (for longitudinal motions) and E_w (for transverse motions) at the strut-fuselage interface are shown in this figure. The corresponding expressions are given by

$$E_u = \frac{|(\hat{u}^o)^2|}{|(\hat{u}^o)^2| + |\hat{w}^2|}\,, \quad E_w = 1 - E_u\,, \tag{46}$$

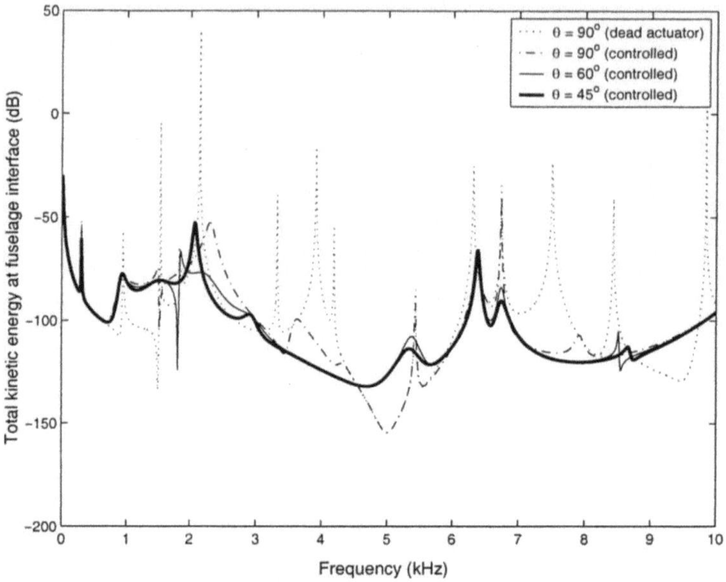

Figure 8. Kinetic energy at strut-fuselage interface in uncontrolled and controlled cases with different inclined orientations. Velocity feedback provided to each actuator based on sensor response measured at $x_s = 0.9\,m$. Inclined actuator is located at $x_a = 0.6\,m$ and and other actuator is located at strut-fuselage interface.

where $\hat{u}^o$ and $\hat{w}$ are the spectral amplitudes of longitudinal and transverse displacements, respectively, at the strut-fuselage interface. From this figure, it can be said that the kinetic energy associated with the significant primary transverse modes is attenuated, except at the frequency locations close to the first transverse resonance mode and the other three modes associated with resonances near $5.2\,kHz$ and $6.8\,kHz$.

In Figure 8, for the same configuration of the closed-loop system as that used to generate the results of Figure 6(b), the total kinetic energy is shown for uncontrolled and controlled cases. From the plots, it is clear that the first mode which is a flexural one, does not contribute much energy compared to the modes at higher resonance frequencies (most of which are axial). Therefore, although good attenuation of the flexural wave component close to the first resonance location is not obtained, because this particular component does not contribute significantly to the kinetic energy, the studied closed-loop scheme can still be an acceptable one. For the particular system configuration studied here, it appears that the results (i.e., in terms of kinetic energy attenuation) are relatively insensitive to the angle of inclination θ of the actuator located at $x_a = 0.6\,m$. Changes in the orientation angle do not affect much the responses at locations close to the primary mode resonances. However, the responses at locations close to the resonances of axial-flexural coupled modes are sensitive to changes in θ. For a different finite-length active strut, depending on the system characteristics, the angle of inclination may have a significant influence on the response of the system.

6. Concluding Remarks

In the present work, the problem complexities associated with the structure-borne noise transmission through helicopter gearbox support strut were addressed. The high frequency behavior of the cylindrical strut was analyzed to study the accuracy of the mechanics model. The application of ASEM for control of multiple waves in a finite-length strut with active materials was demonstrated. A representative closed-loop scheme with constant gain velocity feedback was chosen to carry out this demonstration, and the design of this scheme needs further study. The case studies and results presented here illustrate the possibility of accurate modeling of structure-borne noise transmission in helicopter gearbox support strut. The simulations presented show how a rational design procedure may be developed for such applications to choose appropriate collocated and/or non-collocated actuator-sensor configurations. These simulations can lead to considerable computational savings in complex structural systems where one is interested in coming up with appropriate actuator-sensor configurations. Besides being used to augment analytical work, ASEM based simulations can also be used to complement experimental work.

The case studies show that it is possible to attain considerable attenuation of longitudinal wave transmission and acceptable attenuation of flexural wave transmission. The ease of studying different actuator-sensor configurations for constructing closed-loop feedback schemes, axial-flexural wave coupling in such problems, and distribution of kinetic energy by using ASEM has also been illustrated. Here, in modeling the active finite-length strut, it has been assumed that the sensitivity parameters for sensors and (magnetostrictive) actuators do not change with frequency. But in practice, a large parametric variation may be observed at high frequencies; that is, the parameters α and β can be functions of frequency. This is also true while using piezoceramic or any other acoustic materials. However, effect of such frequency dependent properties and dead band regions in the acoustic response in the materials can be accommodated in the present model, once the sensitivity spectrum for sensors and actuators are known. From the numerical study, it is clear that the identification of feedback gain (constant over the frequency range) may have some limitations for controlling all the resonant mode. Also, for very high transient sensor output signal, the required amplification of the current or voltage may not be available. ASEM has been found suitable for such optimal control parameter estimation through adaptive gain scheduling at different frequencies, subjected to some known upper-bound to the actuator inputs.

References

Staple, A.E. and Wells, D.M. (1990) The Development and Testing of an Active Control of Structural Response System for the EH101 Helicopter, *Proceedings of the 16th European Rotorcraft Forum*, **III.6.1.1–III.6.1.11**.

Vignati, G. (1993) Research on Helicopter Interior Noise: A Survey on Current Activities in Helicopter Industries, *2nd Community Aeronautics RTD Conf., AERO-DAYS '93 (Naples, Italy), (Luxembourg: Commission of the European Communities, Directorate of General Science, Research and Development*, pp. 493–503.

Balachandran, B. and Sampath, A. (1995) Active Interior Noise Control Studies, *Proceedings of the National Technical Specialists' Meeting on Rotorcraft Structures, American Helicopter Society*, Hampton Roads Chapter, Williamsburg, Virginia, October 31-November 2.

Pelinescu, I. and Balachandran, B. (2001) Analytical Study of Active Control of Wave Transmission Through Cylindrical Struts, *Smart Materials and Structures*, **10**, pp. 121–136.

Staple, A.E. and MacDonald, B.A. (1993) *Active Vibration Control System*, US Patent, **No. 5,219,143**.

Sutton, T.J., Elliott, S.J., Brennan, M.J., Heron, K.H. and Jessop, D.A.C.(1997) Active Isolation of Multiple Structural Waves on a Helicopter Gearbox Support Strut, *Journal of Sound and Vibration*,**205**(1), pp. 81–101.

Nelson, P.A. and Elliott S.J. (1992) *Active Control of Sound*, Academic Press, London.

Elliott, S.J. and Billet, L. (1993) Adaptive Control of Flexural Waves Propagating in a Beam, *Journal of Sound and Vibration*, **163**, pp. 295–310.

Brennan, M.J. and Pennington, R.J. and Elliot, S.J. (1994) Mechanisms of Noise Transmission Through Helicopter Gearbox Support Struts, *Transactions of American Society of Mechanical Engineers, Journal of Vibration and Acoustics*, **116**, pp. 548–554,

Yoerkie, Jr C.A., Newington, J., Welsh, W. A., Haven, N. and Sheehy, H. (1994) *Helicopter Active Noise Control System*, US Patent, **No. 5,310,137**.

Millot, T.A., Welsh, W.A., Yoerkie, Jr C.A, MacMartin, D.G. and Davis, M.W. (1998) Flight Test of an Active Gear-mesh Noise Control on the S-76 Aircraft, *54th Annual Forum of the American Helicopter Society* (Washington, DC), pp. 241–249,

Ortel, D. and Balachandran, B. (1999) Control of Flexural Wave Transmission Through Struts, *Proceedings of the SPIE Smart Structures and Materials Conference on Smart Structures and Integrated Systems*, Newport Beach, CA, **3668**(2), pp. 567–577.

Roy Mahapatra, D., Gopalakrishnan, S. and Balachandran, B. (2001) Active Feedback Control of Multiple Waves in Helicopter Gearbox Support Struts: An Active

Spectral Element Model (ASEM), *Smart Materials and Structures* **10**(5), pp 1046-1058.

Doyle, J.F. (1989) *Wave Propagation in Structures*, Springer-Verlag, NY.

Doyle, J.F. (1988) A Spectrally Formulated Finite Element for Longitudinal Wave Propagation,*International Journal of Analytical and Experimental Modal Analysis*, **3**, pp. 1–5.

Doyle, J.F. and Farris, T. N. (1990) A Spectrally Formulated Finite Element for Flexural Wave Propagation in Beams, *International Journal of Analytical and Experimental Modal Analysis*, **5**, pp. 13–23.

Gopalakrishnan, S., Martin, M. and Doyle, J.F. (1992) A Matrix Methodology for Spectral Analysis of Wave Propagation in Multiple Connected Timoshenko Beam, *Journal of Sound and Vibration*, **158**, pp. 11–24.

Miller, D.W. and von Flotow,.H. (1989) A Traveling Wave Approach to Power Flow in Structural Networks, *Journal of Sound and Vibration*, **128**(1), pp. 145–162.

von Flotow, A.H. (1986) Disturbance Propagation in Structural Networks, *Journal of Sound and Vibration*, **106**, pp. 433–450.

Pan, J. and Hansen, C. H. (1991) Active Control of Total Vibratory Power Flow in a Beam. I: Physical System Analysis, *Journal of Acoustical Society of America*, **89**(1), pp. 200–209.

Gardonio, P. and Elliott, S.J. (1996) Active Control of Wave in a One-dimensional Structure with Scattering Termination, *Journal of Sound and Vibration*, **192**, pp. 701–730.

von Flotow, A.H. (1986) Travelling Wave Control for Large Spacecraft Structures, *Journal of Guidance*, **9**, pp. 462–468.

Clark, R.L. and Pan, J. and Hansen, C.H. (1992) An Experimental Study of multiple Wave Types in Elastic Beams, *Journal of Acoustical Society of America*, **89**(1), pp. 871–876.

Pelinescu, I. and Balachandran, B. (2000) Analytical and Experimental Investigations into Active Control of Wave Transmission Through Gearbox Struts, *Proceedings of the SPIE Smart Structures and Materials Conference on Smart Structures and Integrated Systems*, (Newport Beach, CA), **3985**.

Butler, J.L. (1988) *Application Manual for the Design of ETREMA Terfenol-D Magnetostrictive Transducers*, (Ames, IA: EDGE Technologies Incorporated).

IEEE (1991) *Standard on Magnetostrictive Materials: Piezomagnetic Nomenclature*, New York: The Institute of Electrical and Electronics Engineers, **No. 319-1990**.

Song, O. and Liberescu, L. (1997) Structural Modeling and Free Vibration Analysis of Rotating Composite Thin-Walled Beams, *Journal of the American Helicopter Society*, pp. 358–369.

Cowper, G. R. (1996) The Shear Coefficient in Timoshenko's Beam Theory, *Journal of Applied Mechanics*, **33**, pp. 335–340.

ACTIVE NOISE CONTROL IN ACOUSTIC CAVITIES WITH FLEXIBLE WALLS

S. NARAYANAN and C. PADMANABHAN
Department of Applied Mechanics
Indian Institute of Technology Madras
Chennai 600036 India

Abstract A coupled structural-acoustic finite element formulation forms the basis for active control of noise in enclosures bounded by a flexible structure. The primary excitation for the enclosure can be in the form of volume velocity or piston sources as well as excitation acting on the structure. Minimization of acoustic potential energy is employed for deciding the source strength(s) of the secondary sources for active control. The significant feature of this paper is the use of coupled modes for the active control strategy, unlike prior literature. This makes it readily applicable to problems where significant interaction between the structure and acoustic cavity can be expected. Two sample problems are addressed, (a). rectangular enclosure with one flexible wall and (b). flexible cylindrical shell with endplates enclosing an acoustic cavity. The effectiveness of the approach is clearly demonstrated through these examples.

1. Introduction

Active noise control in enclosures has been receiving attention recently due to the need for significant reductions in noise levels of helicopter and turboprop aircraft interiors. In either case the noise in the aircraft cabin is typically periodic in nature and is related to the propeller blade passing frequency and its higher harmonics. These frequencies are usually low and hence traditional passive noise control methods become ineffective. Active control using secondary sources forms an attractive alternative for noise reduction in such situations. Bullmore and his co-workers (Nelson et al., 1987;Bullmore et al., 1987) were among the first to demonstrate the feasibility of active control by developing the required theory and proving it using computational and experimental results. The exercise was carried out for a rigid walled rectangular enclosure. Their main conclusion was that properly placed secondary

M.L. Munjal (ed.), IUTAM Symposium on Designing of Quietness, 129–146.

sources (acoustic pistons) can lead to significant reduction in acoustic potential energy and consequently sound pressures. Since the technique was based on modal information, the need for more modes than those within the frequency range of interest was stressed. Yang et al. (Yang et al., 1994) have used an optimization procedure based on the boundary element method for calculating the optimal secondary source location when single and multiple primary sources are present. Nelson and Elliott (Nelson and Elliot, 1993) have summarized in their book the state-of-the-art (1993) related to noise control research in rigid enclosures. More recently Parkins et al.(Parkins et al., 2000) have proposed an energy density based active control scheme for rectangular enclosure. Unlike Bullmore et al. (Bullmore et al., 1987) they have considered the sum of acoustic potential energy and particle kinetic energy for minimization. Experiments have been carried out to verify the formulation. A special energy density sensor has been developed using six electret microphones. Comparison with purely pressure based measurements seems to suggest that the energy density approach has an advantage for off-resonance control.

Flexible panel backed rectangular enclosure has been the subject of study of a number of researchers. Dowell et al. (Dowell et al., 1977) have developed a technique for coupled structural-acoustic analysis using the uncoupled rigid wall acoustic and *in vacuo* structural modes. Narayanan and Pillai (Narayanan Premachandran pillai, 1985) exploited the gyroscopic nature of the formulation developed by Dowell et al. (Dowell et al., 1977) to reduce the equations to a symmetric form. The coupled system mode shapes and frequencies were obtained directly. A number of investigators have used this formulation to study active noise control for enclosures. Qiu et al. (Qiu et al., 1995) examined the effect of a point force on the structure as a means of control and found that control was effective only for structure controlled modes in the cavity. Pan et al. (Pan et al., 1990) also used a point force actuator and have detailed the control mechanisms in terms of plate and cavity controlled modes. Narayanan and Padmanabhan (Narayanan and Padmanabhan, 1997) demonstrated the effectiveness of active control using minimization of potential energy for a rectangular enclosure with a flexible panel. This was shown for a variety of fluids in the cavity and a new impedance based matrix approach for active control was proposed. Kim and Brennan (Kim and Brennan, 2000) have very recently reported the effectiveness of using both structural and acoustic piston sources for active control of a rectangular enclosure with a flexible plate. The plate is excited by a plane wave incident at an angle of 45^0 and the modeling approach is based on

Dowell et al. (Dowell et al., 1977). Control spillover, due to the effect of modes not included in determining the secondary source strength, has been observed only in the case of actuation using a structural excitation.

From the review of the literature on active control of enclosures it is apparent that most of them are for rectangular shaped cavities. This is mostly due to the fact that analytical modal expansions are readily available for modeling. Also most of the cases reported are for air as a medium in the cavity; this usually implies that the coupling is weak. The assumption that the coupled response can be represented in terms of the uncoupled structure and acoustic modes would work well for cavities with a less dense medium such as air. If one is interested in analysing the effect of active control for a more complex shaped cavity such as a helicopter or aircraft cabin, then a general modeling technique such as finite element method must be used. Hence, in this paper the twin objectives are to develop a finite element model for the structural-acoustic system to calculate coupled modes directly and use these to implement an active control strategy based on acoustic potential energy minimization. Two examples are considered to demonstrate the flexibility and the power of the approach, (i). rectangular box with one flexible wall and (ii). a cylindrical cavity with flexible shell and end plates.

2. Analytical Formulation for Rectangular Enclosure

The governing three-dimensional acoustic equation and the associated boundary conditions for a rectangular cavity with five rigid walls (represented by A_r) and one flexible panel (represented by A_f) is given by:

$$\nabla^2 p_c - \frac{1}{c_0^2}\frac{\partial^2 p_c}{\partial t^2} = -\rho_0 \frac{\partial q}{\partial t} \tag{1}$$

$$\frac{\partial p_c}{\partial n} = \begin{cases} -\rho_0 \frac{\partial^2 w}{\partial t^2} & \text{, on } A_f \\ 0 & \text{, on } A_r \end{cases} \tag{2}$$

In the above equation p_c represents the acoustic pressure inside the rectangular cavity, c_0 is the speed of sound in the medium, ρ_0 the average density of the medium and q represents the effect of both primary and secondary sources. The normal displacement, w, is given by the plate equation (with simply supported boundary conditions) with $D = \frac{Eh^3}{12(1-\nu^2)}$, E the Young's modulus, h the plate thickness, ν the Poisson's ratio and μ the mass per unit area of the plate:

$$D\nabla^4 w + \mu \frac{\partial^2 w}{\partial t^2} = p_c(t) - p_e(t) \tag{3}$$

The plate can also be excited by an external acoustic source which is represented by $p_e(t)$ in equation 3. Equations 1 and 3 can be solved using the modal decomposition method. For the cavity rigid wall modes are used in the modal decomposition and the inx*in vacuo modes* for the structure assuming simply supported boundary conditions is made use of to arrive at:

$$p_c(\mathbf{x_c},t) = \sum_{k=0}^{\infty} p_k(t)\psi_k(\mathbf{x_c}); \quad w(\mathbf{x}_s,t) = \sum_{m=1}^{\infty} w_k(t)\phi_m(\mathbf{x}_s) \qquad (4)$$

where $\mathbf{x_c}$ represents position vector in the cavity and $\mathbf{x_s}$ the position vector for a point on the plate. Using equation 4 and assuming modal damping, ξ_a and ξ_s, for the acoustic and structural modes respectively one arrives at the following differential equations:

$$\begin{aligned} \ddot{p}_k + 2\xi_{ak}\omega_{ak}\dot{p}_k + \omega_{ak}^2 p_k &= \frac{\rho_0 c_0^2}{V_c}\left\{\dot{Q}_k - \sum_{m=1}^{\infty} \ddot{w}_m S_{km}\right\} \\ \ddot{w}_m + 2\xi_{sm}\omega_{sm}\dot{w}_m + \omega_{sm}^2 w_m &= \frac{4}{m}\left\{\sum_{k=0}^{\infty} p_k S_{km} + \frac{Q_{me}}{A_f}\right\} \end{aligned} \qquad (5)$$

The external excitation term Q_{me}, internal excitation term $\dot{Q}_k$ and the structural-acoustic coupling coefficients are given by:

$$\begin{aligned} Q_{me}(t) &= -\int_{A_f} p_e(t)\phi_m dA; \quad \dot{Q}_k(t) = -\int_{V_c} \frac{\partial q}{\partial t}\psi_k dV; \\ S_{km}(t) &= -\frac{1}{A_f}\int_{A_f} \psi_k \phi_m dA \end{aligned} \qquad (6)$$

Assuming a time harmonic dependence for the variables equations 5 and 6 the following matrix equations are obtained:

$$\tilde{\mathbf{p}} = \mathbf{a_p} + \mathbf{Bq_S} + \mathbf{Z_A S \tilde{w}}; \quad \tilde{\mathbf{w}} = \frac{1}{A_f}\mathbf{Z_S \tilde{Q}_e} + \mathbf{Z_S S^T \tilde{p}} \qquad (7)$$

The primary and secondary sources are modeled as acoustic pistons of uniform velocity and the expressions for the components of $\mathbf{a_p}$ and $\mathbf{B}$ are:

$$\begin{aligned} a_{pk} &= \frac{j\omega\rho_0 c_0^2}{V_c\{(\omega_{ak}^2 - \omega^2) + j2\xi_{ak}\omega_{ak}\omega\}} \frac{q_p}{S_P} \int_{S_P} \psi_k dS \\ B_{ml} &= \frac{j\omega\rho_0 c_0^2}{V_c\{(\omega_{am}^2 - \omega^2) + j2\xi_{am}\omega_{am}\omega\}} \frac{1}{S_{Sl}} \int_{S_{Sl}} \psi_m dS \end{aligned} \qquad (8)$$

Here S_p and S_{Sl} represent the primary and secondary source piston areas with $l = 1, 2, \ldots, N_S$, while $\mathbf{Z_S}$ and $\mathbf{Z_A}$ are diagonal receptance matrices with their elements given by:

$$\begin{aligned} Z_{Skk} &= \frac{4}{kA_f\{(\omega_{sk}^2 - \omega^2) + j2\xi_{sk}\omega_{sk}\omega\}} \\ Z_{Ann} &= \frac{\omega^2 A_f \rho_0 c_0^2}{V_c\{(\omega_{an}^2 - \omega^2) + j2\xi_{an}\omega_{an}\omega\}} \end{aligned} \qquad (9)$$

Elimination of $\tilde{\mathbf{w}}$ from the first expression of equation 7 using the second expression leads to the following form for the pressure:

$$\{\mathbf{I} - \mathbf{Z_A S Z_S S^T}\}\tilde{\mathbf{p}} = \mathbf{a_P} + \frac{1}{A_f}\mathbf{Z_A S Z_S}\tilde{\mathbf{Q}}_\mathbf{e} + \mathbf{Bq_S} = \mathbf{d_P} + \mathbf{Bq_S} \quad (10)$$

It appears from equation 10 that simply choosing the secondary source strength for $N_S = 1$ to be $q_S = -\frac{d_{Pk}}{B_k}$ would lead to the cancellation of the kth mode. Hence to cancel m modes then one would simply have to choose m sources of appropriate strengths. However, while one mode may be reduced the secondary source can increase the amplitude of some other mode (control spillover). To prevent this and achieve more global noise reductions it was proposed by Bullmore et al. (Bullmore et al., 1987) to use a metric based on time-averaged acoustic potential energy, $E_P = \frac{1}{4\rho_0 c_0^2}\int_{V_c}\langle p_c^2\rangle_t dV$. The control strategy is to simply choose the secondary source strength vector $\mathbf{q_S}$ so that the E_P is minimized. For harmonic pressure response the expression for the acoustic potential energy is simply given by:

$$E_P = \frac{V_c}{4\rho_0 c_0^2}\tilde{\mathbf{p}}^\mathbf{H}\tilde{\mathbf{p}} = \mathbf{d}_P^H\mathbf{d}_P + \mathbf{d}_P^H\mathbf{B}\mathbf{q}_S + \mathbf{q}_S^H\mathbf{B}^H\mathbf{d}_P + \mathbf{q}_S^H\mathbf{B}^H\mathbf{B}\mathbf{q}_S \quad (11)$$

When we minimize E_P with respect to $\mathbf{q}_S$ one obtains the following expression for the secondary source strengths at a particular excitation frequency:

$$\mathbf{q}_S = -(\mathbf{B}^H\mathbf{B})^{-1}\mathbf{B}^H\mathbf{d}_P \quad (12)$$

The above equation forms the basis for deciding the source strength for reduction of sound pressure in the cavity once the locations of the secondary sources are decided.

3. Finite Element Formulation for Cylindrical Cavity

3.1 Shell Structure Model

Since the structure considered is a cylindrical shell, polar co-ordinates are used for the derivation of various relations. The shell element formulation is based on thin shell theory (Ramalingeswara Rao, 1997). The displacements at any point z away from the shell mid-surface are expressed in terms of mid-surface displacements and rotations and are given by

$$u = u_0 + z\psi_s; v = v_0 + z\psi_\theta; w = w_0 \quad (13)$$

The strain-displacement relations, based on the assumption of linear variation of displacements u, v and w with respect to rotations, are:

$$\begin{aligned}
\varepsilon_{ss} &= k_1\left(\tfrac{\partial u_0}{\partial s} + \tfrac{w_0}{R_\phi} + z\tfrac{\partial \psi_s}{\partial s}\right) \\
\varepsilon_{\theta\theta} &= k_2\{\tfrac{1}{r}\tfrac{\partial v_0}{\partial \theta} + \tfrac{u_0}{r}\cos\phi + \tfrac{w_0}{r}\sin\phi + z(\tfrac{1}{r}\tfrac{\partial \psi_\theta}{\partial \theta} + \tfrac{\psi_s}{r}\cos\phi)\}; \qquad \varepsilon_{zz} = 0 \\
\gamma_{sz} &= k_1\{\psi_s - \tfrac{u_0}{R_\phi} + \tfrac{\partial v_0}{\partial s} + z(-\tfrac{\psi_s}{R_\phi})\} \\
\gamma_{\theta z} &= k_2\{\psi_\theta - \tfrac{v_0}{r}\sin\phi + \tfrac{1}{r}\tfrac{\partial w_0}{\partial \theta} + z(-\tfrac{\psi_\theta}{r}\sin\phi)\} \\
\gamma_{s\theta} &= k_1 k_2\{\tfrac{1}{r}\tfrac{\partial u_0}{\partial \theta} + \tfrac{\partial v_0}{\partial s} - \tfrac{v_0}{r}\cos\phi + z(\tfrac{1}{r}\tfrac{\partial \psi_s}{\partial \theta} + \tfrac{\partial \psi_\theta}{\partial s} - \tfrac{\psi_s}{r}\cos\phi)\}
\end{aligned} \tag{14}$$

where $k_1 = \frac{R_\phi}{R_\phi + z}$ and $k_2 = \frac{R_\theta}{R_\theta + z}$, while ψ_θ and ψ_s represent the rotations of the normals about θ and s axis respectively. A three-noded semi-analytical shell element is used for the study with five degrees-of-freedom per node. Since the shell is axisymmetric, Fourier expansions are used in the circumferential direction. Then the assumed displacement distributions then become:

$$\begin{aligned}
u_0 &= \sum_{m=0}^{\infty}\sum_{i=1}^{3} N_i u_{0i}^m \cos m\theta; \quad v_0 = \sum_{m=0}^{\infty}\sum_{i=1}^{3} N_i v_{0i}^m \sin m\theta; \\
w_0 &= \sum_{m=0}^{\infty}\sum_{i=1}^{3} N_i w_{0i}^m \cos m\theta; \quad \psi_s = \sum_{m=0}^{\infty}\sum_{i=1}^{3} N_i \psi_{si}^m \cos m\theta; \\
\psi_\theta &= \sum_{m=0}^{\infty}\sum_{i=1}^{3} N_i \psi_{\theta i}^m \sin m\theta
\end{aligned} \tag{15}$$

Here N_i, $i = 1, 2, 3$, are the standard quadratic shape functions for a three-noded element:

$$N_1 = \tfrac{1}{2}\xi(\xi - 1); \quad N_2 = 1 - \xi^2; \quad N_3 = \tfrac{1}{2}\xi(\xi + 1) \tag{16}$$

where $\xi = s/l$ is the isoparametric axial co-ordinate and l is the length of the element. The strain vector, $\{\varepsilon\} = [\ \varepsilon_{ss}\ \ \varepsilon_{\theta\theta}\ \ \gamma_{s\theta}\ \ \gamma_{\theta z}\ \ \gamma_{sz}\ ]^{\mathrm{T}}$, can be represented as $\mathbf{Bu}_e$, where $\mathbf{B}$ is the strain-displacement matrix and $\mathbf{u}_e$ is the nodal displacement vector. The element structural stiffness $\mathbf{K}_e$ and mass $\mathbf{M}_e$ matrix can now be obtained, where $\mathbf{D}$ is material property matrix and ρ is the structural density.

$$\mathbf{K}_e = \int_V \mathbf{B}^{\mathrm{T}}\mathbf{D}\mathbf{B}dV; \quad \mathbf{M}_e = \rho\int_V \mathbf{N}^{\mathrm{T}}\mathbf{N}dV \tag{17}$$

Once the element stiffness and mass matrices are obtained the global stiffness and mass matrices can be assembled based on the nodal connectivity data for the structural elements and the undamped free vibration equations are:

$$\mathbf{M}\ddot{\mathbf{U}} + \mathbf{K}\mathbf{U} = \mathbf{0} \tag{18}$$

3.2 Acoustic Model

Axisymmetry of the cylindrical acoustic cavity is exploited in the formulation of the finite element model. Due to axisymmetry cylindrical co-ordinates are preferred for the derivation of relevant matrices. An eight-noded isoparametric element, as shown in Figure 1, with pressure as the nodal degree-of-freedom, forms the basis for the finite element model. From the acoustic potential and kinetic energy expressions the element acoustic stiffness and inertia matrices can be obtained as:

$$\begin{aligned}\mathbf{H}_e &= \int_{V_{ac}} \left(\frac{\partial \mathbf{N}_F}{\partial r}^T \frac{\partial \mathbf{N}_F}{\partial r} + \frac{1}{r^2}\frac{\partial \mathbf{N}_F}{\partial \theta}^T \frac{\partial \mathbf{N}_F}{\partial \theta} + \frac{\partial \mathbf{N}_F}{\partial z}^T \frac{\partial \mathbf{N}_F}{\partial z}\right) dV \\ \mathbf{G}_e &= \frac{1}{c^2} \int_{V_{ac}} \mathbf{N}_F{}^T \mathbf{N}_F dV\end{aligned} \tag{19}$$

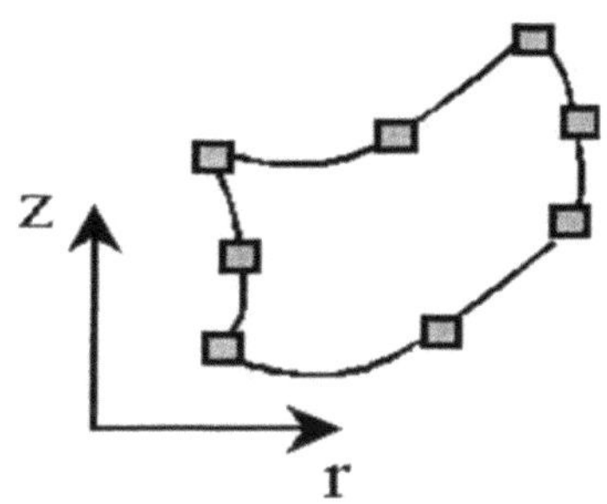

Figure 1. Eight-noded isoparametric element schematic

where $\mathbf{N}_F$ is the shape function vector for the fluid element and c is the speed of sound in the fluid medium. The shape functions in terms of the natural co-ordinates ξ and η are as follows:

$$\begin{aligned}N_{F1} &= -\tfrac{1}{4}(1-\xi)(1-\eta)(1+\xi+\eta); & N_{F2} &= \tfrac{1}{2}(1-\xi^2)(1-\eta) \\ N_{F3} &= -\tfrac{1}{4}(1+\xi)(1-\eta)(1-\xi+\eta); & N_{F4} &= \tfrac{1}{2}(1+\xi)(1-\eta^2) \\ N_{F5} &= -\tfrac{1}{4}(1+\xi)(1+\eta)(1-\xi-\eta); & N_{F6} &= \tfrac{1}{2}(1-\xi^2)(1+\eta) \\ N_{F7} &= -\tfrac{1}{4}(1-\xi)(1+\eta)(1+\xi-\eta); & N_{F8} &= \tfrac{1}{2}(1-\xi)(1-\eta^2)\end{aligned} \tag{20}$$

The pressure in an element can be expanded as a Fourier series and is represented as:

$$p = \sum_{m=0}^{\infty} \sum_{j=1}^{8} N_{Fj} p_j^m \cos m\theta \tag{21}$$

Again following a standard assembly procedure the global matrices **H** and **G** are formed to yield the following acoustic governing equation for no excitations:

$$\mathbf{G}\ddot{\mathbf{p}} + \mathbf{H}\mathbf{p} = \mathbf{0} \tag{22}$$

3.3 Structural-Acoustic Coupling

In situations where the interaction between the fluid and structure is significant, the fluid pressure acts as a load on the structure and the structural acceleration in turn influences the fluid pressure at the interface. This interaction is usually expressed by the following equation:

$$\frac{\partial p}{\partial n} = -\rho_F \frac{\partial v_n}{\partial t} \tag{23}$$

where $\frac{\partial p}{\partial n}$ is the pressure gradient in the direction normal to the interaction surface, v_n is the particle velocity normal to the interface and ρ_F is the fluid density. This leads to the appearance of an interaction matrix, which occurs in both the acoustic and structural equations derived earlier; see equations 18 and 22. The interaction matrix for each element is obtained as follows (Ross, 1994).

$$\mathbf{S}_e = \rho_F \int_{A_{fs}} \mathbf{N}_F{}^T \mathbf{N} dA \tag{24}$$

where $\mathbf{N}_F$ represents the fluid shape function vector and $\mathbf{N}$ is the structural shape function vector. The interaction matrix will be present only for those structural elements in contact with the fluid and vice versa. Equations 18 and 22 now become coupled leading to the form shown below:

$$\begin{bmatrix} \mathbf{M} & \mathbf{0} \\ \mathbf{S} & \mathbf{H} \end{bmatrix} \begin{Bmatrix} \ddot{\mathbf{U}} \\ \ddot{\mathbf{p}} \end{Bmatrix} + \begin{bmatrix} \mathbf{K} & -(1/\rho_F)\,\mathbf{S}^{\mathrm{T}} \\ \mathbf{0} & \mathbf{G} \end{bmatrix} \begin{Bmatrix} \mathbf{U} \\ \mathbf{p} \end{Bmatrix} = \begin{Bmatrix} \mathbf{0} \\ \mathbf{0} \end{Bmatrix} \tag{25}$$

The above equation is the unsymmetrical representation of the coupled structural-acoustic finite element formulation. In order to obtain the response to a harmonic excitation a modal superposition method based on left and right eigenvectors is used. Equation 25 is used to calculate the right eigenvectors by assuming $[\mathbf{U(t)}\ \ \mathbf{p(t)}]^T = [\mathbf{U_O}\ \ \mathbf{p_O}]^T e^{j\omega t}$ which leads to the following eigenvalue problem:

$$\mathbf{A}\mathbf{V}_R = \omega^2 \mathbf{B}\mathbf{V}_R; \qquad \mathbf{V}_R = [\mathbf{U_O}\ \ \mathbf{p_O}]^T$$

$$\mathbf{A} = \begin{bmatrix} \mathbf{K} & -\frac{1}{\rho_F}\mathbf{S}^{\mathrm{T}} \\ \mathbf{0} & \mathbf{G} \end{bmatrix}; \quad \mathbf{B} = \begin{bmatrix} \mathbf{M} & \mathbf{0} \\ \mathbf{S} & \mathbf{H} \end{bmatrix} \tag{26}$$

The adjoint eigenvalue problem is given by the expression $\mathbf{A}^H\mathbf{V}_L = \omega^2\mathbf{B}^H\mathbf{V}_L$. It can be shown (Golub and Van Loan, 1996) that the original eigenvalue problem and its adjoint have the same eigenvalues. For a symmetric matrix the left and right eigenvectors would be identical. The following orthogonality relationships exist for the eigenvectors with respect to $\mathbf{A}$ and $\mathbf{B}$ matrices:

$$(\mathbf{V}_L^m)^H \mathbf{A}\mathbf{V}_R^k = \omega_m^2\delta_{mk}; \quad (\mathbf{V}_L^m)^H \mathbf{B}\mathbf{V}_R^k = \delta_{mk} \tag{27}$$

Using the above orthogonality property matrices and the transformation $\mathbf{V} = \mathbf{V_R}\mathbf{Y}$, equation 25 on pre-multiplying by $\mathbf{V}_L^H$ reduces to a set of uncoupled equations. When there is a harmonic excitation on the right hand side of equation 25 $\mathbf{F}_0 e^{j\Omega t}$ it gets modified as $\mathbf{V}_L^H\mathbf{F}_0 e^{j\Omega t}$. The coupled system steady-state response is then simply obtained, assuming a modal damping ratio of ξ_k, $k = 1, 2, \ldots, M$, where M is the number of included modes, as shown below:

$$\mathbf{V} = \sum_{k=1}^{M} \frac{\left(\mathbf{V}_L^H\mathbf{F}_0\right)_k}{(\omega_k^2 - \Omega^2) + j2\xi_k\omega_k\Omega}\mathbf{V}_R^k \tag{28}$$

The minimization is carried out using the mean square pressures calculated using $\mathbf{V}^H\mathbf{V}$. This would contain both mean square pressures and displacements and only the pressure terms are used for minimization. The procedure adopted is similar to equation 12 in the previous section.

4. Results

Figure 2 shows a rectangular enclosure with one primary and three secondary acoustic piston sources locations. Apart from the internal primary source the box which has a flexible panel on the top is also excited by an acoustic point source located outside as shown. The dimensions of the enclosure are 0.5x0.25x0.752 m while the steel panel on the top has the following material properties: $E = 200$ GPa, $h = 3$ mm, $\nu = 0.3$ and $\rho_S = 7800\ \frac{\text{kg}}{\text{m}^3}$. To generate the results a uniform damping ratio of $\xi_k = 0.01$ is assumed with the source strengths $q_P = q_e = 10^{-4}$. The external source is located at $(0.3, 0.2, 0.85)$ with respect to the origin shown in Figure 9 while the primary internal source (piston of area $0.025m^2$) center is at $(0.2, 0.2, 0)$. The frequencies of the cavity and the structure are given in Table 4 below assuming the medium is air with $c_0 = 343$ m/s and $\rho_0 = 1.23\ \frac{kg}{m^3}$. When the secondary source is placed at S2 with co-ordinates$(0.40, 0.25, 0.30)$ the control is effective for the lower cavity modes, but is not good for structural or higher cavity frequencies as seen in Figure 3.

Table 1. Frequencies below 500 Hz for rectangular enclosure.

Plate: $\frac{\omega_S}{2\pi}, Hz$	Cavity: $\frac{\omega_A}{2\pi}, Hz$
119	0
163	228
285	343
463	411
476	454

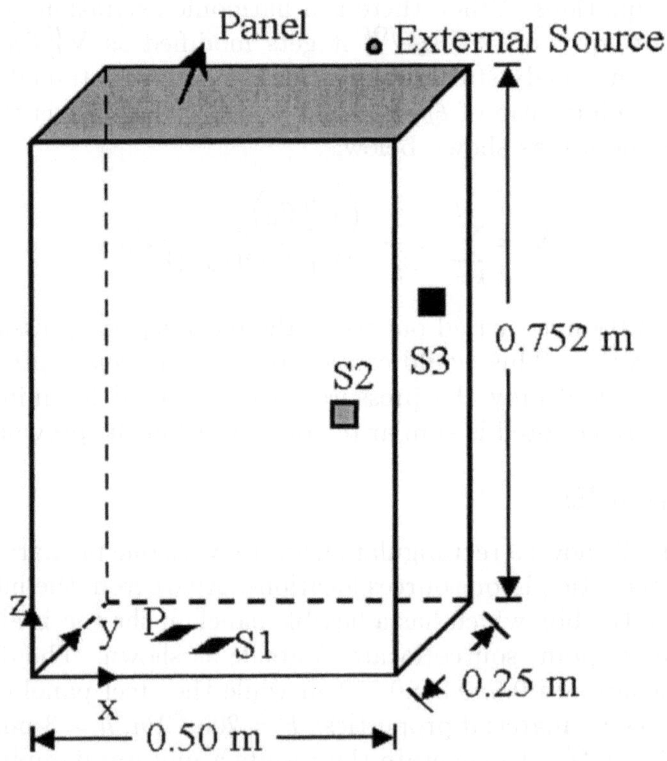

Figure 2. Box with one flexible panel indicating primary and secondary sources.

This is due to the distance of the source from the primary source P. When the source is moved to S1, with co-ordinates $(0.25, 0.125, 0.0)$, the cancellation is significantly better at the higher cavity frequencies and structural modes as can seen from Figure 3. The secondary source at S1 is better for a low modal density cavity. A similar trend can be observed when the secondary source is placed at S3 whose center is located at co-ordinates (0.5, 0.1, 0.5). This behavior is similar to that reported by Bullmore et al. (Bullmore et al., 1987) where it was shown that when the secondary source is separated by significant wavelengths for cavities with low modal density the cancellation is not satisfactory.

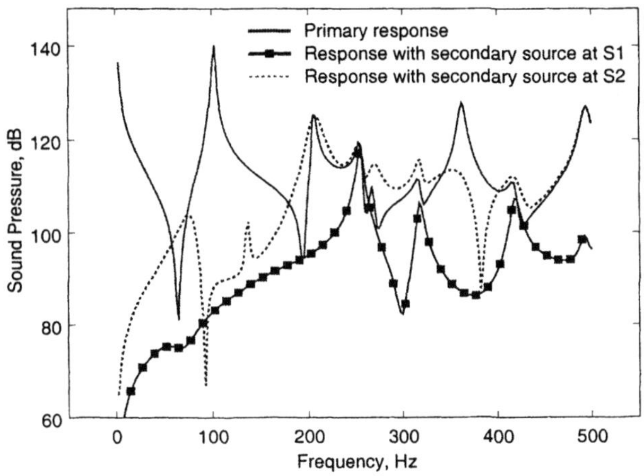

Figure 3. Comparison of sound pressure at location (0.31, 0.12, 0.42) between secondary source at S1 and S2.

Table 2. Cavity frequencies for rectangular enclosure with $c_0 = 155$ m/s and $\rho_0 = 26\ \frac{kg}{m^3}$

0	104	156	
208	260		
312	329	349	375
406	441	453	488

When the modal density is increased by considering a different fluid such as Freon with $c_0 = 155$ m/s and $\rho_0 = 26\ \frac{kg}{m^3}$, the cavity frequencies increase in number as can be seen in the Table 4. For this case if one carries out active control for the same primary and external excitation with secondary sources at S1 and S3, it is found that the behavior is the same as the low modal density case, but at higher frequencies both sources are becoming ineffective. This would imply that one secondary source may be insufficient for active control of high modal density cavities.

Figure 6 shows the schematic of the shell-cavity system with the locations of the primary and secondary sources. The acoustic cavity is the space between the outer cylinder and the inner cylinder shown in dark color. In this case the primary source is a monopole point source while the secondary sources are (piston type) placed on the wall. The excitation for this case is only due to an internal acoustic source. Since a semi-analytical formulation is used and the load is localized in space, Fourier expansion of the sources is carried out in terms of the circumferential modes to determine the excitation term for each circumferential mode. The response for each mode is obtained using the unsymmetrical mode superposition technique outlined in the earlier section. It is assumed that the cavity fluid has a density, $\rho_F = 5.3\ kg/m^3$ and speed of sound $c = 170\ m/s$. The shell is conventional steel material and the monopole source strength i s assumed to be 0.10. Figure 7 shows a comparison of the response with and without the secondary sources for circumferential mode, $m = 1$. It is clear that substantial reductions in pressure have been achieved near resonances. However, at off-resonance regions the sound pressure levels have increased. When only one secondary source, at S1, is used instead of three, the control is ineffective for certain modes as seen in Figure 7. Using three sources it is found that the sound pressure in the cavity is reduced even for higher order circumferential modes of the structural-acoustic systems. This is demonstrated in Figures 8 and 9. Again when only one source is used, at S1, the response is not very good for the $m = 2$ case as seen in Figure 8.

5. Conclusions

In this paper a finite element model for the structural-acoustic system to calculate coupled modes directly has been presented. An active control strategy based on acoustic potential energy minimization has been shown to be effective for the finite element model. A cylindrical cavity with flexible shell and end plates has been used to demonstrate the power of the approach. This approach makes it more amenable for examining the effectiveness of active control strategies in practical situations such as aircraft cabins. Future research issues such as optimization of placement of secondary actuators are being addressed.

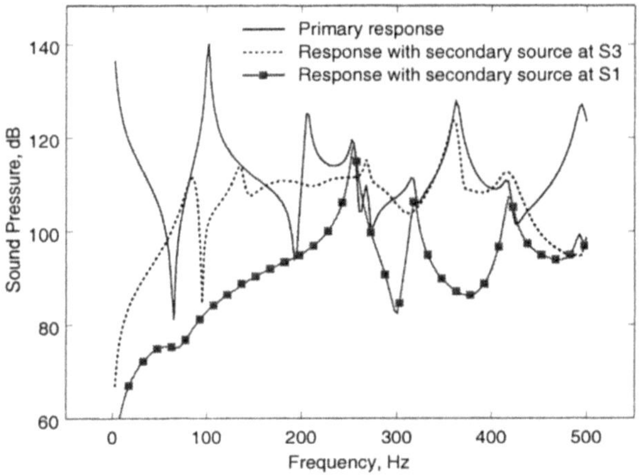

Figure 4. Comparison of sound pressure at location (0.31, 0.12, 0.42) between secondary source at S1 and S3.

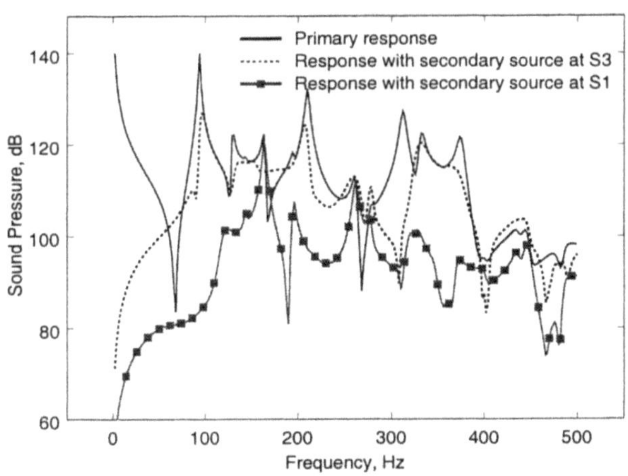

Figure 5. Comparison of sound pressure at location (0.31, 0.12, 0.42) between secondary source at S1 and S3 for high modal density case.

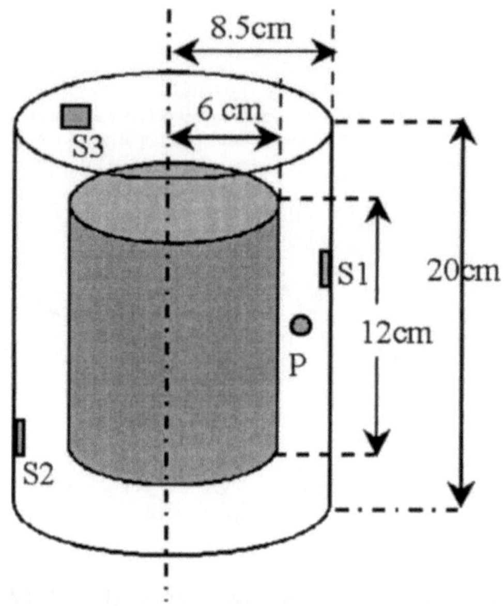

Figure 6. Shell-cavity system showing primary and secondary sources.

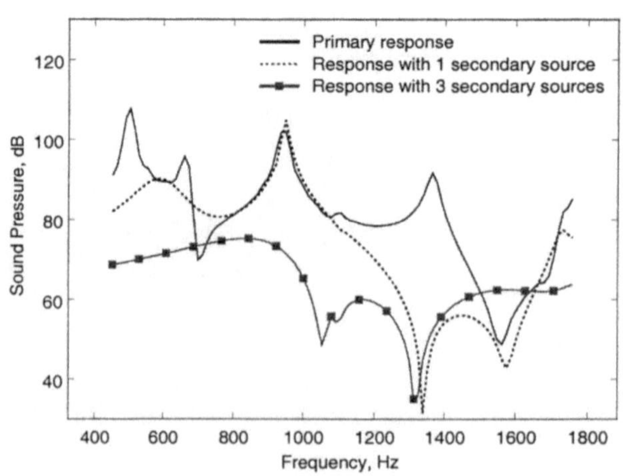

Figure 7. Active control for circumferential mode $m = 1$.

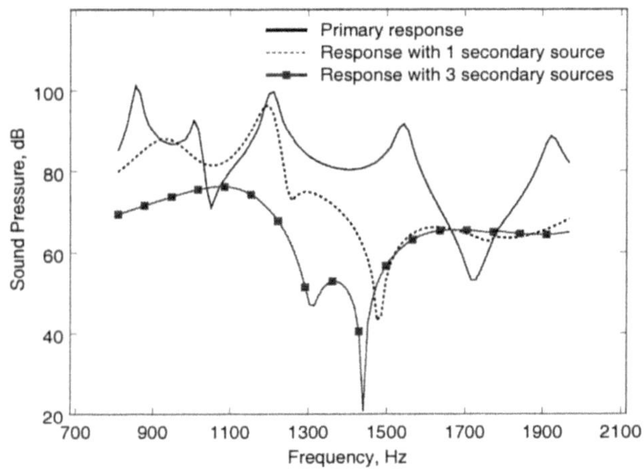

Figure 8. Active control for circumferential mode $m = 2$.

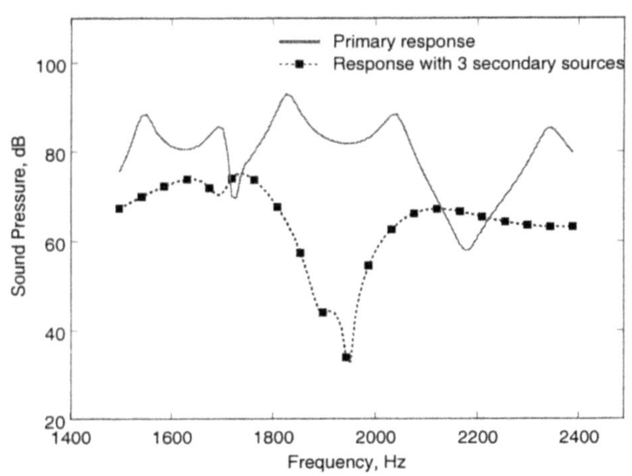

Figure 9. Active control for circumferential mode $m = 4$.

References

Bullmore, A. J., P. A. Nelson, A. R. D. Curtis, and S. J. Elliott: 1987, 'The Active Minimization of Harmonic Enclosed Sound Fields, Part II: Computer Simulation'. *Journal of Sound and Vibration* **117**, 15–33.

Dowell, E. H., G. F. Gorman, and D. A. Smith: 1977, 'Acoustoelasticity: General Theory, Acoustic Modes and Forced Response to Sinusoidal Excitation, Including Comparisons with Experiment'. *Journal of Sound and Vibration* **52**, 519–542.

Golub, G. H. and C. F. Van Loan: 1996, *Matrix Computations*. Baltimore: Johns Hopkins University Press, 3rd edition.

Kim, S. M. and M. J. Brennan: 2000, 'Active Control of Harmonic Sound Transmission Into an Acoustic Enclosure Using Both Structural and Acoustic Actuators'. *Journal of Acoustical Society of America* **107**, 2523–2534.

Narayanan, S. and C. Padmanabhan: 1997, 'Active Noise Control of Enclosed Sound Fields'. *Journal of Acoustical Society of India* **XXV**, II–3.1–3.9.

Narayanan, S. and D. Premachandran Pillai: 1985, 'Acousto-Elasticity of a Panel Backed by a Rectangular Cavity'. In: G. Comte-Bellot and J. E. Ffowcs-Williams (eds.): *Proceedings IUTAM Symposium on Aero- and Hydro-Acoustics*. pp. 393–402, Kluwer.

Nelson, P. A., A. R. D. Curtis, S. J. Elliott, and A. J. Bullmore: 1987, 'The Active Minimization of Harmonic Enclosed Sound Fields, Part I: Theory'. *Journal of Sound and Vibration* **117**, 1–13.

Nelson, P. A. and S. J. Elliott: 1993, *Active Control of Sound*. San Diego: Academic Press.

Pan, J., C. H. Hansen, and D. A. Bies: 1990, 'Active Control of Noise Transmission Through a Panel Into a Cavity, I: Analytical Study'. *Journal of Acoustical Society of America* **87**, 2098–2108.

Parkins, J. W., S. W. Sommerfeldt, and J. Tichy: 2000, 'Narrowband and Broadband Active Control in an Enclosure Using the Acoustic Energy Density'. *Journal of Acoustical Society of America* **108**, 192–203.

Qiu, X. J., J. Z. Sha, and J. Yang: 1995, 'Mechanisms of Active Control of Noise Transmission Through a Panel Into the Cavity Using a Point Force Actuator On the Panel'. *Journal of Sound and Vibration* **182**, 167–170.

Ramalingeswara Rao, S.: 1997, 'Static and Dynamic Problems in Laminated Beams and Axisymmetric Shells'. Ph.D. thesis, Indian Institute of Technology, Chennai, India.

Ross, C. T. F.: 1994, *Pressure Vessels under External Pressure: Statics and Dynamics*. London: Elsevier Applied Science.

Yang, T. C., C. H. Tseng, and S. F. Ling: 1994, 'Constrained Optimization of Active Noise Control Systems in Enclosures'. *Journal of Acoustical Society of America* **95**, 3390–3399.

SEA FOR DIESEL ENGINE TRANSFER PATH

Application of SEA for Diesel Engine Noise Reduction through Transfer Path Analysis

DHANESH N. MANIK
Associate Professor
IIT Bombay, India
dnm@me.iitb.ac.in

PARAG H. MATHURIA
Section Engineer
Mahindra (R&D), Nasik, India
mathuria@usa.net

Abstract The main objective of the paper was to evolve a methodology for minimising vibration transmission through a reciprocating mechanism, which is the dominant path of vibration transmission in a diesel engine. Vibration transmission in this path in a diesel engine is in the range of 800 to 3000 Hz. Since statistical energy analysis is applicable in this frequency range, it was used for the transfer path analysis, from the viewpoint of minimising vibration transmission. The transfer path, consisting of piston,piston pin, connecting rod and crank pin was constructed using simple elements like beam, plate and shell, which enabled the use of analytical results for modal density. Damping loss factors and coupling loss factors were measured experimentally. Modal density of all the subsystems was computed analytically and then compared with the experimental measurements. Coupling loss factor was analytically determined for a beam-beam connection, which enabled fine tuning of the experimental method of determining coupling loss factor between the subsystems of the transfer path model. After evaluating all the SEA parameters, the energy ratio between the piston and crank pin was computed. A total of eight different transfer path models were constructed with some variation in their dimensions and materials used for construction. Based on these results, certain recommendations were made for designing such paths for minimum vibration transmission.

Keywords: SEA, Diesel Engine, Transfer Path

M.L. Munjal (ed.), IUTAM Symposium on Designing of Quietness, 147–165.

Introduction

Diesel engine is widely used in many applications due to its high thermal efficiency and fuel economy. The main applications of diesel engine are trucks and railways for transportation. In addition, diesel engine is used in power generation, earth moving vehicles, ships, tractors, irrigation pumps, etc. Even though diesel engine has such wide applications, it is not commonly used in passenger cars because, its noise and vibration are higher than a gasoline engine.

Diesel engine noise can be minimised in three ways-source, path and receiver. Noise control at the source is preferable in any type of system. A number of researchers have done extensive work to identify sources of diesel engine and its transfer path within the engine by using different techniques. Modification of the external structure of the engine, and providing local acoustic enclosures for auxiliary systems like valves, fuel pumps is also common. If any of the above methods are not possible, acoustic enclosures can be provided. A well designed muffler is of course necessary to ensure lower noise levels from a diesel engine. In this paper, attention is focused upon the reciprocating mechanism, which is the dominant transfer path of the diesel engine noise, by using the principle of SEA.

Noise and vibration inside a diesel engine travel along three different paths as shown in Figure 1. In case of piston slap, there are two vibration paths, one through the liner(path 2) and the other through the piston-piston pin-connecting rod-crank pin (path 3). On the other hand, combustion noise is transmitted along the following three paths, namely-cylinder head (path 1), liner (path 2) and piston-piston pin-connecting rod-crank pin (path 3). From experimental observations it has been observed that the connecting rod path is the one through which most of the vibration and noise are transmitted to the exterior of the engine. Therefore, modification of this path is expected to reduce vibration transmission and the accompanying noise radiated by external surfaces of the engine.

Main objective of the work presented here is to analyze the transfer path for vibration transmission through the connecting rod and suggest suitable modifications. These modifications can be accounted for designing the components of the reciprocating mechanism at the drawing board stage. Statistical energy analysis (SEA) is a powerful tool for predicting response of vibro-acoustic systems at the drawing board stage. Since the excitation frequency of combustion noise and piston slap forces are in the range of 800 -3000 Hz, SEA can be advantageously used for designing such reciprocating mechanisms used in diesel engines.

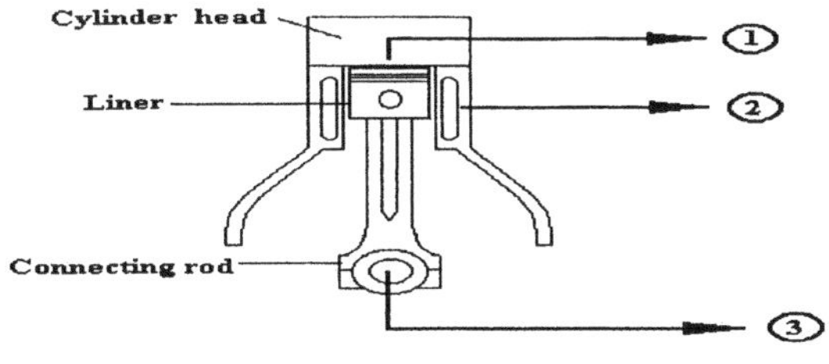

Figure 1. Transfer paths in a diesel engine

1. Literature Survey

Lyon [1] has developed Statistical Energy Analysis (SEA) in 1960. SEA is an useful tool for predicting vibration and noise in the high frequency range. Initially SEA was used to design rocket launch vehicle structures and later in aircraft (Miller and Faulkner [2], automotive vehicles (Steel [3]) and ships(Jensen [4]). Alpini et al. [5], have analysed combustion noise and vibration in a diesel engine. They have found that piston slap dominates the region between 1200 Hz - 1500 Hz and combustion forces dominate around 2500 Hz and 3000 Hz. Okuba et al. [6], have analyzed the piston slap using simulation techniques. They have suggested that the modification of the piston profile as well as piston pin center offset and reduction of piston-liner clearance will reduce piston slap noise. Lalor et al. [7] have analyzed engine noise due to mechanical impacts at pistons and bearings. They have underscored the importance of clearances in these systems. Okamura et al. [8] have analysed noise and vibration of engine structure and crank shaft vibration. They have found correlation between crankshaft vibration, engine structure vibration and engine noise. Kanda [9] has analysed the source and transmission path of diesel engines. He has concluded that more than 80% of the engine radiated noise is due to combustion and piston slap and that the reciprocating mechanism is the dominant path for vibration and noise transmission. Manik [10] has presented an elegant technique for determining coupling loss factors.

Surprisingly, SEA has not be used thus far to design the reciprocating mechanism of a diesel engine, from the viewpoint of minimizing vibration transmission.

2. Theory

The power flow between two systems is given by

$$\pi_{12} = 2\pi f(\eta_{12}E_1 - \eta_{21}E_2) \tag{1}$$

where,
π_{12}=net power flow between system 1 to system 2 ,
η_{12} coupling loss factor from system 1 to system 2,
η_{21} coupling loss factor from system 2 to system 1,
E_1 is the total energy of system 1,
E_2 is the total energy of system 2.

The reciprocal relationship between coupling loss factors and modal density is given by

$$n_1(\omega)\eta_{12} = n_2(\omega)\eta_{21}, \tag{2}$$

where $n_1(\omega)$ is the modal density of system 1 and $n_2(\omega)$ modal density of system 2.

In terms of average frequency spacing, the reciprocal relationship between coupling loss factors and average frequency spacing of systems 1 and 2 is given by

$$\eta_{12}\delta\bar{f}_2 = \eta_{21}\delta\bar{f}_1 \tag{3}$$

Equation (1) can be expressed in terms of modal coupling coefficient $\beta_{12} = \beta_{21}$ and modal energies Φ_1 and Φ_2 as follows:

$$\pi_{12} = \beta_{12}\left\{\Phi_1 - \Phi_2\right\} \tag{4}$$

where,

$$\Phi_1 = \frac{E_1}{n_1(\omega)}, \tag{5}$$

and

$$\Phi_2 = \frac{E_2}{n_2(\omega)}, \tag{6}$$

where,

$$\beta_{12} = 2\pi f\eta_{12}n_1(\omega) = f\eta_{12}/\delta\bar{f}_1 \tag{7}$$

and

$$\beta_{21} = 2\pi f\eta_{21}n_2(\omega) = f\eta_{21}/\delta\bar{f}_2 \tag{8}$$

Due to the reciprocity relationship defined by equation (2) or equation (3), the modal coupling coefficients defined by equations (7) and (8) are equal. That is

$$\beta_{12} = \beta_{21} \tag{9}$$

The coupling loss factors in general will have to be measured. But in many simple cases they can be calculated.

The above equations can be extended to a general system of connected subsystems. By index notation the modal coupling coefficient between systems i and j is given by

$$\beta_{ij} = \beta_{ji} = 2\pi f \eta_{ij} n_i(\omega) = f\eta_{ij}/\delta\bar{f}_i \tag{10}$$

and the modal overlap factor is given by

$$\beta_{ii} = f\eta_i/\delta\bar{f}_i \tag{11}$$

The above modal coupling coefficients and modal overlap factors can be combined into a matrix A, whose diagonal elements are defined by

$$A_{ii} = \sum_{j=1}^{m} \beta_{ij}, \tag{12}$$

where β_{ij} are given by equation (10) and β_{ii} are given by equation (11).

The off-diagonal symmetric elements are given by $-\beta_{ij}$. If the power input to each system is defined by the vector $\tilde{\pi_{in}}$, the modal power potential - similar to equations (5) and (6) is given by

$$\tilde{\Phi} == \begin{bmatrix} E_1/n_1(\omega) \\ E_2/n_2(\omega) \\ \vdots \\ E_m/n_m(\omega) \end{bmatrix} \tag{13}$$

The modal energy potential vector $\tilde{\Phi}$ can be obtained from the following equation.

$$\tilde{\Phi} = [A]^{-1}\tilde{\pi_{in}} \tag{14}$$

3. Application of SEA

The general procedure of SEA is as shown in Figure 2. The procedure outlined in this Figure is applicable to any vibro- acoustic system. It can discussed under the following headings.

3.1 System Model

The SEA model is based on the balance between dynamical energy and power flow among groups of natural modes in a system. A complex system can be modeled as a set of coupled mode groups, that are associated with the physical components of the system. The general

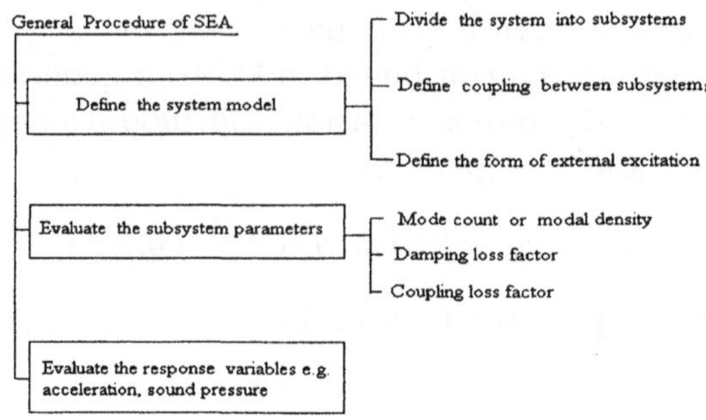

Figure 2. General Procedure of SEA

steps used in developing the SEA model of a dynamical system are as follows:

1 Define subsystems containing groups of natural modes within each of the system components.

2 Define the physical coupling between subsystems and

3 Define the form of external excitation to the subsystems

3.1.1 Subsystem parameters.

1 Modal density

2 Damping loss factor (DLF)

3 Coupling loss factor (CLF)

Modal density is defined as the number of modes within a frequency band of interest. It can be computed theoretically for a number of simple systems. For complex systems, experimental methods will have to be used. Basically there are two types of experimental methods for determining modal density: (1) Modal count technique (2) Point mobility method. Modal count technique consists of counting the number of resonant peaks within a frequency band and it is applicable when the resonant peaks are distinct without any modal overlap. Point mobility method consists of measuring point mobility and then relating it to modal density. This method is a direct method of measuring modal density, irrespective of the degree of modal overlap.

Damping Loss Factor (DLF), which is a measure of power dissipation, is a major source of uncertainty in the estimation of dynamic response. It has to be experimentally determined.

Coupling Loss Factor (CLF), is a measure of the power flow between connected systems. CLF can be evaluated in many cases using theoretical and empirical methods. Some times, CLF can be related to power transmission parameters like transmission loss and radiation efficiency, for which information is already available in the literature. Coupling loss factors can be experimentally measured.

Input power is one of the most important parameter in SEA modeling, since response predictions are directly proportional to input power. The power input to a structure by a point excitation is equal to the time averaged product of force and velocity.

3.2 SEA Model

The transfer path of the reciprocating mechanism of a diesel engine, considered for the present study, has four components: piston, piston pin, connecting rod and crank pin. It is shown in Figure.3. Additional components after the crank pin could have been added for the study. However, based on the earlier work of many researchers, it was decided to restrict to the above components, which are expected to play a significant role in vibration transmission. Each of the above compo-

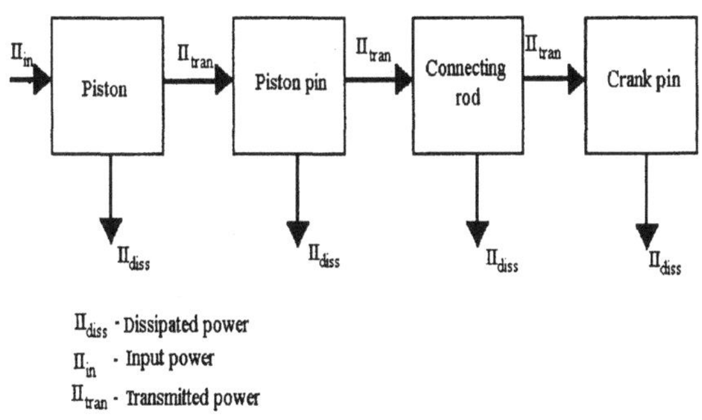

Figure 3. SEA model of the transmission path

nents mentioned above were constructed using simple elements, so they could be easily fabricated. In addition, for some of the elements, modal density data are readily available.

Piston was modeled as a cylinder with one end closed with a circular plate. Connecting rod was modeled as shank connected with two rings. Piston pin was modeled as a hollow cylinder and the crank pin was modeled as a solid cylinder. Details of the construction of the above subsystems is presented in Appendix. Eight different models were constructed by varying some parameters of the connecting rod and the material of the piston pin as shown in Table A.1 of the Appendix.

4. Experimental Setup

4.1 Modal density

Modal density was evaluated by measuring the point mobility of the structure for broadband excitation. The space-averaged point mobility was obtained by measuring point mobility at various locations of the structure and averaging it. Modal density and space-averaged point mobility of the structure are related by the following equation.

$$n(f) = \frac{1}{\Delta f} \int_{f_\ell}^{f_u} 4M_s < \Re(Y) > df \tag{15}$$

where Δf is the frequency band between frequencies f_ℓ and f_u, M_s is the mass of the structure and Y is the mobility. The experimental setup for measuring modal density is as shown in Figure 4. In this setup measuring the modal density of a connecting rod is shown. The same setup was used for measuring the modal density of other subsystems. It consists of an electro-magnetic exciter driven by a random signal generator. The exciter was connected to the force transducer and the response of the system was measured as close to the force transducer as possible, to obtain drive point mobility. Output of the force transducer and accelerometer were connected to the charge amplifier and then to the FFT analyzer, wherein the ratio of velocity and force in the frequency domain was measured to obtain drive point mobility. The data were then transferred to a computer through a GPIB interface. The drive point mobility of the structure was measured at a number of locations to obtain the space-averaged drive-point mobility and then related to modal density using equation. Modal density of various subsystems were measured from 500 Hz to 5000 Hz, which is the frequency range of interest in diesel engines.

4.2 Damping loss factor

The damping loss factor of each subsystem was computed by measuring the input power and the space-averaged response of the system, given by the following equation.

$$\eta(f) = \frac{1}{f_u - f_l} \int_{f_l}^{f_u} \frac{\overline{F(t)v(t)2\pi f}}{\langle a(\omega)^2 \rangle \, M_s} df. \tag{16}$$

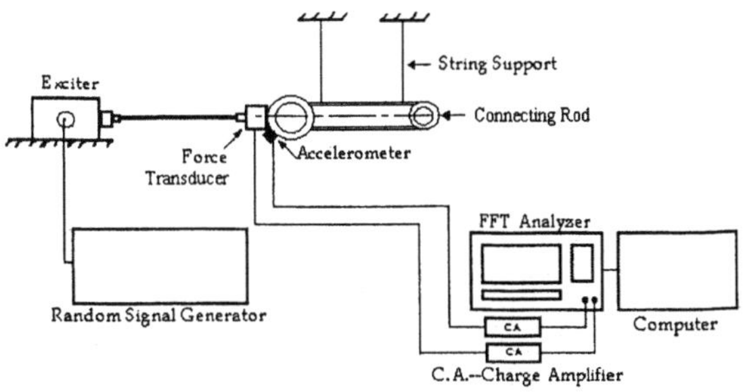

Figure 4. Mobility Measurements of the Connecting Rod

In addition, damping loss factor was measured based on modal density, by using the following equation.

$$\eta(f) = \frac{1}{f_u - f_l} \int_{f_l}^{f_u} \left\langle \frac{\overline{F^2(t)}}{\overline{\langle v^2(t) \rangle}} \right\rangle \frac{n(f)}{8\pi f M^2} df. \tag{17}$$

The experimental setup is as shown in Figure 5. The exciter receives random signal from the random signal generator and it drives the structure through a force transducer. An accelerometer was mounted very close to the force transducer to measure the input power and another accelerometer was used to measure the space- averaged response of the system. By using equation (16) and equation (17), the damping loss factor in 1/3 octave was computed for each of the subsystems.

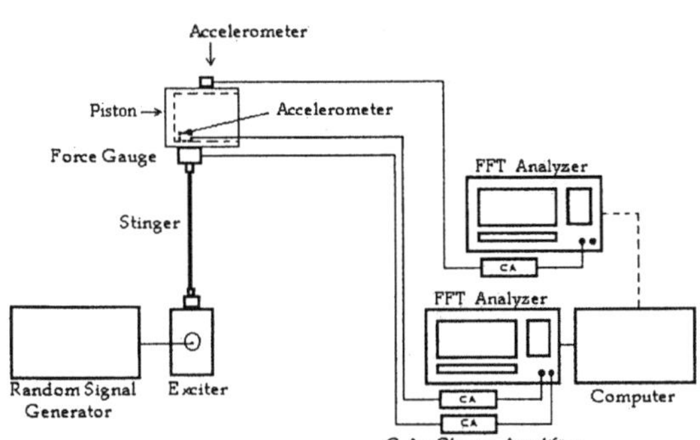

Figure 5. Damping loss factor measurement using power balance method.

4.3 Coupling loss factor

The coupling loss factor between two systems connected across a junction can be expressed by the following equation.

$$\eta_{12} = \frac{\eta_2}{\frac{m_1}{m_2}\left(\frac{v_1^2}{v_2^2} - \frac{v_1}{v_2}\right)} \tag{18}$$

Therefore with the knowledge of damping factor of the second system, the coupling loss factor between the first system and the second system can be obtained based on the velocity response of both the systems and their masses.

The experimental setup for measuring coupling loss factor is as shown in Figure 6. It consists of a random signal generator, which is connected to the electromagnetic exciter. The exciter directly excites the first system and the space averaged response of the first and the second system were measured to obtain the coupling loss factor between them at various 1/3 octave frequencies. Extreme care has to be taken while measuring the coupling loss factors, especially when the subsystems are lightly coupled, since CLF is an order of magnitude lower than the corresponding damping loss factors.

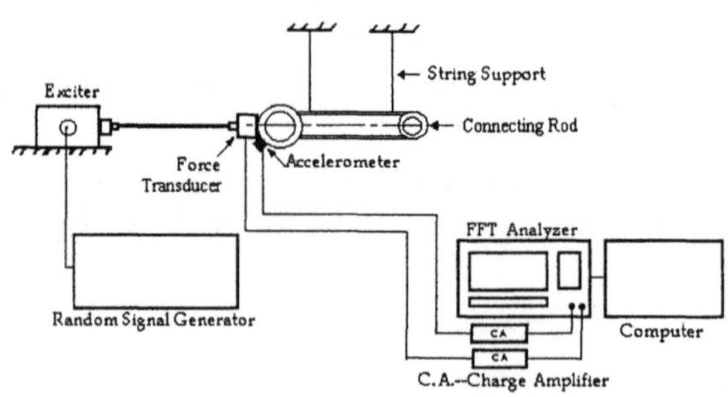

Figure 6. Typical coupling loss factor measurement

5. Results and Discussion

A large data base was generated for modal density, damping loss factor and coupling loss factor of all the subsystems of the reciprocating mechanism, for all the models shown in Table A.1. In case of modal density, theoretical predictions were also made for all the subsystems. Regarding coupling loss factor, experiments were conducted on some connected elements for which theoretical coupling loss factor was available and then compared with the experimentally measured coupling loss factor. This enabled fine tuning the experimental procedure for determining coupling loss factor between the subsystems of the reciprocating mechanism. Only relevant portion of the above results will be presented in this section.

The modal density of simple elements like a beam, plate, etc. can be theoretically determined. In case a complex structure is built from simple elements, the modal density of the complex structure can be taken as the modal density of the sum of all the elements comprising the structure. Modal density of all the dominant mode groups were also considered corresponding to each basic element. For example, the connecting rod was modeled as a shank connected to a small ring at one end and a big ring at the other end. Only radial modes of both the rings were considered, whereas torsional, longitudinal and bending modes of the shank were accounted for.

The theoretical modal density, expressed as a sum of the modal density of its elements and the experimental modal density of a connecting rod of shank length 200 mm is shown in Figure 7.

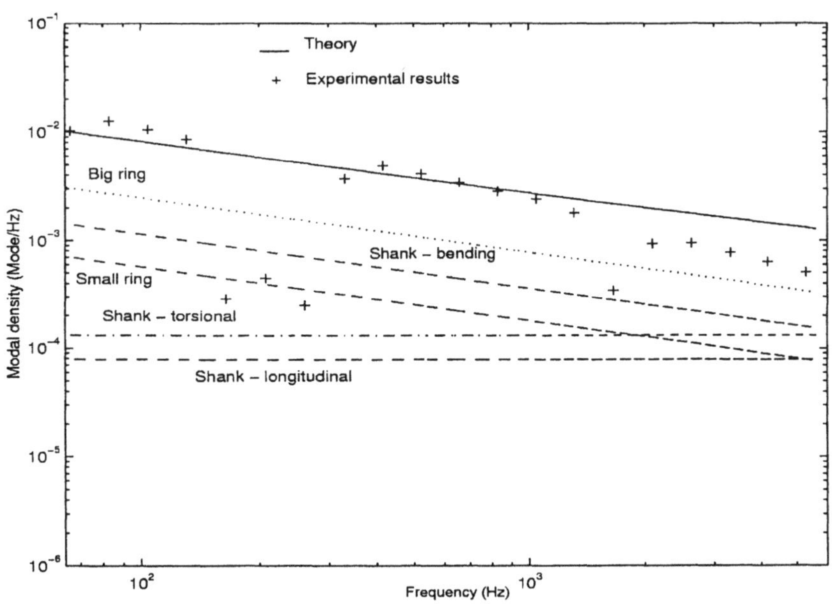

Figure 7. Modal density of connecting rod (shank length - 200 mm)

Damping loss factor of all the subsystems was measured using the power balance method and modal density method. Although there is a discrepancy in the damping loss factor between these two methods at lower frequencies, there is good agreement in the higher frequency range. The damping loss factor of a connecting rod of shank length 100 mm, measured using both the above methods, is shown in Figure 8.

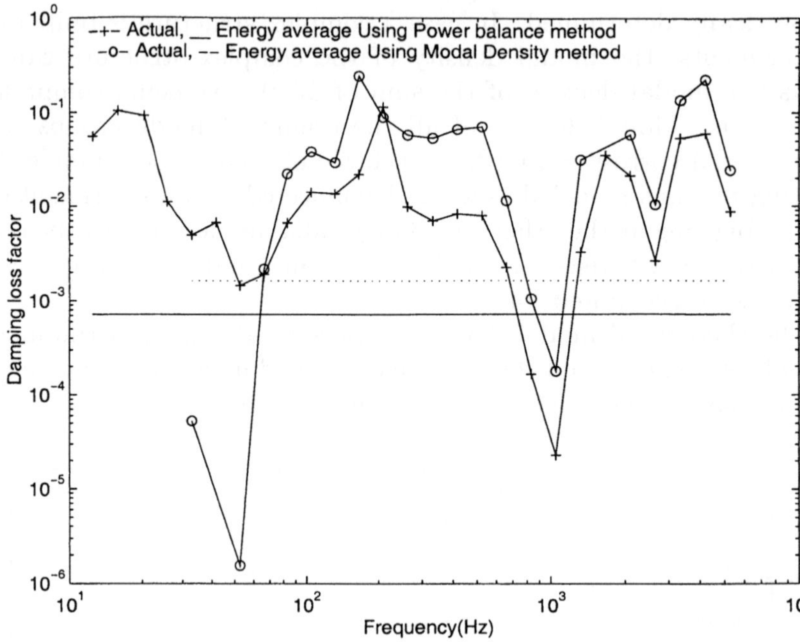

Figure 8. Damping loss factor of connecting Rod (shank length 100mm)

Coupling loss factor between various subsystems was experimentally evaluated using equation (18). A typical plot of coupling loss factor for various types of connecting rods and crank pin is shown in Figure 9.

Once the data base of modal density, damping loss factor and coupling loss factor of all the transmission path models were determined, they were used in determining the energy ratio between the crank-pin and the piston. A typical plot is shown in Figure 10. One can see from the plot a significant variation in the energy ratio between various types of connecting rods. Since energy ratio determines the extent of vibration transmission through the reciprocating mechanism, it serves a good purpose in designing such mechanisms for minimum vibration transmission. Although other considerations like fatigue strength, buckling, balancing, etc. determine the design of the connecting rod, minimum vibration

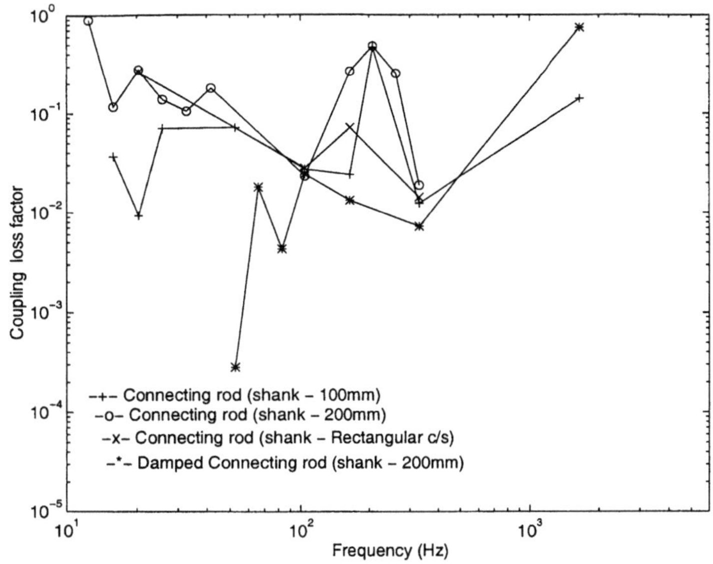

Figure 9. Coupling loss factor between connecting rods and crank pin

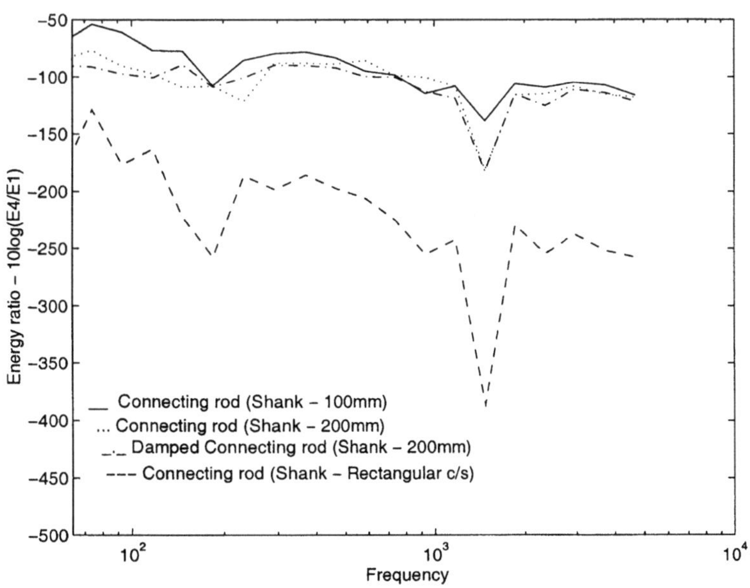

Figure 10. Comparison of energy ratio using aluminium pin and four different types of connecting rod

transmission will have to be given due consideration for reducing diesel engine noise.

6. Conclusions

Among the eight models of varying dimensions and material that were considered for application of SEA, the model with Aluminium as the piston -pin material and the connecting rod shank with rectangular cross section (Table A.1, Appendix, model no. 4) gave the best results. For this model, the energy ratio between crank pin to piston is about 5-20 dB lower than other models in the frequency range of 800 - 3000 Hz. Modal density of the connecting rod, maximum contribution of which comes from the shank of the connecting rod, significantly influences the energy ratio. Higher the modal density of the connecting rod, lower the energy ratio. Therefore while giving consideration for other parameters like buckling, fatigue strength, etc., due consideration must be given to increasing the modal density of the connecting rod.

Damping loss factor of the piston-pin has some influence on the energy ratio. Model number 4 with Aluminium piston pin was found to have lower energy ratio than model number 8 with steel pin. However, since the manufacturing tolerances could not be kept the same for both the pins, it is not clear whether damping could have played any role on the energy ratio. Damping ratio of the connecting rod has some influence on the energy ratio. Additional damping was introduced on connecting rods corresponding to model number 3 and 7.

Although the coupling loss factor between the subsystems of the models were measured for the eight different models, they were almost similar in the frequency range of 800-3000 Hz. This concept will be useful at the drawing board stage of designing the reciprocating mechanism, where coupling loss factors of existing similar systems can be used to predict the response at the crank pin.

Nomenclature

$a(t)$	Acceleration
f	Frequency, Hz
$\bar{\delta f}$	average separation frequency, Hz,
$n(f)$	modal density, Hz,
m_1, m_2	masses,
$v(t)$	velocity,
E	Energy,
$F(t)$	Force,
M_s	Mass,

Y	Mobility,
β_{ii}	Modal overlap factor,
β_{ij}	Modal coupling coefficient,
η	Damping factor,
η_{ij}	Coupling loss factor,
π_{ij}	Power flow,
Φ	Modal energy.

Appendix

There are four subsystems in the transmission path. Piston is considered as a cylinder with one end closed with a circular plate. Piston dimensions are shown in Figure A.1. Piston pin is considered as a beam with hollow circular cross section. The dimensions of piston pin are shown in Figure A.2. Connecting rod consists of a two rings connected by a rib (called shank). The cross section of the rib is an I section which is shown in Figure A.3. The fourth component is a crank pin of circular cross section which is shown in Figure A.4. All components are modified for simplicity of the analysis. Since modal density depends upon shape of the structure, it was attempted to maintain the shape similar to the actual shape of the transfer path. Eight models of varying dimensions and materials, as shown in Table A.1, were considered for applying SEA. Piston and crankpin remain the same for all the models.

Table A.1. List of models for the transfer path of a diesel engine

No	Piston pin	Connecting rod cross section	Connecting rod shank length
1	Aluminium	I	100 mm
2	Aluminium	I	200 mm
3	Aluminium	I,Damped	200 mm
4	Aluminium	Rectangular	200 mm
5	Mild steel	I	100 mm
6	Mild steel	I	200 mm
7	Mild steel	I,Damped	200 mm
8	Mild steel	Rectangular	200 mm

There are four subsystems in the transmission path. Physical coupling between the subsystems is as follows : Piston is coupled with piston pin, area of the coupling is twice the thickness of piston multiplied by outer circumference of the piston pin. Piston pin is also connected to small end of the connecting rod. Big end of connecting rod is connected to the crank pin. Both ends of the connecting rod are push fit.

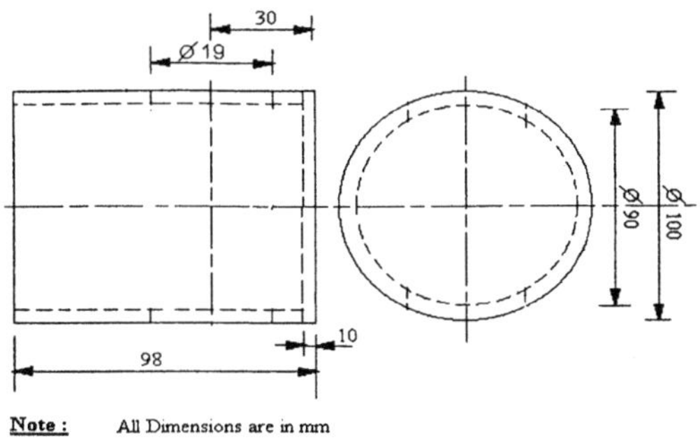

Figure A.1. Piston

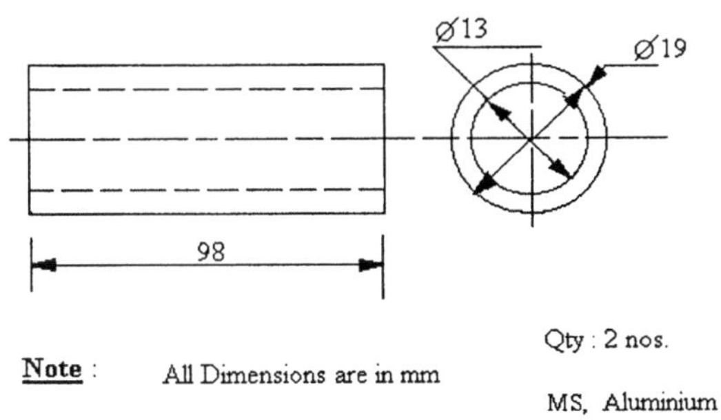

Figure A.2. Piston pin

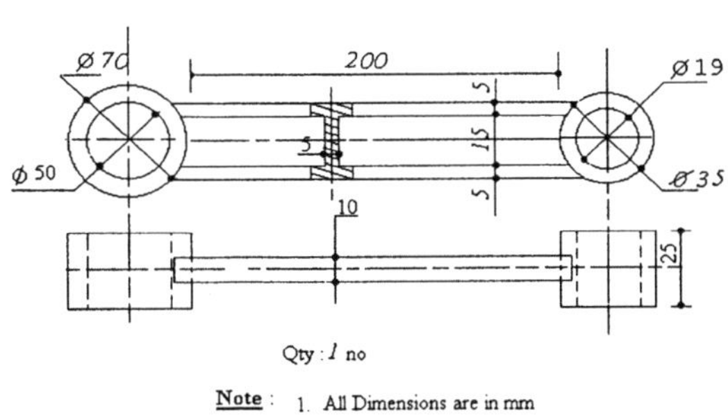

Figure A.3. Connecting rod

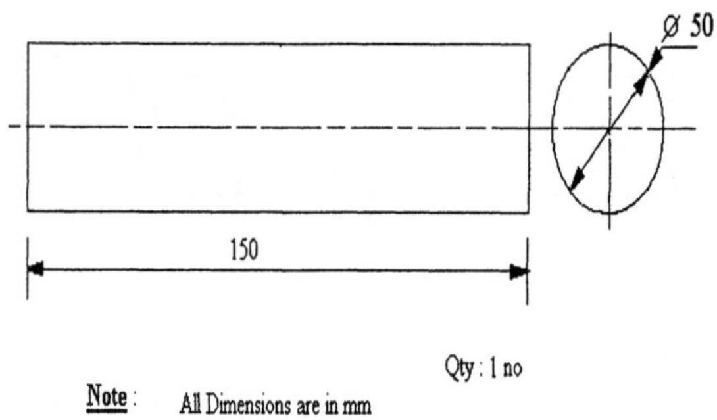

Figure A.4. Crank pin

References

R.H.Lyon, and R.G.DeJong. *Theory and Application of Statistical Energy Analysis*, Butterworth- Heinmann, Second Edition (1995)

V.R.Miller and L.L.Faulkner. *Prediction of aircraft interior noise using the Statistical Energy Analysis Method*, Journal of Vibration, Acoustics, Stress and Reliability Design, 105, pp. 512-518 (1983)

J.A.Steel. *The prediction of structural vibration transmission through a motor vehicle using Statistical Energy Analysis*, Journal of Sound and Vibration, 197, pp.691-703 (1996)

J.O.Jensen and J.H.Jensen(Editors). *Calculation of structure-borne noise transmission in Ships using Statistical Energy Analysis*, Proceedings of the International Symposium on Ship Board Acoustics (1977).

Alpini et al. *Analysis techniques of combustion noise and vibration in diesel engines*, Society of Automotive Engineers, 800406 (1980).

M.Okuba et al. *Analysis and reduction of piston slap noise in diesel engines*, Society of Automotive Engineers, 890127 (1989)

Lalor et al. *Engine noise due to mechanical impacts at pistons and bearings*, Society of Automotive Engineers, 800402 (1980).

Okamura et al. *Experimental study of the correlation between crank shaft vibrations, engine-structure vibrations and engine noise in high speed engines*, Society of Automotive Engineers, 951290 (1995).

H.Kanda et al. *Analysis of noise sources and their transfer paths in diesel engines*, SAE Transactions on Engines, 900014, pp. 34-41 (1990)

D.N.Manik. *A new method for determining coupling loss factors for SEA*, Journal of Sound and Vibration, 214, pp. 261-267 (1998)

AN ENVIRONMENTAL IMPACT ASSESSMENT OF A PROJECTED AIRPORT NOISE

K. KONISHI
Ken-on Engineering Inc. 1-17-1 Shofudai, Tarumi, Kobe, 655-0871 JAPAN

Z. MAEKAWA
Environmental Acoustics Laboratory, Kobe University Faculty of Engineering, Rokko, Nada, Kobe, 657-8501 JAPAN

Abstract

In order to obtain the data for environmental impact assessment of a new planned airport on an artificial island, long term (14 months) continuous measurement of long range (5 km) sound propagation over a sea surface was carried out with the MLS correlation method. Although, clear relationships could not be found between the measured noise reduction data and the meteorological conditions observed simultaneously, useful statistical data of excess attenuation for propagation of estimated aircraft noise was invented with no meteorological condition but using only noise reduction data themselves.

1. Introduction

When a new airport planned on an artificial island, the aircraft noise at the residential area on the land about 5 km far from the airport had to be predicted. Long range sound propagation can be expected to be unstable with atmospheric condition. Our major interest is to know the varying patterns of excess attenuation at the site, because when they deviate to negative and the arriving noise level becomes high, noise nuisance must happen. At what time and for what percentage of time does noise exceed the acceptable level? If it could be forecast that the noise level would be higher than acceptable, the operation of aircraft might be controlled in order not to incur social noise problems. For this purpose, some simulation programs of sound propagation such as the First Field Program (FFP) [1],[2] and Parabolic Equation (PE) [3]

M.L. Munjal (ed.), IUTAM Symposium on Designing of Quietness, 167–179.

were tried to apply. And, application of the Integrated Noise Model [4] was also carefully examined. However, there was no program applicable to the assessment, because the necessary meteorological data could not be obtained, which would represent local conditions. Furthermore, the propagation distance was too long to assume that the boundary layer on the land could be extended in its stratified condition to the sea [5]. Therefore, long term continuous measurement of sound propagation had to be carried out at the real site.

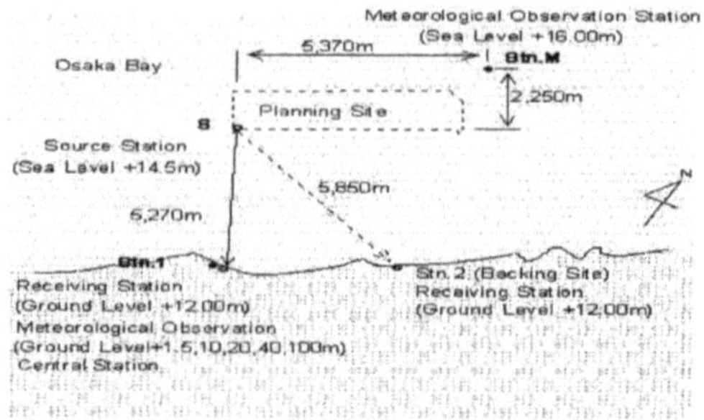

Figure 1. Measurement site

2. Outline of Measurement of Noise Propagation

2.1 Measurement Site And The Measuring System

Figure 1 shows the measurement site. The opposite shore of the artificial island is a suburb of Osaka City. A horn-type loud speaker source was installed on a revetment on the artificial island 14.5m above the sea level. Two receiving points were settled on the opposite shore at 12m above the ground more than 5 km far from the source. Stn.1 was the main station for data acquisition, while Stn.2 was a supplement. A parabolic reflector for the microphone was adopted in order to increase sensitivity for the received signal and to shield the noise coming from the rear side, where high level of traffic and city noise existed. A fully automated measuring device using the MLS correlation method was applied [6][7], in order not to annoy the people near the site, since the method can obtain the sound propagation data with signal levels lower than the background noise level. Details of the measurement system were published already [8]. As regarding the meteorological conditions, they were observed on a 100m tower near Stn.1, and also at Stn.M, 16m above sea level. These two meteorological stations were constructed according to ICAO instruction not for noise propagation but for operating aircraft [9]. So, a lot of meteorological conditions observed here were not used for predicting long range noise propagation, even though every possible means of statistical analysis were performed as shown in the Appendix.

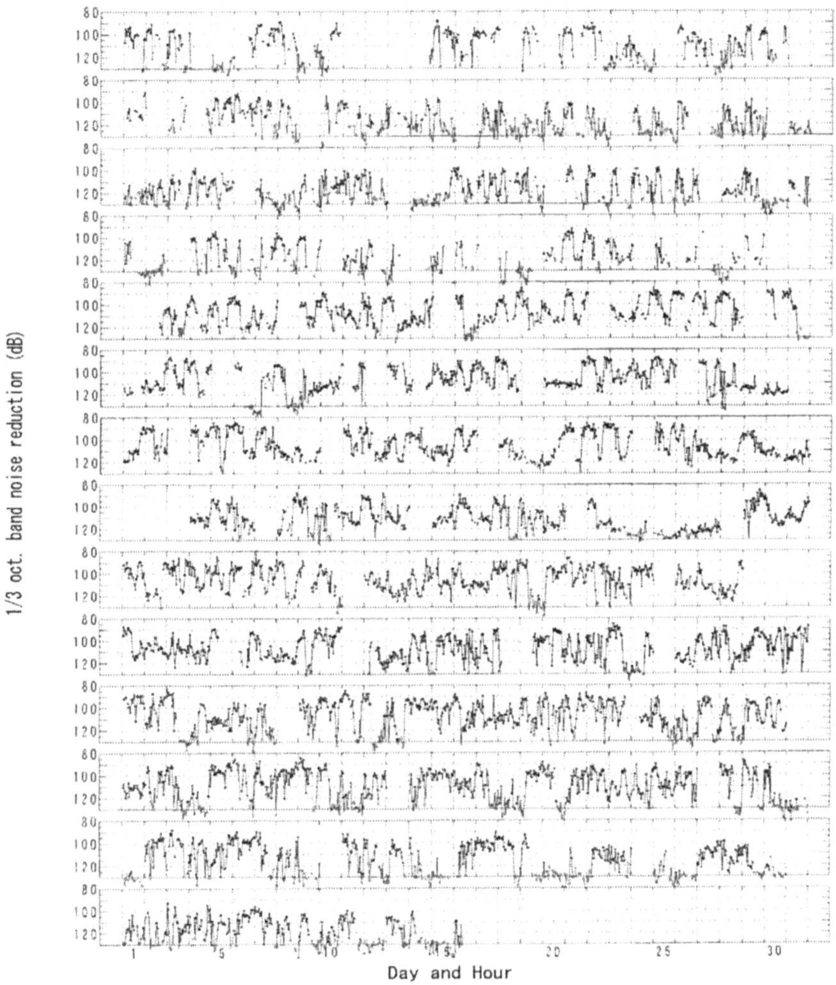

Figure 2. Noise reduction data of 1000 Hz measured at Stn. 1.

2.2 Measured Results For Noise Reduction

Measurement was carried out for 10 minutes at every hour, every day except holidays, during 14 months. Noise reduction was measured 8334 times and the measurements were successful for 62at Stn.1. An example of the measured data of noise reduction is shown in Figure 2.

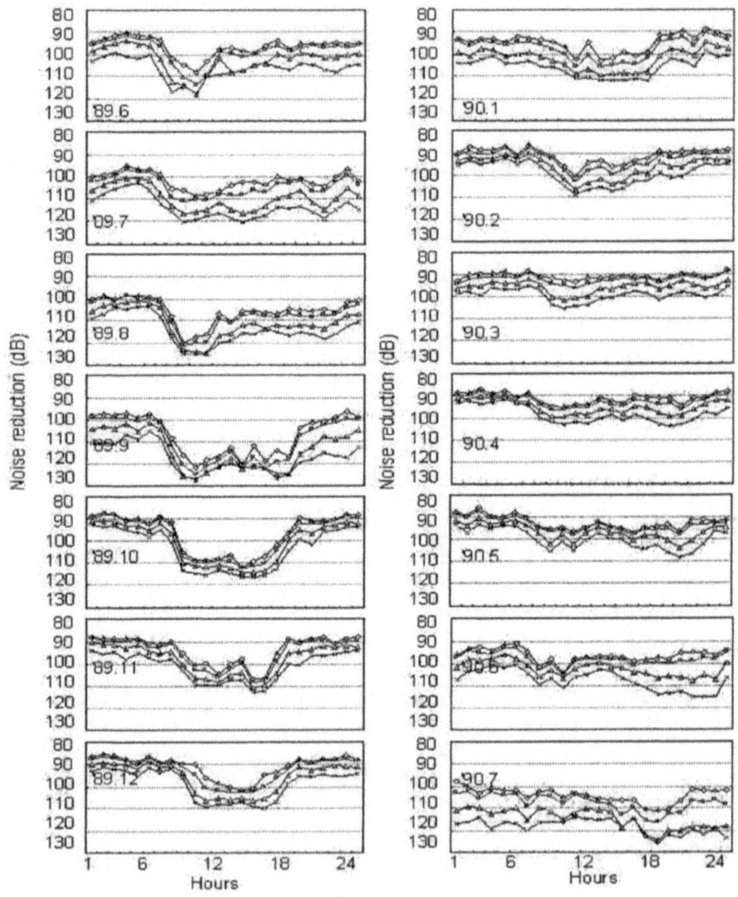

Figure 3. Monthly averaged hourly noise reduction for accrately and estimated data of 1000 Hz at Stn.1. Number of data from lower to heigher noise reduction is 3(◇), 5(□), 10(△) and 15(×)

They are including geometrical divergence of 74.4 dB where the distance is 5270m from the source. All values of noise reduction also include the effect of air absorption. It can be seen that the noise reduction seems to be fluctuating wide range with one day period. Figure 3 shows monthly average of noise reduction at every hour. It is clear that the noise reduction is larger in daytime than in nighttime especially in autumn.

3. Excess Attenuation Measured In 1/3 Octave Band Noise

Noise reduction data were divided into 20 ranks, with every rank including 5% of the data in 500 Hz at Stn.1. And all data, including also 250 Hz and 1 kHz measured simultaneously put into the ranks, were shown in Fig.4(a). The figure shows the distribution of the frequency characteristics of measured noise reduction. Then, the excess attenuation can be obtained by subtracting the geometrical divergence and the effect of air absorption from the noise reduction data. The necessary values for temperature and humidity were adopted from the mean values of those at Stn.1 and Stn.M in the sea [10]. The frequency characteristics of the excess attenuation is shown in Fig.4(b), with the same manner as Fig.4(a). Figure 5 shows the time distribution of these frequency characteristics of the same data averaged for every hour through 14 months.

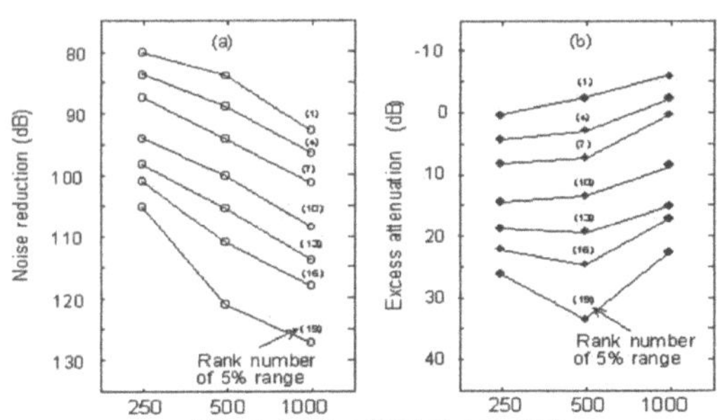

Figure 4. Distribution of frequency characteristics of (a) measured noise reduction and (b) measured excess attenuation at each 5% rank
Total number of noise reduction values at 250 Hz, 500 Hz and 1000 Hz is 4822, 6566 and 6220 respectively.

4. Estimated Excess Attenuation For Aircraft Noise

4.1 Estimation of Aircraft Noise

As shown in Fig.6, the A-weighted sound power level of an aircraft is approximated to be 145 dB based on a sound level of 91dB at a distance of 500m. Its spectrum is estimated from 5 octave band levels measured from 125 Hz to 2kHz for an L-1011 aircraft, at 64m far from an engine on the ground.

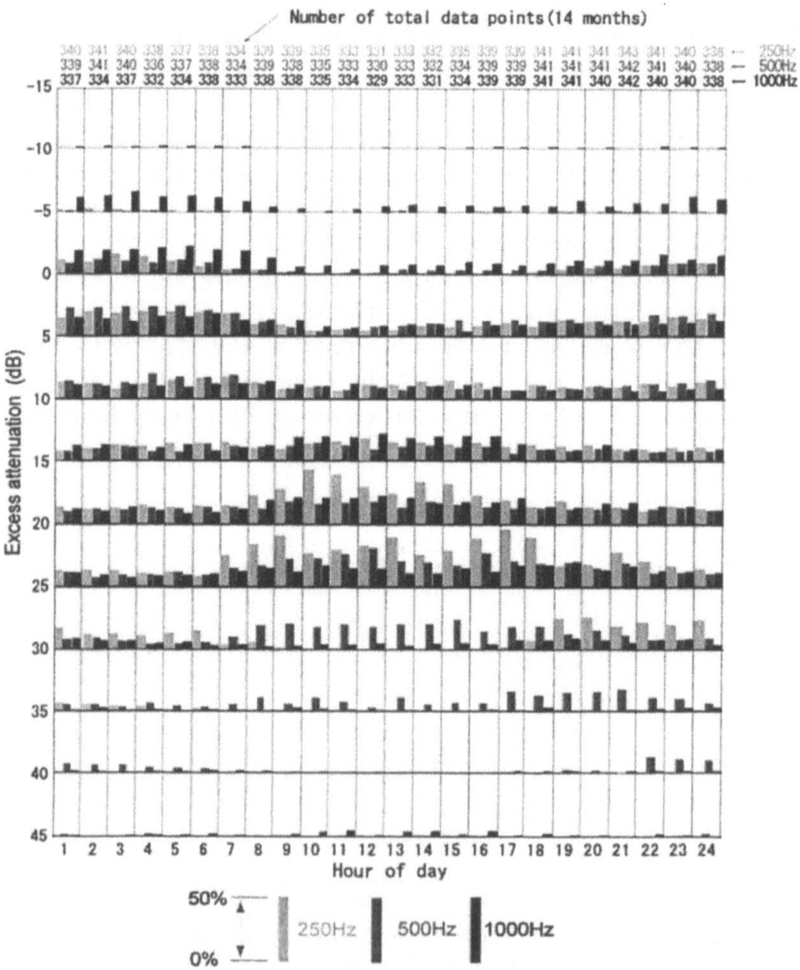

Figure 5. Frequency characteristics of excess attenuation for 3 oct. bands measured at Stn. 1. Excess attenuations are shown as percent distribution in 5dB steps every hour for three measured frequencies.(For example, for hour 10, an excess attenuation of 25dB occured about 25% of the time at 250Hz.)

Excess attenuation of the aircraft noise in each band is calculated using measured excess attenuation values shown in Fig.5. Since there are no data of noise reduction in two bands, 125 Hz and 2kHz, their noise reductions are assumed to be the same as 250 Hz and 1 kHz

respectively, and the excess attenuation are calculated with air absorption data for their own frequencies.

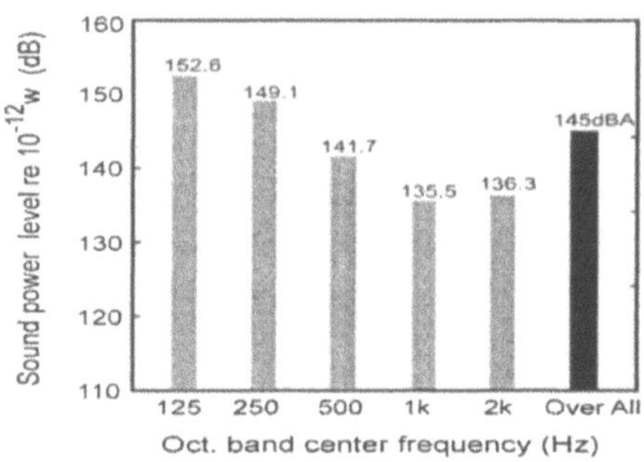

Figure 6. Sound Power level of an aircraft.
A-weighted sound power level 145 dB is approximated from the data measured at a 500m slant distance at low angle of elevation.Octave band power levels are measured by the measured spectrum.

4.2 Time Distribution of Excess Attenuation For Estimated Aircraft Noise

The probability of occurrence for estimated excess attenuation for aircraft noise is obtained by combining the data shown in Fig.5 and Fig.6; the result is shown hourly in Fig.7. Negative values of excess attenuation in this figure might be a possible problem for the environment in the nearby community. These negative values do not appear in the daytime but do appear in the nighttime, though it is a rare observation.

4.3 Percentage Distribution of Estimated Excess Attenuation For Aircraft noise

Figure 8 shows the hourly percent distribution and histograms of the occurrence of excess attenuation for aircraft noise. The number of total data is almost uniform for every hour as also shown at the top of Fig. 7. The distribution of the excess attenuation is wider and more uniform in nighttime than in daytime.

5. Discussions

5.1 Long Range Noise Propagation and Meteorological Conditions

It seems that there is no factor to affect the sound propagation in the atmosphere except meteorological condition.

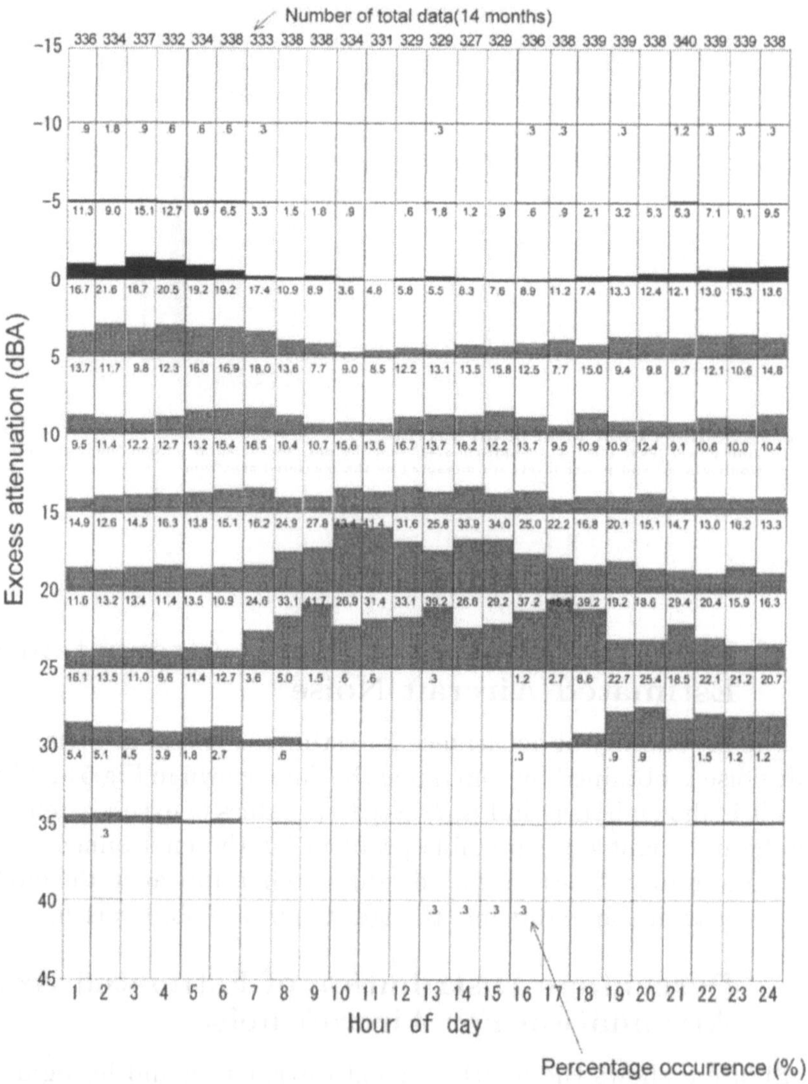

Figure 7. Probability of occurence of excess attenuation for aircraft noise estimated at Stn.1.
A-weighted sound power level of the aircraft is approximated by 145 dB, and its spectrum is etimated from the 5 oct.bands data shown in Fig.55

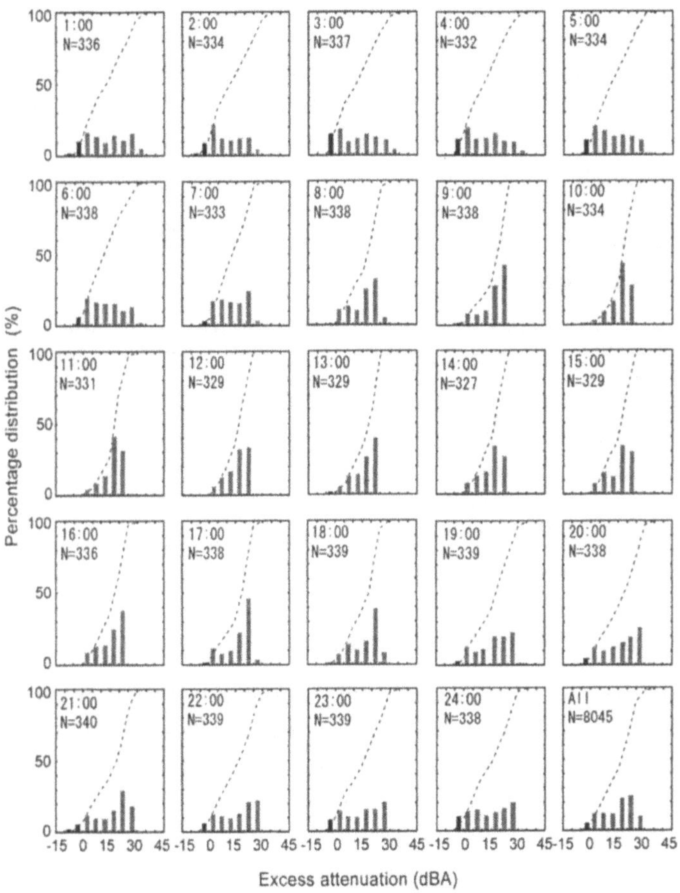

Figure 8. Hourly distribution (brs) and Cumulative values (dotted line) of excess attenuation for aircraft noise, same as shown in Fig. 7. N: number of total data(14 months)

However, it is a disappointing fact that the many results obtained by the statistical analyses could not explain explicitly the relation between noise reduction and meteorological conditions observed simultaneously as shown in the Appendix. This fact might be due to unavoidable errors with spatial sampling for the meteorological data over the propagation path, and to the turbulent eddies that always appear randomly in the convective boundary layer [11]. It might be possible to obtain the fine data corresponding to the sound propagation, if the meteorological observation could perform at more suitable positions. Unfortunately, however, a prediction method for the noise environment was required using conventional ordinary data of meteorological conditions observed at Stn.M more than 5 km apart from the Source stations.

5.2 Computation Of The Long Range Noise Propagation

In order to predict the excess attenuation, it is not practicable to use the FFP or the PE computation programs, even if they can take into account the existence of turbulence in the atmosphere, because the turbulence is an unpredictable phenomenon. In other words, there is no method to obtain the meteorological data sufficient for computing the sound propagation. Therefore, development of a new proper method to obtain the useful local meteorological data for predicting the sound propagation is essential for environmental noise impact assessment.

5.3 Concluding Remarks

In spite of the fact that the relationship between noise and meteorological data had not been clear, long range and long term measurement of noise propagation was still useful in this assessment, using a probability method of noise data analysis. The excess attenuation of aircraft noise estimated at a receiving point is almost always positive. And it has become clear that the excess attenuation is usually negative only at nighttime infrequently. This fact will be very useful for the environmental impact assessment of noise from the planned airport.

Appendix

Some Relations Between Noise And Meteorological Conditions[12]

The meteorological conditions were observed on a 100m tower near Stn.1 in Fig.1, i.e. temperature, humidity, wind speed/direction at the elevations 1.5m, 10m,

20m, 40m and 100m above ground. And also at Stn.M, 16m above sea level, temperature, humidity, wind speed/direction were observed.

Figure A.1 shows examples of simple relation between noise reduction at 500 Hz and temperature difference between 100m-10m at Stn.1. Figure A.2 shows the same between noise reduction and wind direction. These meteorological conditions were expected to show strong effects on the sound propagation, but the results show such disappointing facts.

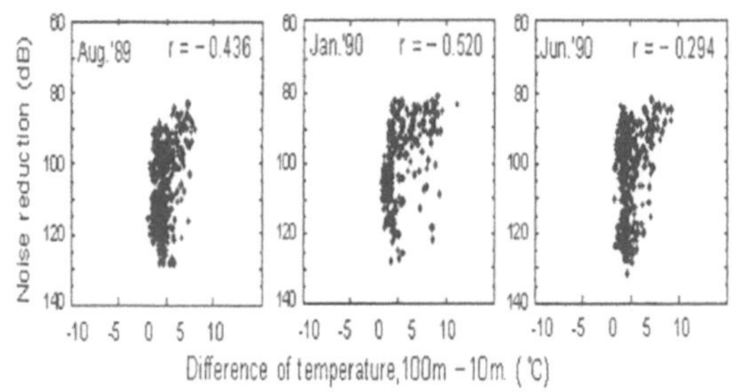

Figure A.1. Examples of simple relation between noise reduction data at 500Hz and difference of temperature from 100m to 10m at Stn.1

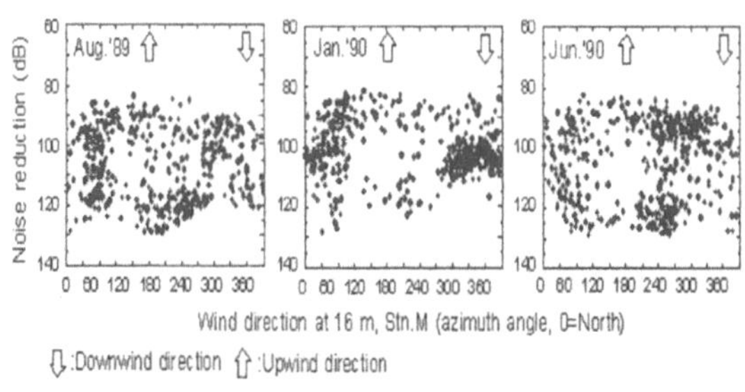

Figure A.2. Examples of simple relation between noise reduction and wind direction

Figure A.3 shows the correlation between noise levels measured and estimated ones using the results of the multiple regression analysis of 500 Hz at Stn.1.

For the analysis selected 5 meteorological data were used, such as temperature at 10m, temperature difference 100m-10m, relative humidity, incoming-outgoing radiation and vector wind speed at 100m above ground. This results show that this method is not sufficient for the prediction.

Figure A.4 shows the similar result of Fig. A.3 using the multi-dimensional quantification(1st kind) analysis[13]. For this analysis ten meteorological factors were used as dependent variables while the noise reduction as the independent variable. But this figure shows only slightly better result than that of Fig.A.3, and reveals the fact that reliable prediction of noise reduction with using the meteorological data obtained here is almost impossible.

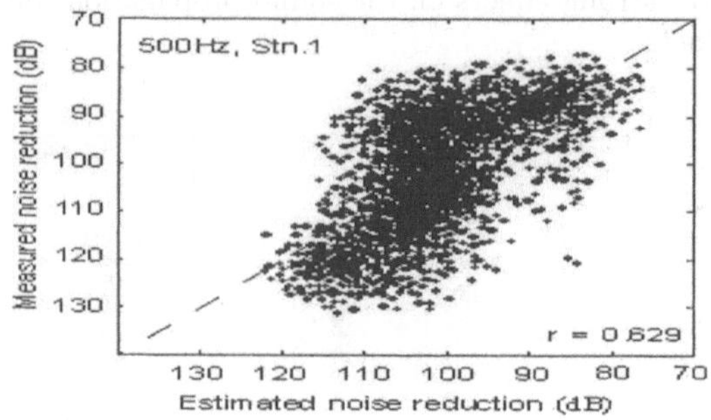

Figure A.3. Measured and estimated noise reduction obtained by using the multiple regression analysis.

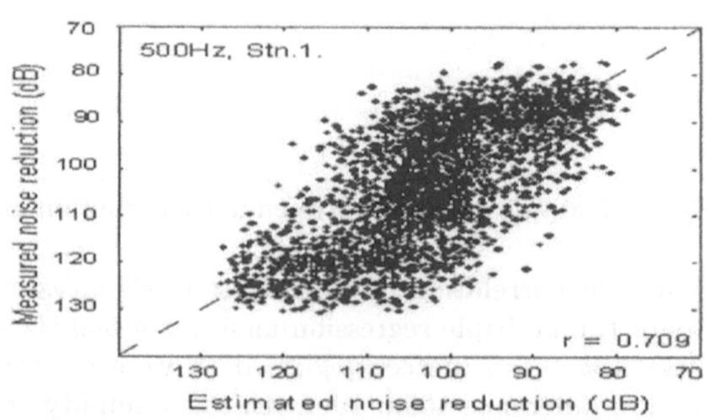

Figure A.4. Measured and estimated noise reduction obtained by using the result of multi-dimensional qualification (1st. kind) analysis

References

Raspet, R., Lee, S. W., Kuester, E., Chang, D.C. , Richards, W. F. , Gilbert, R. and Bong, N. (1985) Fast-field program for a layered medium bounded by complex impedance surfaces, *J. Acoust. Soc. Am.*,**77(2)**,p .345-352.

L'Esperance, A., Niclas, J. , Wilson, D. K. , Gabillet, Y. and Daigle, G. (1993) Sound Propagation in the Atmospheric Surface Layer; Comparisons of Experiment with FFP Predictions, *Applied Acoustics,*,**40, No 4**,p .325-346.

Gilbert, K. E. and White, M. J. (1989) Application of the parabolic equation to sound propagation in a refracting atmosphere,*J. Acoust. Soc. Am.*,**85(2)**, p .630-637.

Gados, R.G. and Aldord, J.M. (1979) FAA Integrated noise model validation-phase1, *Analysis of integrated noise model calculation for air carrier fly-over*, **U.S. Department of transportation.**,

Embleton, T. F. W. , Piercy, J. E. and Olson, N. (1976) Outdoor sound propagation over ground of finite impedance, *J. Acoust. Soc. Am.*, **59(2)**, p .267-277.

Aoshima, N. and Igarashi, J. (1968) The M-sequence Signal Applied to the Correlation Measurement of the Wave Propagation,*Proc. 6th I.C.A.*, F-1-5,

Konishi, K. , Aoki, M. and Maekawa, Z. (1979) Measurement of the long distant noise propagation over ground, *Proc. Inter-Noise*,**79**, p .761-764.

Konishi, K. , Tanioku, Y. and Maekawa, Z. (2000) Long time measurement of long range sound propagation over an ocean surface, *Applied Acoustics*,**61(2)**,p .149-172.

Annexes 14-1 to the Convention on International Civil Aviation, (Jul. 1990)"Aerodromes. Vol. 1 Aerodrome Design and Operations", *I.C.A.O. Publications*

International Standard, Acoustics-Attenuation of sound during propagation outdoors-Part 1: Calculation of the absorption of sound by the atmosphere", **ISO 9613-1:1993(E)**

Kaimal, J. C. (1988) The Atmospheric Boundary Layer - Its Structure and Measurement, Lecture Note, *Indian Institute of Tropical Meteorology, P.O. Box 913, Shivajingar, Pune 411005*,

Konishi, K. and Maekawa, Z. (2001) Interpretation of long-term data measured continuously on long range sound propagation over sea surfaces , *to be published in Applied Acoustics*,

Hayashi, C (1952) On the Prediction of Phenomena from Qualitative Data and the Quantification of Qualitative Data from the Mathematico-Statistical Point of View,*Ann. Inst. Stat. Math.*,**3**,p .69-98.

DYNAMICS OF HEARING - SENSITIVITY TO NOISE

ALBRECHT EIBER and WERNER SCHIEHLEN
Institute B of Mechanics
University of Stuttgart
ae@mechb.uni-stuttgart.de

Introduction

For communication between humans sound events are widely used to transmit information. Also for pleasure like music or fireworks, sound waves as pressure variations in the ambient medium are most important. On the other hand, these cases show negative aspects of sound, too: Depending on the frequency distribution sound will result in discomfort and high sound pressure levels causing pain or even injuries to the human body. There are many unintended sound events in daily life due to machines and other devices: in working (forging, riveting, punching), in relaxing (music, shooting), in travelling (cars, air bags, railways, air planes). The design for quietness is intended to reduce such sound events generally denoted as noise.

In this paper the effects of sound to the auditory system are considered based on mechanical models. The spatial motion of the middle ear structure is excited via the ear drum, and transfers the sound to the inner ear. The analysis is based on three-dimensional mechanical models of the middle ear and its adjacent structures. These models provide the spatial motion of the members of the middle ear which are related to the hearing impression as well as to the hazard of damage to the hearing by intensive noise.

Due to its dangerous nature and the high risk of hearing loss transient sound events are considered in detail. The time history of the resulting motion is essential since global measures like energy of the transient impulse allow only a very rough estimation of probable hearing loss.

M.L. Munjal (ed.), IUTAM Symposium on Designing of Quietness, 181–200.

1. Sound Perception and Hearing Process

The human body is able to perceive sound events in different manners. The common and most important path is via the hearing organ, but some additional feeling is achieved by organs like skin, stomach, and lung. The organ of Corti is located in the inner ear and produces an electrical signal in the auditory nerves. This voltage is generated by an electro-chemical process from mechanical stimulation of the cilia of the outer and inner hair cells.

There are different sound transmissions to the inner ear, resulting from different mechanisms of excitation:

- structural vibrations of the skull, in audiology called bone conduction;
- mechanical vibrations of the ossicular chain in the middle ear, called air conduction and to some extent
- air borne sound in the tympanic cavity.

Figure 1 shows schematically the transduction of the sound pressure p applied at the external ear canal to the spatial motion of the ossicular chain in particular the piston-like inward/outward motion of the stapes footplate y_S and the rocking motions α_S, γ_S around its long and short axis resulting in a hearing impression h.

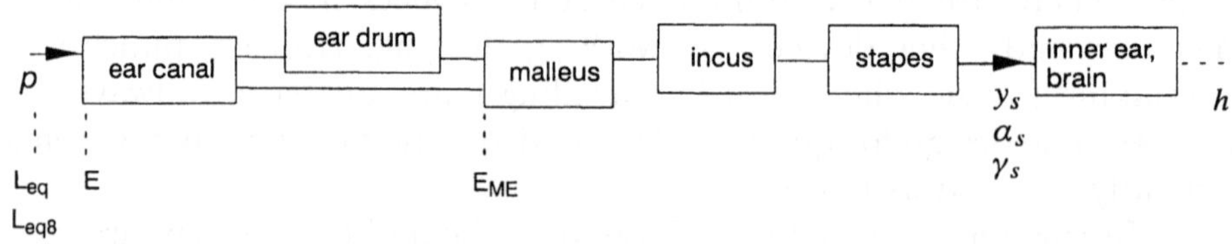

Figure 1. Path of sound pressure p through the ossicular chain (air conduction) to the hearing impression h. Transmitted energy is denoted by E, L means equivalent sound pressure levels and y_S, α_S and γ_S are the motions of the stapes

The hearing organ is a very powerful sensor covering several orders of magnitude but it is very sensitive against singular overload and against exposure to high noise level over time. To understand the damage caused by noise, it is necessary to design appropriate models of the hearing organ.

Classical models are often based on electric circuits using electro-acoustical and electro-mechanical analogies. These models are mostly dealing with scalar entities in a global sense and do not describe the local three-dimensional dynamical behavior of the ossicles [1, 2, 3, 4].

Mechanical models allow the consideration of spatial motions of the specific ossicles due to a prescribed pressure or force function based on the laws of dynamics. Then, the motion follows directly from the applied forces.

2. Mechanical Model of the Middle Ear

Mechanics generally deals with the relation between forces and motion or strength in distinct components of a system. This is also the case for investigation of the hearing process. Moving particles in the air connected with pressure variations form air borne sound. A sound event reaching the concha is focused and brings the air column in motion. Consequently the ear drum and the middle ear structure are brought in motion. The spatial displacement of the stapes acts as the excitation of the fluid in the inner ear which deforms the membranes and the organ of Corti. The whole process is a mechanical one therefore mechanical model should be used to describe it [5, 6, 7].

The air column in the outer ear canal is modeled as a continuous system in a rigid cylindrical tube. Derived from that a lumped mass system was chosen to represent the dynamics of the air column in the audible frequency range.

For the ear drum with its conical shape the finite element approach is used to describe the nonuniform membrane- and bending stiffness as well as the nonuniform thickness. From these studies a lumped mass model was derived and coupled with the ossicular chain.

The ossicles are considered as rigid bodies and the ligaments as massless spring/damper combinations. Position and orientation of each ligament is taken from human anatomy. The stiffness and damping properties of the ear drum and the annular ring are introduced by full matrix relations. Using a multibody systems formalism [8] a model was generated consisting of three bodies viscoelastically hinged in the tympanum by various ligaments as shown in Fig. 2. Normally the joint between malleus and incus is a fixed connection but for high force intensity it is articulated to decouple the incus from the motion of malleus to some extent. The joint between incus and stapes has an important role. It is modeled as a inxsix-dimensional coupling with three forces and three moments.

The model of the ossicles has 9 degrees of freedom (dof) which are enlarged by adding the subsystems for the ear drum, the air column and the fluid column of the inner ear to f = 25 dof. Due to small displacements in the hearing region, up to 80 dB SPL linear kinematics and linear constitutive equations of the compliant materials are assumed, and the incudo-malleal joint is considered as fixed. For describing the normal hearing process dealing with moderate sound pressure levels (< 100 dB) and even for impulsive noise the nonlinear equations of motion can be linearized as long the deflections are below certain limits [9]. Presently the nonlinear form of the models contain a lot of rough assumptions and hypotheses but the investigations are ongoing. In Fig. 2 the multibody system model of the entire model is sketched. The general form of the equation of motion reads as

$$\mathbf{M}\ddot{\mathbf{y}} + \mathbf{k} = \mathbf{q}. \tag{1}$$

Having f degrees of freedom, y(t) denotes the time dependent f×1 position vector containing the f generalized coordinates, $\dot{\mathrm{y}}$ and $\ddot{\mathrm{y}}$ are the derivatives with respect to time describing the velocities and accelerations. The f×f mass matrix **M** contains the mass and inertia properties of the multibody system's members, the vectors **k** and **q** describes the generalized coriolis forces and the generalized applied forces, respectively. After linearization these equations have the form

$$\mathbf{M}\ddot{\mathbf{y}} + \mathbf{P}\dot{\mathbf{y}} + \mathbf{Q}\mathbf{y} = \mathbf{h}, \tag{2}$$

where the forces dependent on the velocity are described by the matrix **P** and those dependent on the displacement by the matrix **Q**. The external excitation of the multibody system is summarized in the vector **h**.

Using today's powerful computational tools for modeling and generating the governing equations of motion for complex mechanical systems, it is no big problem to establish a detailed model of the hearing organ. The main problem however is the determination of the belonging parameters. After estimations exploiting literature and previous simulations, series of measurements on cadaver specimen and living subjects have been carried out [10, 11].

Based on Laser Doppler vibrometry the inxmechanical transfer functions were derived from the motions of different points of a temporal bone preparation driven acoustically or by a permanent magnet and an electrical coil. In Fig. 3 the transfer function G between the stimulation at the manubrium and the stapes footplate is shown for different driving voltages corresponding to sound pressure levels of about 64 and 74 dB SPL showing the frequency-dependent behavior of the middle ear. The cochlea was removed to get a perpendicular sight to the footplate. Therefore, the measured data have to be compared with a corresponding model without cochlea.

Some data have been derived using the concept of multifrequency tympanometry. A multi frequency probe tone consisting of 400 sinusoids equidistant with Δf =15 Hz in the range of 15 Hz and 6 kHz was applied

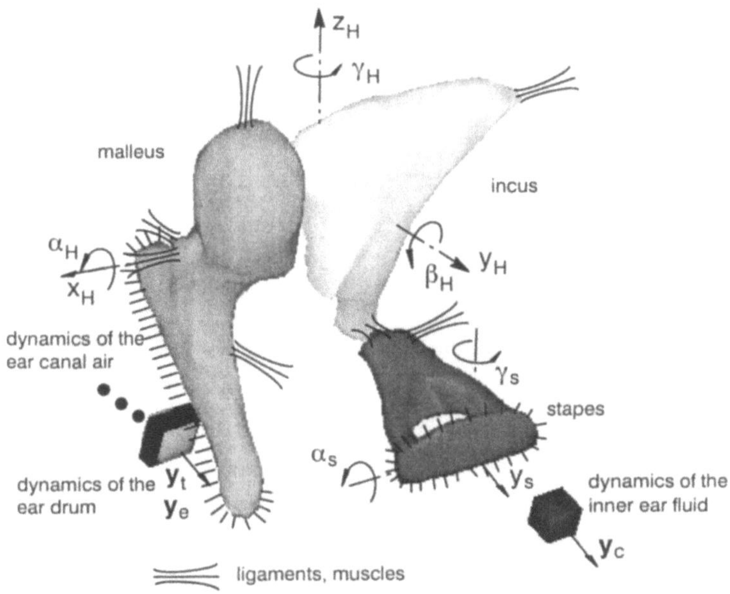

Figure 2. Mechanical model of the middle ear and its adjacent structures. Multibody system with translations x_H, y_H and z_H and rotations α_H, β_H and γ_H of the incudo-malleal block, translation y_S and rotations α_S, and γ_S of the stapes. The ligaments are indicated as black lines

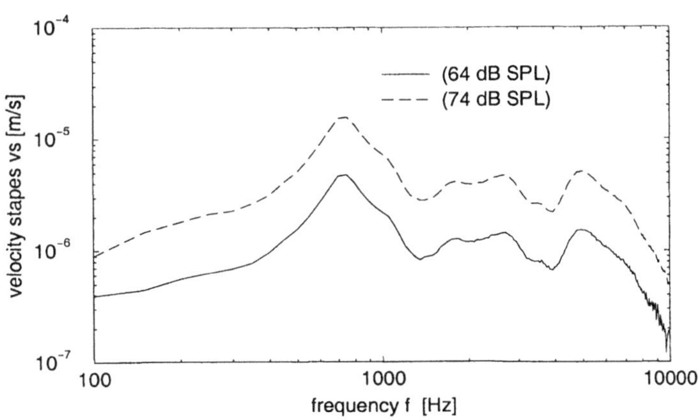

Figure 3. Measured transfer function between magnetical stimulation at the malleus handle and the velocity of stapes footplate. Excitation with different voltages corresponding to 64 and 74 dB SPL

to the ear canal. The superposed static pressure varied from -300 daPa to +250 daPa. By means of Laser Doppler vibrometry the velocities of different points on the ossicles and the ear drum have been measured. A shift of natural frequencies can be seen due to the nonlinear behavior of the ear drum and the middle ear ligaments [12]. In Fig. 4 the velocity of the tip of manubrium (umbo) is indicated by different colors and plotted depending on the static pressure in the ear canal and the frequency of excitation.

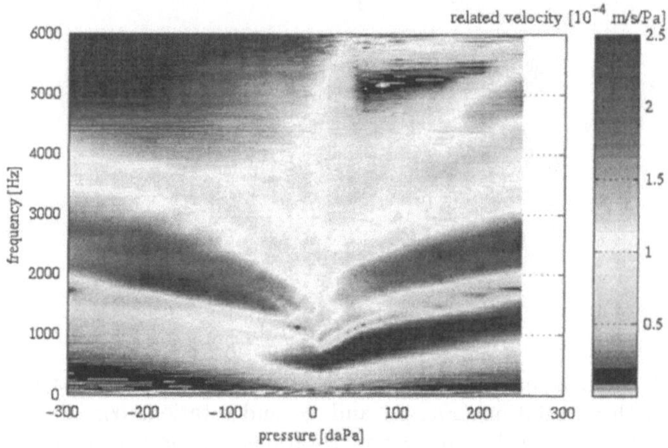

Figure 4. Multifrequency tympanogram derived from the measured velocity of umbo

Connected with the shift of natural frequencies a reduced sensitivity to the applied sound can be observed. This fact is known from situations with a reasonable pressure difference between ear canal and tympanic cavity, it was proposed for the design of a hearing protector.

The established three-dimensional mechanical models can be used to simulate the spatial motions of the middle ear structure due to different sound events $p(t)$. Transient motions occur for a short time when a steady state motion starts or in the case of transient excitations like inximpulsive noise. A steady state motion, mostly a superposition of harmonical motions, is the response to a stationary excitation.

Concerning the loads to the system components maximal values of deformations and of velocities of deformation are determining the forces in the system and thus the stresses. Particularly the spatial motion of the stapes consisting of a piston-like motion y_S and rocking motions α_S, γ_S around the long and short axis of the stapes footplate, respectively the corresponding velocities are the input to the inner ear and determine the excitation and load.

In the case of transient solution the peak values of the displacements and the velocities are the entities to assess the hazard of a particular sound event. For this the time history $y(t)$ has to be calculated by time integration of the equations of motion in the nonlinear or even in the linearized form.

Dealing with a steady state situation, for a linearized consideration the elegant concept of inxtransfer functions can be used working in the frequency domain where a time integration is avoided.

Another application of such simulations is the development of passive and active middle ear implants. Using extended models the effect of different surgery procedures and different types of implants to the hearing can be studied and serve for optimization of prostheses [13].

Generally, simulations using such mechanical models can be considered as numerical experiments carried out at the computer to reduce the cumbersome physical experiments in the lab on temporal bones and living patients. Here there are no ethical restrictions and distinct parameters can be varied keeping the others in a defined form. This advantage becomes obvious when the different criteria for assessment of impulsive noise have to be compared.

3. Risks, Evaluation, Assessment

Exposure to an intense noise may cause a noise induced hearing loss (NIHL) which appears as a temporary threshold shift (TTS) or in a permanent form as a permanent threshold shift (PTS).

It is thought that TTS is due to the changed electro-chemical process in the organ of Corti caused by a reduced number of open ion channels in the hair cells. The consequence is a reduced gain of the so-called cochlear amplifier, the active element in the inner ear. This effect may explain the observation [14] that the hearing organ shows a kind of self protecting mechanism of reduced sensitivity when it is exposed to intense permanent noise. Another hypothesis supposes a reversible change of the mechanical properties of the ear drum and the ligaments and/or an active tension of the middle ear's muscles. Recovering from a TTS needs several hours and is facilitated by a silent ambience. The method of optimal medical treatment is still under discussion, generally it is supposed that improving the blood flow and the oxygen supply will support the recovering process.

The permanent loss of hearing power PTS has its origin in a mechanical destruction of parts of the hearing organ or in irreversible changes of its mechanical parameters.
Damages of the hearing organ may occur in different ways:

a) A high static load to the tympanic membrane is caused e.g. during diving, flying or by diagnostical investigations like tympanometry. Here the difference in pressure Δp between the ear canal and the tympanic cavity is about ± 300 daPa causing sometimes pain to the patient. In normal conditions a static pressure difference is limited by the function of the Eustachian tube. In the case of an overpressure the tuba opens around +400 daPa, a vacuum must be reduced actively e.g. by swallowing. Static pressure causes ruptures in the tympanic membrane. Different values for the critical pressure p_{rupt} can be found in literature based on different conditions of excitation [15, 16, 17]. An applied pressure of 35 kPa, 100 kPa or 280 kPa leads to a probability of rupture of 10%, 50% or 100% respectively. These values are not very confident because experiments cannot carried out easily.

b) Steady state noise causes damages of outer and/or inner hair cells of Corti organ. An intense touching of the tectorial membrane may destroy a reasonable number of hair cells. In the human ear there are about 16000 hair cells. Destroyed hair cells cannot be reconstructed.

c) Transient sound events at first cause high stresses in the mechanical structure and may also damage the outer and/or inner hair cells of Corti organ. A weak element in the mechanical structure is the ear drum. Dynamic load by a pressure wave may cause ruptures in the ear drum due to high stress at a certain location. This stress is the consequence of large deformations. Therefore the time history of deformation and the corresponding velocities as the answer to a particular excitation have to be considered. Looking at the structures of the middle or inner ear, again the strength in particular elements is determined by the deformation and the time rate of deformation which are the dynamic answer of the mechanical system to the excitation. This fact must be regarded particularly in the case of transient excitation. Such impulsive noise events are very dangerous because dynamical overshots of deformation depending on the particular time history of the excitation may play the essential role in the load process. Moreover, a self protection like in the case of permanent noise exposure does not exist. The only protection is a rupture of the ear drum or the articulation of the incudo-malleal joint.

Criteria for assessment of noise:

In the static or quasistatic case the maximum of applied pressure load is characterizing the strength of the hearing organ.

A permanent noise can be characterized by its effective mean level, the peak values and the frequency content. Currently the iso-energy principle is used to define a measure of noise impact or noise dose related to the exposure time T. The energy-equivalent mean value of sound pressure level is defined as

$$L_{eq} = 10 \cdot \lg(\frac{1}{Tp_0^2} \int_0^T p^2(t)dt). \tag{3}$$

To get a valuation of the noise dose, rating levels related to a rating time of 8 hours were introduced as

$$L_{eq8} = L_{eq} + 10 \cdot \lg(T/28800s). \tag{4}$$

These levels can be calculated using the measured sound pressure without regarding the reaction of the hearing organ. It is a well known fact, that the hearing impression is strongly dependent on the frequency of excitation which is roughly described by the different, normalized weighting curves A, B and C.

Calculating the equivalent levels via the spectrum of the applied sound allows to regard the frequency-dependence of the ear by weighting, [18, 19, 20], e.g.

$$L_{Aeq} = 10 \cdot \lg(\frac{1}{Tp_0^2} \int_{F_{min}}^{F_{max}} (\, A(f)p(f)\,)^2 \, df). \tag{5}$$

Particularly in the case of impulsive noise the time history or form of the excitation determines the dynamical response, i. e. the displacements $y(t)$ and the velocities of the system's members. A damage may be caused by exceeding the ultimate breaking stress or the ultimate fracture point leading to a permanent hearing loss. In this case only the time history $y(t)$ can be used to describe the situation [4, 9]. If these limits were not exceeded, a temporal threshold shift may be caused. In this case the noise load should be computed also from the time history $y(t)$. To evade this calculation and to get a rough guess of hazard different criteria have been stated based on the data of the impulse.

To assess the noise of different weapons the equivalence of energy is used [21, 22, 23]. Defining a time of duration, the noise dose is considered to be the peak pressure p_{max} multiplied by the duration time and number of sound events per day. This rating level is limited by a straight line. Fig. 5 shows the time of duration and the criteria of Pfander and Smoorenburg. Both do not use any information of the hearing organ itself. This drawback is lessened by calculating the energy equivalent level from the weighted spectrum of the impulse, e.g. using A-weighting. A limit for such level L_{Aeq8} of 85 dB was proposed in [24].

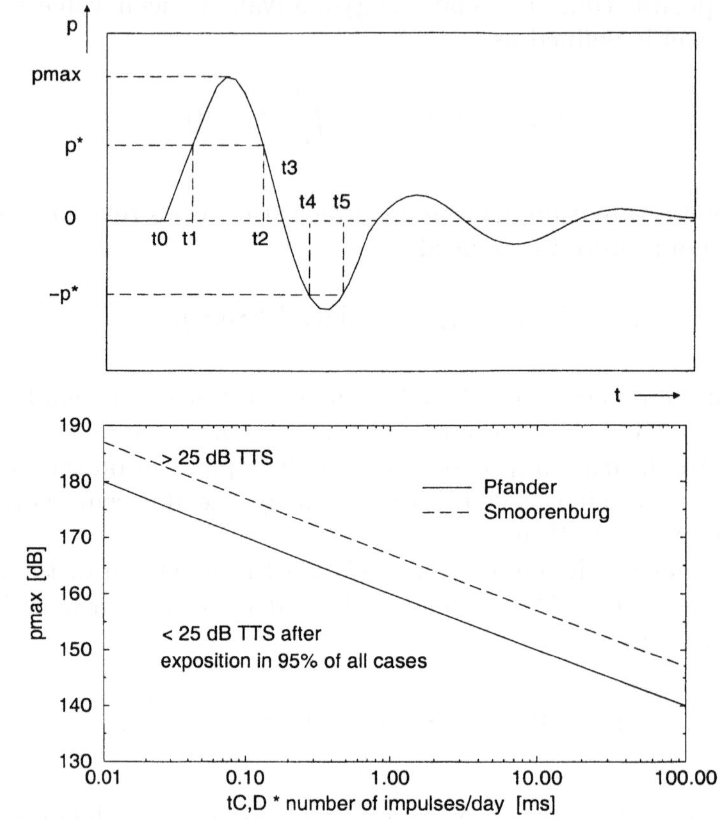

Figure 5. Definitions of duration of impulse: $p^* = p_{max} - 10dB = 0.316\ p_{max}$. A-duration $t_A = t_3 - t_0$; C-duration (Pfander) $t_C = (t_2 - t_1) + (t_5 - t_4)$; D-duration (Smoorenburg) $t_D = t_5 - t_1$. Pfander and Smoorenburg criterion

4. Applications

4.1 Transient sound

A shot of a 9 mm pistol was taken from [21] and modified first to a single triangular pulse taking only the first peak, second to a full shot consisting of two triangular pulses in the positive and in the negative pressure range and third to two positive pulses as shown in Fig. 6. This is to demonstrate that the response of the ear is strongly dependent on the particular form of the excitation. Especially the negative branch may have an important contribution to the system's response. This is the reason for window glasses falling outside of a building even due to an external detonation.

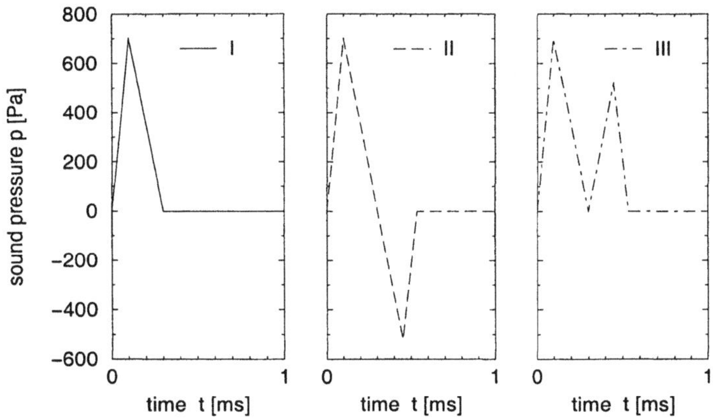

Figure 6. Sound pressure of different impulses derived from a pistol shot. I: triangular pulse; II: two triangular pulses, two-sided; III: two triangular pulses, one-sided

From the resulting response of the stapes piston-like motion y_S and rocking motions α_S, γ_S the displacements are shown in Fig. 7. In all three cases the maximum peak is caused by the first positive branch of the of the shot. In contrary the velocity peaks reaches their maximum later and in the negative direction. The high danger of the two-sided pulses becomes obvious because the stapes velocity is the primary excitation of the inner ear, Fig. 8. It should be mentioned that significant rocking motions of the stapes take place. It is postulated that they have a reasonable influence to the inner ear.

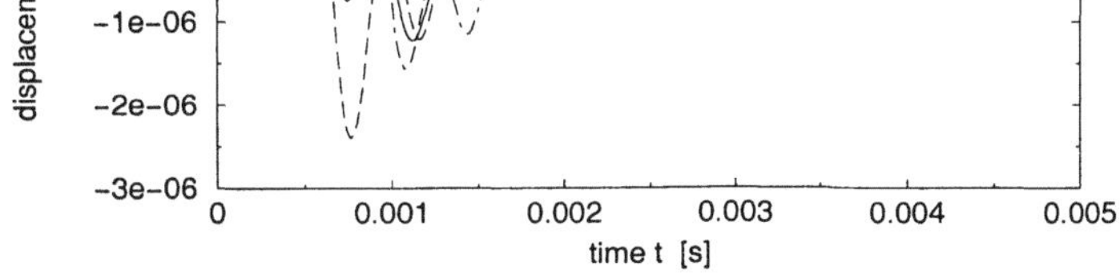

Figure 7. Stapes displacement y_S due to different impulses

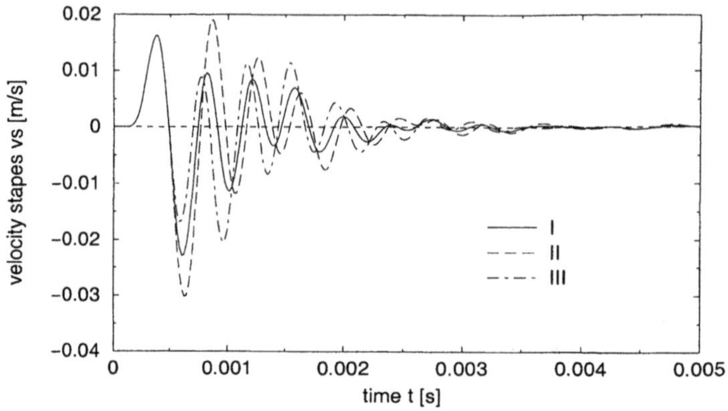

Figure 8. Stapes velocity v_S due to different impulses

Very often the energy of the sound event or the energy transmitted to the ear is used to characterize the danger of a sound event. Taking only the sound pressure, the energy and the corresponding rating levels are not strictly related to excitation of the inner ear. Even the energy transmitted to the air in the ear canal is not an appropriate measure, more likely seems the energy transmitted to the middle ear. The time fluctuations in Figs. 9 and 10 show that the transmitted energy for the considered sound events are different time dependent functions. A flow back of energy due to the internal dynamical behavior of the ear can be seen particularly in Fig. 10 at the entrance to the middle ear. Only a part of the energy introduced to the air column at the entrance of the ear canal is transmitted to the middle ear structure. In Table 1 the characteristic values of the different impulses and the corresponding reactions of the stapes are put together.

By evaluation of the different assessment criteria the number of admissible rounds are fairly different as it can be seen in Table 2.

From this example it becomes clear that for assessment of impulsive noise it is not sufficient to look on the energy delivered by the sound source. It is rather a question of energy transmitted to the inner ear which is determined by the spatial motion of the stapes. Therefore the most appropriate assessment of noise, in particular impulsive noise is based on the time history of the spatial stapes motion.

For other sound events measured by Prof. Fleischer, Giessen, the response of the middle ear have been calculated as shown in Figs. 11 and 12. Similar to the idealized impulses above, the time history of the impulses determine the stapes motion and the negative branches sometimes plays an important role.

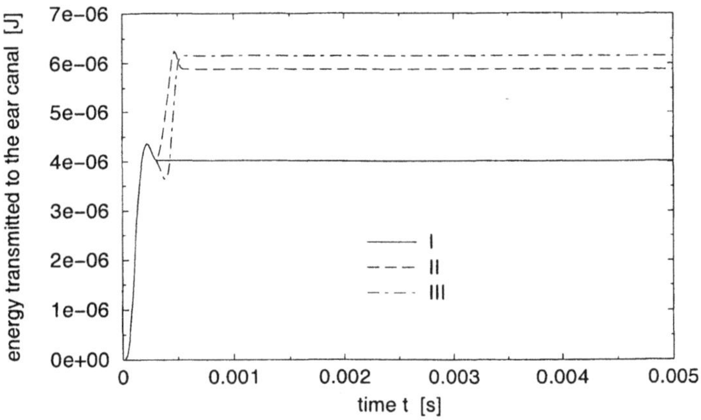

Figure 9. Transmitted energy from free field to the entrance of the ear canal

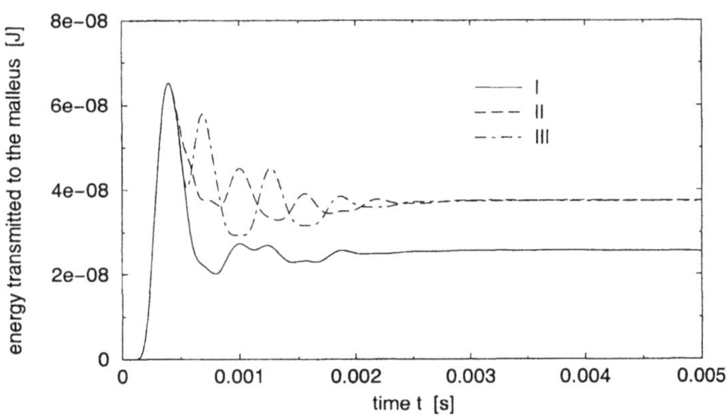

Figure 10. Transmitted energy from the air column and the ear drum to the malleus

4.2 Harmonic sound

The dynamical behavior of linear systems can be characterized by means of transfer functions. Applying a steady state pure tone at the entrance of the outer ear canal, the response of the middle ear structure is strongly dependent on the frequency of excitation. In Fig. 13 the displacement of the stapes is plotted showing roughly a decay of 40 dB/decade above around 1 kHz but also several relative peaks resulting from resonances of the air in the ear canal and the middle ear structure. Multiplying this curve by the excitation frequency ω the transfer function between the sound pressure p and the stapes velocity v_S is received.

Table 1. Characteristic values of excitation with different impulses and the reaction of the middle ear. I: triangular pulse; II: two triangular pulses, two-sided; III: two triangular pulses, one-sided

	I	II	III
p_{max}	700 Pa (151 dB)		
T	0.301 ms	0.535 ms	
t_A	0.301 ms	0.301 ms	
t_C	0.206 ms	0.340 ms	
t_D	0.206 ms	0.467 ms	
E_p L_{eq}; L_{eq8} L_{Aeq}; L_{Aeq8}	0.178 J/m^2 146.1 dB; 66.3 dB 145.6 dB; 65.7 dB	0.255 J/m^2 145.2 dB; 67.8 dB 145.6 dB; 68.3 dB	0.255 J/m^2 145.2 dB; 67.8 dB 143.9 dB; 66.5 dB
E; E_{ME}	4.03 μJ; 2.55 μJ	5.87 μJ; 3.73 μJ	6.14 μJ; 3.71 μJ
y_{Smax}, peak no. positive negative	 2.85 μm, first -1.26 μm, second	 2.85 μm, first -2.43 μm, first	 2.85 μm, first -1.52 μm, first
v_{Smax}, peak no. positive negative	 9.7 mm/s, second -22.8 mm/s, first	 19.3 mm/s, second -30.4 mm/s, first	 16.1 mm/s, first -20.6 mm/s, second

Table 2. Admissible rounds due to different criteria

	I	II	III
Pfander	38	23	23
Smoorenburg	242	147	147
$\Sigma L_{Aeq8} = 85$ dB	85	46	70
$\Sigma L_{Aeq8} = 90$ dB	269	147	223

Knowing the spectrum of a steady state excitation the transfer functions can be easily used to calculate the spectrum of the response according to the superposition principle for linear systems.

4.3 Reconstructed ears

The sound transfer through the normal, pathological and reconstructed middle ear has to be investigated to develop optimal passive or active implants. Theoretical investigations based on mechanical models allow an insight into the dynamical behavior [13, 25] and can be used to study different types of prostheses and situations of incision.

The dynamics of the reconstructed middle ear is very sensitive to the points where and how the prosthesis is coupled with the ossicles, to the spatial direction in which the prosthesis is oriented and in particular

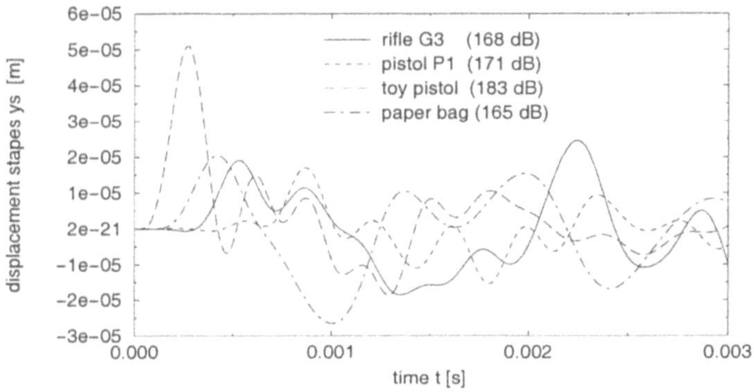

Figure 11. Stapes displacement y_S due to shots of rifle, pistol, toy pistol (fired into the ear) and blasting paper bag. The peak pressure p_{max} is indicated in dB

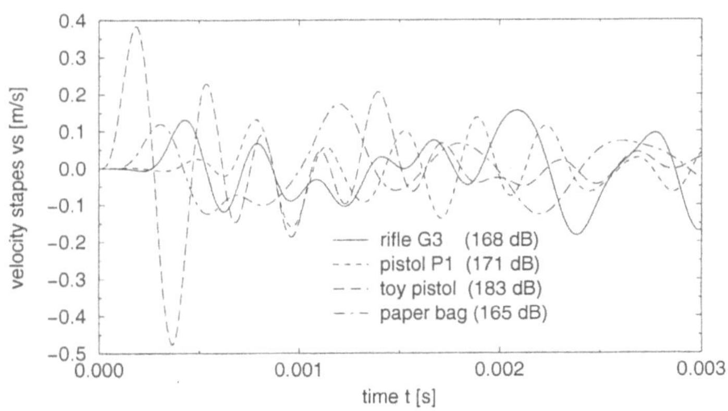

Figure 12. Stapes velocity v_S due to shots of rifle, pistol, toy pistol (fired into the ear) and blasting paper bag

to the type of coupling. Moreover, the surgical manipulation on the remaining ossicular chain is of influence.

5. Conclusions

The hearing is a marvelous organ most important for the daily life. It should be treated well and carefully protected since destroyed receptor cells of the organ of corti can not be rebuilt.

A high technical challenge is the reduction of noise levels of which humans and animals are exposed. This concerns the working environment as well as the leisure activities and private conditions. Reliable and appropriate modeling techniques and criteria for assessment of sound

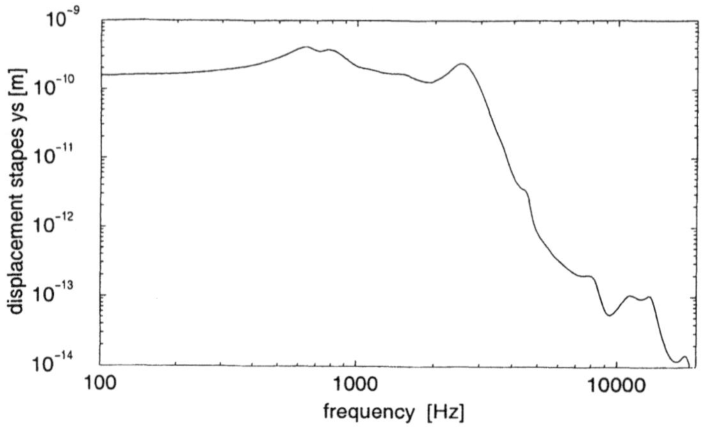

Figure 13. Transfer function between a sound pressure p applied at the entrance of the ear canal and stapes displacement y_S for a normal ear

events are necessary. They can be derived from the dynamical behavior of the hearing organ using computer software.

Because of the spatial orientation and suspension, the ossicles of the normal and the reconstructed ear carry out a three-dimensional motion in the tympanum due to a mechanical excitation by the sound pressure or by active hearing implants, respectively. Therefore mechanical models should be used to get a description close to the physical hearing process. For this purpose, all the powerful and well developed tools and approaches known from engineering disciplines can be applied.

To investigate the dynamical load to the hearing organ and the susceptibility to noise induced hearing loss, the mechanical load of the parts of the organ have to be taken into account. However, the sound pressure of a particular sound event influences the stress of a particular part like ear drum, ligaments and inner ear only secondary. The mechanical load is primarily determined by the mechanical stress and stress rate derived from a dynamical analysis using the equations of motion.

In the case of steady state noise the frequency transfer function through the middle ear can be used. This leads only to a rough guess for the case of transient noise. For the evaluation of peak values an integration in the time domain s necessary.

Impulsive noise is very dangerous since self-protection mechanisms are missing. In contrary to steady state noise, the pain is not immediately felt and one can not escape. Measurements are difficult due to uncertain parameters like incidence angle, integration time and filtering as well as the current criteria for assessment are not very reliable. Therefore, computer simulation is most important to evaluate the progress in designing for quietness.

Acknowledgments

This work was partially granted by the German Research Council under Ze 149/7-1 and Schi 119/27-1.

References

Zwislocki, J.: Analysis of the middle ear function. Part I Input Impedance. J Acoust Soc Am 1962; 34:1514-1523

Shaw, E. A. G., Stinson, M R.: The Human External and Middle Ear: Models and Concepts. In: De Boer, E.; Viergever, M. A. (eds.), Mechanics of Hearing. Delft: University press, 1983: 3-10.

Hudde, H.; Weistenhöfer, C.: A three-dimensional circuit model of the middle ear. Acustica Acta Acust 1997; 83: 535-549.

Price, G. R.; Kalb, J. T.: Insights into hazard from intense impulses from a mathematical model of the ear. JASA, 1991; 90: 219-2227.

Wada, H.; Metoki, T.; Kobayashi, T.: Analysis of dynamic behaviour of human middle ear using a finite element model. J Acoust Soc Am 1992; 24:319-327.

Beer, H.-J.; Bornitz, M.; Hardtke, H.-J.; Schmid, R.; Hofmann, G.; Vogel, U.; Zahnert, T.; Hüttenbrink, K.-B.: Modelling of Components of the Human Middle Ear and Simulation of Their Dynamic Behaviour. Audiol Neurootol 1999; 4:156-162.

Eiber, A.: Mechanical Modeling and Dynamical Investigation of Middle Ear. Proc. of the International Workshop on Middle Ear Mechanics in Research and Otosurgery, Dresden. K-B Hüttenbrink, (ed.). Dresden: Techn. Universität, 1997.

Schiehlen, W.(ed.): Advanced Multibody Systems Dynamics. Dordrecht: Kluwer, 1993.

Eiber, A.: Mechanical Modeling and Dynamical Behavior of the Human Middle Ear. Audiol Neurootol 1999; 4:170-177.

Eiber, A.; Kauf, A.; Maassen, M.; Burkhardt, C.; Rodriguez Jorge, J.; Zenner, H.-P.: Vergleich von Laservibrometriemessungen und Computersimulationen der Gehörknöochelchenbewegung. HNO 1997; 45:538-544.

Rodriguez Jorge, J.; Zenner, H. P.; Hemmert, W.; Burkhardt, C.; Gummer, A.: Laservibrometrie. HNO 1997; 45: 997-1007.

Eiber, A; Freitag, H.-G.; Hocke, T.: On the Relationship between Multifrequency Tympanometry Patterns and the Dynamic Behavior of the Middle Ear. In: Rosowski, J.; Merchant, S. (eds.), The function and Mechanics of Normal, Diseased and Reconstructed Ears. The Hague: Kugler, 2000.

Eiber, A.; Freitag, H.-G.: Middle Ear Reconstructions - A Mechanical View. Machine Dynamics Problems 2000; 24: 63-77

Hoffmann, E.: Hörfähigkeit und Hörschäden junger Erwachsener. Heidelberg: Median-Verlag, 1997.

Hedegaard Jensen, J; Bonding, P.: Experimental Pressure Induced Rupture of the Tympanic Membrane in Man. Acta Otolaryngol 1993; 113: 62-67.

Messerschmidt, O.: Auswirkungen atomarer Detonationen auf den Menschen. München: Karl Thiemig Verlag, 1960.

White, C. S.; Bowen, I. G.; Richmond, D. R.: The relation between eardrum failure and blast induced pressure variations. Space Life Science 1970;2: 158-205.

VDI 2058 Part 2: Assessment of noise with regard to the risk of hearing damages. Part 3: Assessment of noise in the working area with regard to specific operations.

ISO 1999: Acoustics - Determination of occupational noise exposure and estimation of noise-induced impairment.

DIN 45645: Determination of rating levels from measurement data - Part 1: Noise immission in the neighbourhood; Determination of rating levels from measurements - Part 2: Noise immissions at the work place.

Pfander, F.: Das Knalltrauma. Schriftenreihe Präventivmedizin-PM1. Bonn: Bundesministerium für Verteidigung - Referat Hygiene, Arbeits-, Umweltmedizin, 1994.

Smoorenburg, G.: Damage risk criteria for impulsive noise. In: Hamernik, R. P.; Henderson, D.; Salvi, R. (eds.), Noise Induced Hearing Loss, New York: Raven Press, 1982.

Dancer, A.; Henderson, D.; Salvi, R.; Hamernik, R. (eds.): Noise-Induced Hearing Loss. Saint-Louis: Mosby Year Book, 1992.

Dancer, A.; Franke, R.: Hearing Hazard from Impulsive Noise: a Comparative Study of Two Classical Criteria for Weapon Noises (Pfander Criterion and Smoorenburg Criterion) and the L_{Aeq8} Method. Acta Acustica 1995; 3: 539-547.

Eiber, A.; Freitag, H.-G.; Burkhardt, C.; Hemmert, W.; Maassen, M.; Rodriguez Jorge, J.; Zenner, H.-P.: Dynamics of Middle Ear Prostheses - Simulations and Measurements. Audiol Neurootol 1999; 4:178-184.

DESIGN OF A MEMS PRESSURE SENSOR FOR ACOUSTIC APPLICATIONS

RUDRA PRATAP and YASH K. DUNGERPURIA
Facility for Research in Technical Acoustics
Department of Mechanical Engineering
Indian Institute of Science
Bangalore 560 012, INDIA

Abstract In this work,we propose a novel design for the main structural element of a MEMS acoustic pressure sensor and carry out a detailed analysis for its performance characteristics. The design incorporates the basic principle ofconventional condenser microphones but uses innovative boundary conditions to eliminate one of the major shortcomings of the conventional sensor — the widely varying displacement field across the spatial extent of the microphone membrane. The new design consists of a square diaphragm made of polysilicon. The diaphragm is suspended by four cantilever beams, one on each side of the diaphragm, that are fixed to the substrate. This removes the zero displacement boundary condition on the edges of the diaphragm and, thus, makes the displacement of the diaphragm more uniform under uniform acoustic pressure. Combdrives are used on the edges of the diaphragm to sense the displacement.The combdrives provide capacitive sensing through the change in the overlap area of the fixed and the movable fingers, thus eliminating the effect of transduction.

Keywords: Pressure sensor, MEMS, microphone, sensor design

1. Introduction

Sound sensors form an important class of transducers. When sound sensors operate in air they are called *microphones*. Various types of microphones are available — carbon microphones, condenser microphones, piezoelectric microphones, etc. Of these,the condenser microphones are probably the most popular ones. They offer several advantages over the other kinds of microphones such as adequate acoustic sensitivity,low self noise and low sensitivity to mechanical vibrations [1].In a conventional condenser microphone [2], the sound pressure is sensed through displacement of the diaphragm. This displacement $w(t)$ causes

M.L. Munjal (ed.), IUTAM Symposium on Designing of Quietness, 201–213.

variation in the capacitance C as per the relation,

$$C(t) = \frac{\epsilon_0 A}{d - w(t)} \tag{1}$$

where ϵ_0 is the permittivity of the free space, A is the area of overlap, and d is the stationary gap between the plates. Since $w(t) \ll d$, using binomial expansion and neglecting higher order terms $C(t)$ is written as,

$$C(t) = \frac{\epsilon_0 A}{d} + \frac{\epsilon_0 A}{d}\left(\frac{w(t)}{d}\right) \sin \omega t \tag{2}$$

This relationship leads to a contradiction. On one hand, it is desirable to increase the response of the microphone for a given sound pressure and on the other hand, with this increase the assumption for getting eqn. (2) from eqn. (1) becomes invalid. Here, neglecting higher order terms introduces error in measurements as large as 2% at high sound pressure levels (above 120 dB) [3].To retain the characteristics of a condenser microphone at microlevel one would need very sophisticated fabrication techniques as a very thin diaphragm (nm thick) would be required. These features in the conventional designs often hinder the integration of sound sensors in microsystems.In this work, a new design for the main structural element of a sound sensor is proposed using the principle of the condenser microphone. The major advantages offered by the new design are the simplicity and the compatibility with the micromachining technology. The design employs polysilicon whose properties are available in the literature [4]. A complete analysis has been carried out on the proposed design and various damping phenomena present have also been considered. Through this innovative design it is emphasized that the approach towards the microsensors should not be just that of a scaled down prototype of their macro counterpart.

2. Sensor Design

For developing the new design the following objectives were set:

- Reduce the errors that creep i n due to inverse relation between the capacitance and the plate gap,
- Reduce the resistance due to the electrostatic force (transduction phenomenon) [2],
- Retain the characteristics of the microphone which are available at the macro level,
- Employ low cost silicon micromachining to form the diaphragm.

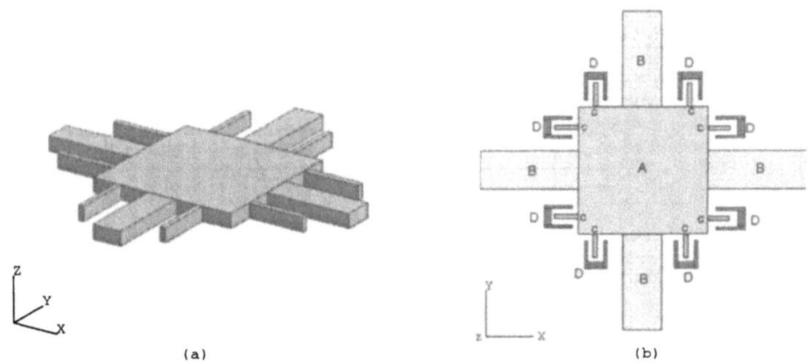

Figure 1. Sensor design: a central diaphragm A suspended by four beams B. The diaphragm A is $200\mu m \times 200\mu m$, beams B are $150\mu m \times 60\mu m$ each. The thickness of each element is $5\mu m$.

The proposed design is shown in fig. 1. The 3D view shown in fig. 1(a) is obtained from the 3D-modeller of MSC-Nastran [5]. The planar view in fig. 1(b) shows various components of the design schematically.As shown in the fig. 1(b), the main part is A — a diaphragm exposed to the sound pressure. The element A is suspended by four thin (same thickness as A) rectangular plates B, also made of polysilicon. One end of B is attached to the diaphragm and the other is attached tothe supports.This is done to get better displacement response for a given sound pressure. This innovation changes the boundary conditions at the edges of the diaphragm. Instead of having boundary condition of fixed-fixed type at the edges of the diaphragm, now dynamic boundary conditions exist at the edges in the proposed design.Another distinguishing feature of this design is the beam element C which is an integral part of the diaphragm. The capacitor for sensing purpose is formed between the moving combdrive C and the stationary combdrive D. There is no capacitor between the main diaphragm A and the back plate. The length of C is kept low so that its movement is dependent on the movement of the diaphragm and the deflections due to beam flexure are minimal.Integration of the combdrives into the diaphragm design shifts the focus from the gap-variation driven capacitance change to the area-of-overlap-variation driven capacitance change for detecting the sound signals. Combdrives also bring other advantages along with them. Now, there exists a linear relationship between the capacitance and the displacement response which was an inverse relationship without the combdrives. The resistance offered to the electrode motion is also reduced. It is now offered only by the fringing fields which are quite low in magnitude compared to the normal electrical fields present in conventional designs.

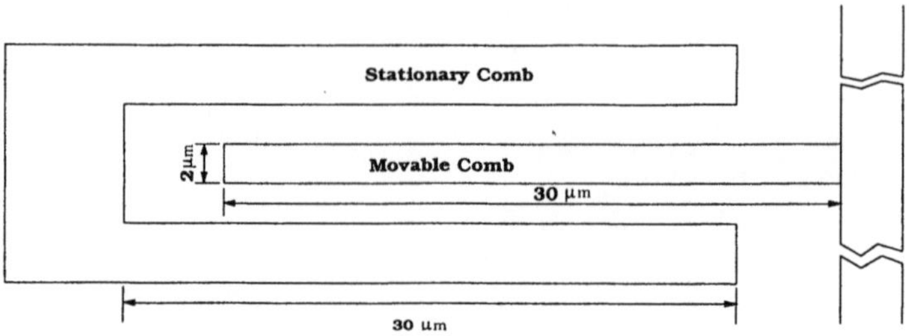

Figure 2. Detailed view of a combdrive.

A detailed view of the combdrive is shown in fig. 2. These combs form capacitor plates in xz and yz planes. When the diaphragm vibrates in response to the sound signal, these combs move up and down in the z-direction. Thus the area of overlap between the capacitor plates varies and a fluctuating voltage wave about a mean voltage is obtained in response to a sound pressure.

The natural frequency of the diaphragm can be increased by increasing the stiffness which can be achieved by increasing the thickness. The ratio of the lateral dimensions to the thickness is kept around 40 to 100. Further, no tension is applied to the diaphragm. Due to these features the diaphragm can be modeled as a thin plate. The governing equation for free vibrations of thin plates [6, 7] is given by,

$$D\nabla^4 w + \sigma\ddot{w} = 0 \tag{3}$$

where, $D = \frac{Eh^3}{12(1-\nu^2)}$, E is the Young's modulus, h is the thickness,ν is the Poisson's ratio, and $\sigma = \rho h$ is the mass density of the plate per unit area. The natural frequencies of the plate are given by,

$$\omega_{mn} = \sqrt{\frac{D}{\sigma}}\left[\left(\frac{m\pi}{a}\right)^2 + \left(\frac{n\pi}{b}\right)^2\right] \qquad m = 1, 2, 3\ldots \quad n = 1, 2, 3\ldots \tag{4}$$

3. Damping in the design

Damping in any system is associated with various phenomena that account for energy dissipation,usually in the form of heat.In microstructures,the various phenomena responsible for energy dissipation are [8]:

1 Internal friction,

2 Squeeze film damping,and

3 Sound radiation while the diaphragm vibrates,i.e.,loss due to sound radiation.

3.1 Structural damping

Structural damping describes the energy loss due to internal dissipation in the structure. It is due to mechanisms such as elastic-plastic hysteresis loss and structural joint friction. The usual model taken is the *Rayleigh* or *proportional* damping model [9]. However, with reduction in size the contribution of this mechanism reduces [8].

3.2 Squeeze film damping

Microstructures can be greatly influenced by the fluid that surrounds them. In the context of fluid damping in the proposed design, the system can essentially be modeled as having relative motion in the normal direction between two rectangular plates — one being the diaphragm and the other being the back plate. When the diaphragm moves towards the back plate, it is forced to squeeze the fluid trapped in between the plates. Therefore, the fluid offers resistance to the motion. Thus the fluid, which is air in this case, functions as a damping medium and the phenomenon is termed as *squeeze film damping.*As the width of the suspensions (overhangs B) is small, the air between the suspension plates B and the back plate can interact freely with the ambient and therefore the pressure there remains essentially equal to the ambient. Thus the design is modeled as a square plate of dimensions equal to that of element A in fig. 1. The dimensionless damping and the stiffness coefficient for rectangular plates vibrating normal to each other are given as [10],

$$f_s = \frac{64\sigma^2\epsilon}{\pi^8} \sum_{m,n,odd} \frac{1}{(mn)^2} \left[\frac{1}{\left[\{m^2 + (n/\beta)^2\}^2 + \sigma^2/\pi^4 \right]} \right] \tag{5}$$

$$f_d = \frac{64\sigma\epsilon}{\pi^6} \sum_{m,n,odd} \frac{m^2 + n^2}{(mn)^2} \left[\frac{1}{\left[\{m^2 + (n/\beta)^2\}^2 + \sigma^2/\pi^4 \right]} \right] \tag{6}$$

where σ is the squeeze number, ϵ is the normalized film thickness perturbation amlitude,and $\beta = b/a$ is the plate aspect ratio. The conditions in the film are assumed to be isothermal and the surfaces may be rigid or continuous [11]. The solutions given above were verified by Andrews *et al.* for microstructures through experiments [12].

Loss due to radiation of sound

This phenomenon is not one which occurs due to resistance offered to the motion of the diaphragm but is due to the vibration of the diaphragm. When a structure vibrates in response to a forcing function then it also functions as a source emitting signals.There is always some energy that is lost due to this phenomenon. In the diaphragm case,this radiated sound from one infinitesimal area acts as a pressure wave on another infinitesimal area. The radiation resistance which is the force per unit velocity exerted by the vibrating surface on the adjacent medium is given as [2],

$$R_r = \rho_0 c \pi a^2 R_1(2ka) \tag{7}$$

where,ρ_0 is the density of the medium,c is the velocity of sound in the medium,a is the radius of the radiating circular surface, k is the wave number, and R_1 is a polynomial function given by

$$R_1(x) = 1 - \frac{J_1(x)}{x} = \frac{x^2}{2^2.1!.2!} - \frac{x^4}{2^4.2!.3!} + \cdots \tag{8}$$

Neglecting higher order terms, the above equation can be written as,

$$R_r = \frac{\rho_0 c k^2}{2\pi} S^2 \tag{9}$$

where $S = \pi a^2$ is the radiating area.It is assumed while analyzing this mode of loss that the diaphragm moves as a rigid piston in an infinite baffle. This assumption is valid as the amplitude of vibration is small. Further, the diaphragm is assumed circular. This assumption is justified since the wavelength of the sound wave so produced is much larger than the diaphragmdimensions.

4. Results and Discussion

The dynamic analysis of the design was carried out using finite element method in MSC-Nastran [5]. The complete geometry was divided into several 4-noded-2D-shell elements. The combs were modeled as beams and were divided into 2-noded-1D-bar elements. The material properties are taken from James *et al.* [4]. The material is assumed to be a linear elastic material. The thickness of the 2D-shell elements is varied from $3\mu m$ to $5\mu m$. Consequently, the cross-section of the beam is taken as $2\mu m \times 3\mu m$, $2\mu m \times 4\mu m$ and $2\mu m \times 5\mu m$. The first type of analysis carried out in Nastran is the normal-mode analysis. The results are shown in fig. 3.

As can be seen from the figure, the modal frequency varies linearly with the thickness. This variation

can be explained by eqn. (4). For a given structure and a given mode (the selection of mode fixes the values of m and n), the natural frequency depends on the factor $\sqrt{D/\sigma}$. For a given material$D$ is proportional to h^3 and σ is proportional to h. Thus effectively the natural

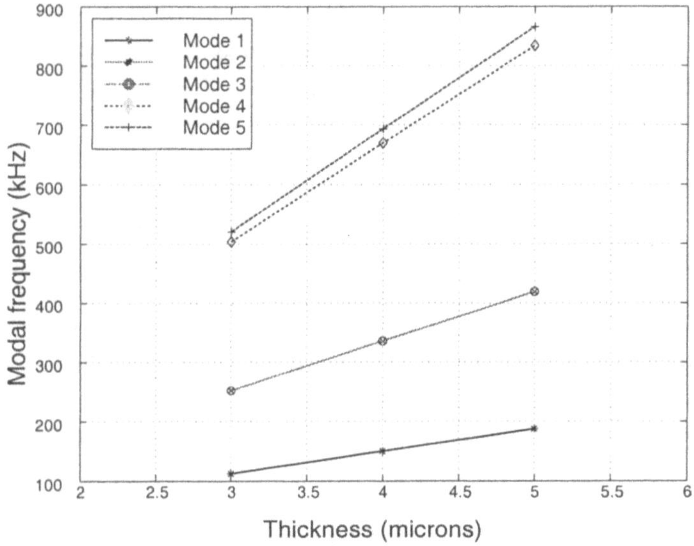

Figure 3. Natural frequencies in various modes for different thickness in the design.

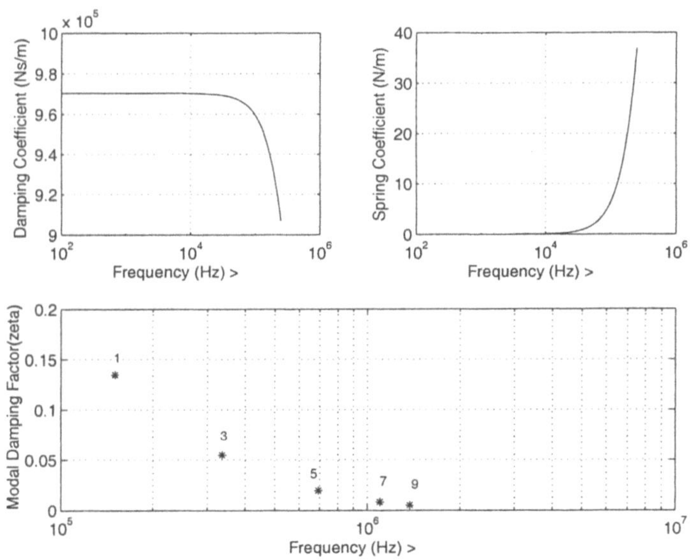

Figure 4. Squeeze film damping in the design (diaphragm thickness=4 μm)

frequency in a mode becomes proportional to $\sqrt{h^2}$ or$h$. The effect of the squeeze film damping is shown in fig. 4. This figure illustrates variation of damping and stiffness coefficients with frequency of vibration for the proposed design. Similar analysis was done for other thicknesses. The results show that the damping coefficient is invariant with respect to thickness. It is to be noted that the thickness of the diaphragm is varied but the gap between the diaphragm and the back plate is kept the same. Thus the volume of air trapped remains same. This explains the observed phenomenon. Further, we also find a sudden drop in damping coefficient when the excitation frequency exceeds a limiting value called the *critical frequency.* Above this frequency, stiffness effect of the air dominates. This is due to the inability of air to escape at high frequencies. It can be seen that the critical frequency for the design is near 50 kHz mark. This also is one of the criteria to decide the frequency limit of the sensor as beyond 50 KHz the assumption of isothermal conditions may break down [12]. Fig. 5 illustrates the effect of sound radiation. The figure shows that the value of the damping coefficient due to sound radiation is less than that due to squeeze film damping. Thus for further analysis only squeeze film damping is considered. However, at high frequencies the contribution of this phenomenon increases, especially because at high frequencies damping due to squeeze film reduces. In the

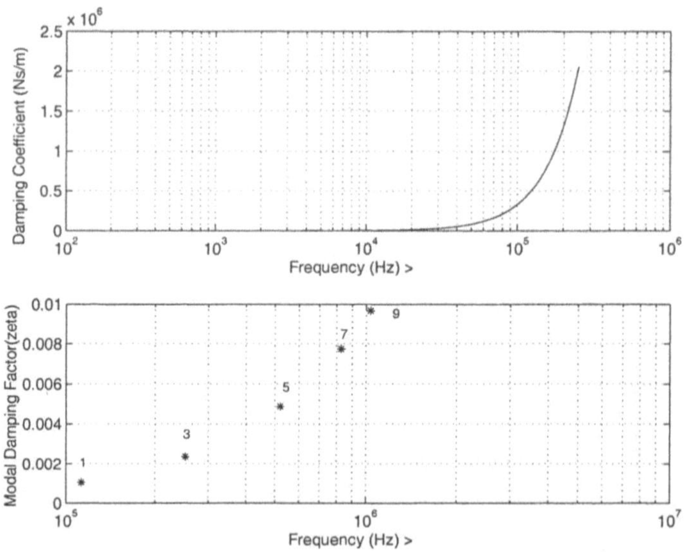

Figure 5. Sound radiation losses in the design (diaphragm thickness=4 μm)

analyses carried out in Nastran, sound pressure was varied from 40 dB to 160 dB.One feature that came out from these analyses is the decrease in response to a sound pressure at a particular frequency with increasing thickness. This feature can be attributed to the increase in stiffness with increasing thickness which reduces the static displacement of the diaphragm in response to a particular sound pressure. The invariance of response with respect to the variation in the frequency of excitation is referred to as the *flatness of response.* In any frequency response curve, the amplitude of the response increases as the frequency of excitation approaches the natural frequency of the system. However, this increase is not uniform and is considerably more near the resonant frequency. In the regions where frequency of excitation is much less than the resonant frequency, the variation with frequency is small. In this region the system is said to have a *flat frequency response.*This characteristic is very important as it decides the frequency limit till which the linearity is exhibited and consequently the frequency sensing range of the sensor.The deviation of the response from the flat frequency response in the design is shown in fig. 6. A tangent parallel to the horizontal axis is drawn to illustrate this deviation. It can be seen from the figure that higher

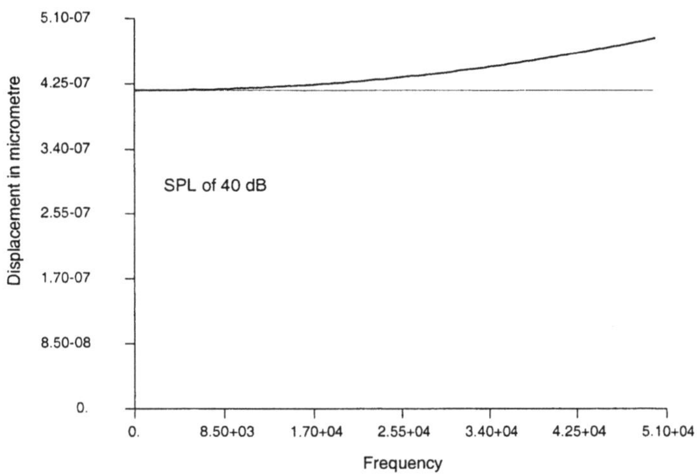

Figure 6. Flatness in the sensor design (diaphragm thickness=3 μm)

the first natural frequency higher is the limiting frequency for flat response. The permissible error limit is taken as ±0.5dB for deciding the frequency limit for flat response. The frequency limit for various thicknesses is given in Table 1 which shows that even for the most sensitive design the frequency limit extends up to 17 kHz.

The linearity is a characteristic that decides the range of operation of the sensor. It describes the relation between the input and the output. However, the linear response holds only till the diaphragm response remains invariant with respect to frequency of excitation (flat frequency response). Figure 7 shows that a linear relationship exists between the plate displacement (measured by the output voltage) and the pressure amplitude. The curves appear almost too linear to be true. However, if

Table 1. Frequency range for various designs

Thickness	Frequency
3 microns	17 kHz
4 microns	30 kHz
5 microns	45 kHz

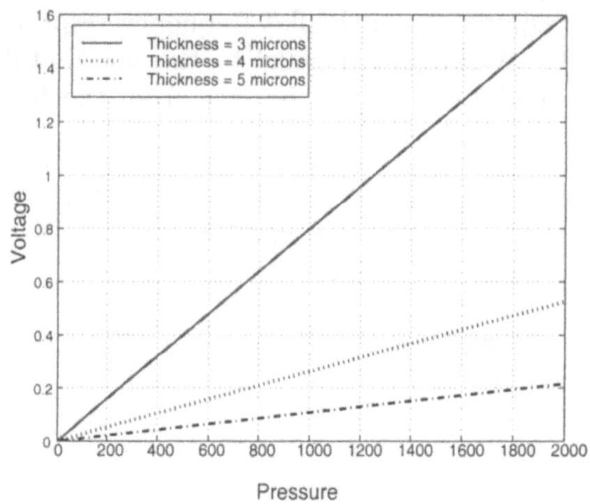

Figure 7. Sensitivity curve for different thicknesses of the design

the plate displacement equation for a linear elastic material is analyzed, it can be seen that displacement amplitude for a given plate depends on frequency and pressure amplitude. If the frequency is kept constant then the only variable is the pressure amplitude.Based on the analyses done for various thicknesses of the diaphragm, the sensitivity of the new design is shown in Table 2. The polarization voltage is taken as 12 V. The values in dBV are computed using 1 V/Pa as the reference.

5. Conclusions

This work focuses on the design of a sound sensing device using MEMS technology. Since at microlevel it is easy to implement a capacitor, a capacitive approach is considered.The major shortcomings in the existing designs have been pointed out and as an alternative a new design is proposed and analysed. Further, a huge reduction in size has been identified as a major advantage that can be exploited to extend the

Table 2. Sensitivities of the design

Thickness (μm)	Sensitivity (mV/Pa)	Sensitivity (dBV)
3	0.95	-60
4	0.26	-71.86
5	0.108	-79

integration of sound sensing devices with microsystems.The sensitivity achieved ranges from -60 dBV to -79 dBV for different thicknesses in the proposed design. A frequency range of 17 kHz is obtained for very sensitive sensors which will increase for lower sensitivities. These ranges have been defined keeping the error in measurements within the permissible limit of $\pm 0.5\, dB$. It has been demonstrated that for better sensitivity one need not design a very thin diaphragm. Higher sensitivity can always be achieved with innovative variations in the design as has been illustrated in the present work without opting for very thin diaphragms. These designs do not require sophisticated fabrication techniques. In all the designs the thickness has not been taken lower than $3\mu m$. The designs suggested can be fabricated from the presently available surface micromachining techniques. In the present work, to reduce the error due to the inverse relation between the capacitance and the electrode gap, a small modification in the form of combdrives extending from the diaphragm is made. This removes the above error which can be as high as 2%. Further, this modification reduces the resistance which is offered by the electrostatic forces between the charged plates. The resistance in conventional designs is offered by the normal electric field whereas in the proposed design, it is offered by the fringes emerging from combdrive edges which are lower in strength.

Acknowledgements

This work was partly supported by the Department of Science and Technology (DST),Government of India, through their grant to the Facility for Research in Technical Acoustics (FRITA), Indian Institute of Science, Bangalore.

References

Beranek, L. L., *Noise and Vibration Control*, McGraw-hill BookCompany, New York 1971.

Frey, A. R., Coppens, A. B., Sanders, J. V. and Kinsler, L. E., *Fundamentals of Acoustics* , Third edition, Singapore: John Wiley & Sons, 1982.

Dungerpuria, Y. K., "Design of a MEMS Pressure Sensor for Acoustic Applications," M.E. dissertation, Indian institute of Science, Bangalore, 2000.

Bustillo, J. M., Howe, R. T. and Muller, R. S. "Surface Micromachining for Microelectromechanical Systems," *Proceedings of the IEEE*,**vol. 86(8)**, (1998), pp 1552-1574.

MSC/NASTRAN Quick Reference Guide, version 70.5, The MacNeal-Schwendler Corporation, Colorado, USA, 1998.

Chen, Y., *Vibrations:Theoretical Methods*, Massachusetts, USA:Addison-Wesley Publishing Company, Inc., 1966.

Geradin, M. and Rixen, D., *Mechanical Vibrations - Theory and Applications to Structural Dynamics*, 2nd ed. England: John Wiley & Sons, Ltd.,1997.

Hosaka, H., Itao, K. and Kuroda, S., "Evaluation of Energy Dissipation Mechanisms in Vibrational Microactuators," *Proceedings IEEE on Microelectromechanical Systems, January 25-28, 1994, Osio, Japan*, pp 193-198, 1994.

Clough, R. M. and Penzien, J., *Dynamics of structures*,Second Edition, Singapore: McGraw-Hill, Inc., 1993.

Blech, J. J., "On Isothermal squeeze Films," *Journal of Lubrication Technology*, **vol. 105**, pp 615-620, October 1983.

Langlois, W. E., "Isothermal Squeeze Films," *Quart. Appl.Maths.*, **XX(2)**, pp 131-150, 1962.

Andrews, M., Harris, I. and Turner, G., "A comparison of squeeze-film theory with measurements on a microstructure," *Sensors and Actuators A*, **vol. 36** pp 79-87, 1993.

REDUCING RADIATED SOUND POWER BY MINIMIZING THE DYNAMIC COMPLIANCE

C. S. JOG
Facility for Research in Technical Acoustics (FRITA)
Dept. of Mechanical Engg.
Indian Institute of Science
Bangalore 560012
India
jogc@mecheng.iisc.ernet.in

Abstract Since the input power due to a periodic load on a structure is equal to the sum of the power dissipated in the structure and the acoustic power radiated to the surroundings, it is not obvious that redesigning the structure so as to minimize the input power results in a reduction of the radiated sound power. In this work, we present numerical evidence to show that dramatic reductions in sound power radiated from a structure immersed in a light fluid such as air can be achieved by using topology optimization techniques to minimize the input power due to the periodic loading. Minimizing the input power moves the natural frequencies of the structure away from the driving frequencies thereby reducing the vibrations, and hence the radiated power. The numerical examples show that in the case of baffled plates, the reduction obtained in the emitted sound power obtained by minimizing the input power is almost as much as that obtained by minimizing the sound power directly; however, the proposed technique circumvents the need to solve the acoustical equations, thereby resulting in considerable saving in computational cost.

Keywords: Topology optimization, Acoustic power

1. Introduction

Topology optimization involves distributing a given amount of material in a domain subjected to certain loading and boundary conditions, in a configuration of macroscopic solid and void regions so as to optimize a given performance functional. This technique has been applied extensively for various kinds of problems ranging from designing the stiffest

M.L. Munjal (ed.), IUTAM Symposium on Designing of Quietness, 215–236.

structure under static loading to reducing vibrations of structures subjected to transient loading (see, e.g., [3]).

For dynamics problems, attention has mostly focused on optimizing the natural frequencies of structures (see, e.g., [2],[7],[11],[13],[14]. However, from the viewpoint of acoustics, it is important to consider the case of periodic loading since a major source of vibration, and hence noise, in a machine is a periodic force due to rotating components. A vast majority of applications involve 'light' fluid loading, e.g., a structure vibrating in air. In this case the structural-acoustic analysis problem can be treated as an uncoupled one, i.e., the effects of fluid loading are ignored while carrying out the structural analysis, and the structure surface velocities are used as boundary conditions in the acoustic analysis code. We shall focus on this case throughout this work.

Ma et al. [14] briefly discuss the problem of reducing the vibrations of structures subjected to periodic loading. They formulate a 'dynamic compliance', the minimization of which results in reduced vibration of the structure. However, it was shown in [9] that in some cases, this measure can drive a system towards resonance instead of away from it–to overcome this shortcoming, a new definition was proposed for the dynamic compliance, namely, the average input power over a cycle. It was shown by means of numerical examples that minimizing this measure is effective in moving the natural frequencies of a structure away from the driving frequencies. In this work, we show that for a light fluid-loaded structure, the radiated sound power has a very close correlation with the input power, and minimizing the input power using the methodology in [9] results in dramatic reductions in the radiated sound power as well; in fact we show that in the case of baffled plates, the reductions are comparable to those obtained by minimizing the sound power directly.

Before discussing the formulation, we shall briefly discuss the previous work that has been done in optimization of structures with a view towards optimizing the acoustic response. Lamancusa [12] discusses various choices of design variables and objective functions that have been used–he concludes that acoustic power when used as an objective function produces the most consistently improved designs. However, he also points out the disadvantage of using acoustic power as an objective function, namely, that it is costly to evaluate (except in special cases such as baffled plates, where the Rayleigh integral yields a simple expression for the power), especially when the loading has several frequency components. A recent and comprehensive review of optimization of coupled and uncoupled structural-acoustic systems is the one by Christensen et al. [4]. This review also notes that several researchers have used acoustic power as an objective function. For example, Salagame et al.[16] discuss the computation of the analytical sensitivity of the acoustic power radiated from plates–this in turn is used in [17] for minimizing the radiated sound power from plate structures by adding optimally sized discrete masses. Constans et al. [6] minimize the sound power from vibrating shell structures; however in contrast to the work in [17] they use a non-gradient based approach. This is natural, since the sensitivity analysis of the radiated acoustic power in the case of shell structures as opposed to plates in infinite baffles is quite complicated. Even if one were able to compute analytical sensitivity expressions, the computational cost involved would be quite prohibitive, since both the structure and acoustic equations have to be solved (an idea of the considerable difficulties in writing a general purpose code for evaluating the sound power can be obtained from [10]. The problem with non-gradient based approaches as used in [6] is that a lot of function evaluations are required to find the minimum, thereby restricting the number of design variables. In contrast, in topology optimization, one typically has hundreds of design variables, so that one would ideally like to pursue a gradient-based strategy. In [9], analytical sensitivity expressions for the input power valid for *any* structure have been presented. In addition, once the finite element analysis for evaluating the input power has been carried out, the incremental cost of computing the sensitivities is almost negligible. Minimizing the input power moves the natural frequencies of the structure away from the driving frequencies, and as we show in the following sections, indirectly brings about a great reduction in the radiated sound power as well. Thus, significant reductions in radiated sound power for 'light' fluid-loaded structures can be obtained simply by optimizing the structural input power, without the need to solve the acoustic equations.

We would like to note, however, that the proposed technique would obviously not be appropriate to use when fluid loading on the structure is significant (e.g., the structure is immersed in a heavy fluid as discussed in [5], or when one is interested in amplifying or reducing sound intensity in certain directions (see, e.g., [18],[19]–in such a case, of course, a structural-acoustic analysis needs to be carried out to find the optimum structure. Since, however, in the vast majority of cases, we are simply interested in reducing the noise radiated in air, the technique proposed in this work could prove to be useful.

In section 2, we present a brief summary of the results presented in [9]. Section 3 presents some example problems to show the efficacy of the proposed method, and Section 4 presents the conclusions.

2. The Dynamic Compliance

In [9], the average input power over a cycle has been proposed as a global measure of vibrations of the structure. This measure has been termed as the 'dynamic compliance' in analogy with the static compliance which is typically used as an objective function in topology optimization in order to find the stiffest structure under a given set of static loading conditions. It has been shown that this objective function is positive definite in the sense that it is positive in the presence of periodic loading, and zero if and only if the structure is stationary. Thus, an optimization strategy that minimizes the dynamic compliance tries to redesign the structure so as to reach as close to the 'zero vibration' state as possible. Of course, the choice of starting design does have a significant influence on the optimum design due to the large number of local optima, a fact also noted by Lamancusa [12] in his review paper.

For a continuum structure, the expression for the dynamic compliance (assuming that the prescribed displacement is zero) is

$$J_d = \frac{1}{T} \int_0^T \left[\int_{\Gamma_t} \bar{t} \cdot v d\Gamma + \int_{\Omega} \mathbf{b} \cdot v \mathbf{d\Omega} \right] dt,$$

where v is the velocity field, $\mathbf{b} = \mathbf{b_0 cos\hat{\omega}t}$ is the prescribed body force on Ω, $\bar{t} = \bar{t_0} cos\hat{\omega}t$ is the prescribed traction on Γ_t, and $T = 2\pi/\hat{\omega}$ is the time period. Note that the proposed dynamic compliance is a 'global' measure of vibrations, in the sense that it tries to minimize vibrations over the entire structure in an overall manner, even though the loading might be concentrated in a small region of the structure. We are interested in finding the distribution of material within a domain Ω that minimizes the dynamic compliance of the resulting structure, and such that the total volume of material equals a specified volume $\bar{V}$.

In the context of the finite element method, the governing equations for the case when the loading is harmonic and given by $\mathbf{f_0} \cos \hat{\omega}\mathbf{t}$, are

$$\left[\mathbf{K} - \hat{\omega}^2 \mathbf{M} + \mathbf{i}\hat{\omega}\mathbf{C} \right] (\mathbf{x_r} + \mathbf{i}\mathbf{x_s}) = \mathbf{f_0}, \quad (1)$$

where $\mathbf{M}$, $\mathbf{C}$ and $\mathbf{K}$ are the mass, damping and stiffness matrices, respectively. The displacement $\mathbf{u}$ is given by the real part of $(\mathbf{x_r} + \mathbf{i}\mathbf{x_s})\mathbf{e}^{\mathbf{i}\hat{\omega}\mathbf{t}}$. Separating the real and imaginary parts of the above equation and simplifying, we get

$$\bar{K}\mathbf{x_s} = -\hat{\omega}\mathbf{f_0},$$

where

$$\bar{K} = (\mathbf{K} - \hat{\omega}^2\mathbf{M})\mathbf{C}^{-1}(\mathbf{K} - \hat{\omega}^2\mathbf{M}) + \hat{\omega}^2\mathbf{C}.$$

Since $\mathbf{M}$, $\mathbf{C}$ and $\mathbf{K}$ are symmetric and positive definite, we see that $\bar{K}$ is also symmetric and positive definite. Hence, the problem of determining

$\mathbf{x_s}$ can also be stated as a minimization of a potential function defined by

$$\Pi(\mathbf{y}) = \frac{1}{2}\mathbf{y} \cdot \bar{\mathbf{K}}\mathbf{y} + \hat{\omega}\mathbf{y} \cdot \mathbf{f_0}.$$

At the solution $\mathbf{x_s}$, the value of Π is

$$\Pi(\mathbf{x_s}) = -\frac{1}{2}\mathbf{x_s} \cdot \bar{\mathbf{K}}\mathbf{x_s}.$$

The displacement and velocity vectors are given by

$$\mathbf{u} = (\mathbf{cos}\hat{\omega}\mathbf{t})\mathbf{x_r} - (\mathbf{sin}\hat{\omega}\mathbf{t})\mathbf{x_s},$$

$$\mathbf{v} = -\hat{\omega}\left[(\mathbf{sin}\hat{\omega}\mathbf{t})\mathbf{x_r} + (\mathbf{cos}\hat{\omega}\mathbf{t})\mathbf{x_s}\right].$$

Hence, the dynamic compliance is given by

$$\begin{aligned} J_d &= \frac{\hat{\omega}}{2\pi}\int_0^{2\pi/\hat{\omega}} (\mathbf{f_0}\mathbf{cos}\hat{\omega}\mathbf{t}) \cdot v\mathbf{dt} \\ &= -\frac{1}{2}\hat{\omega}\mathbf{f_0} \cdot \mathbf{x_s} \\ &= \frac{1}{2}\mathbf{x_s} \cdot \bar{\mathbf{K}}\mathbf{x_s}. \end{aligned}$$

Since

$$J_d = -\Pi(\mathbf{x_s}),$$

the problem of minimizing the dynamic compliance can be written as Find the vector of optimum design variables, ρ^*, and the associated vector $\mathbf{x_s}$, that solves

$$\max_{\rho} \min_{\mathbf{y}} \Pi(\rho, \mathbf{y}),$$

subject to appropriate constraints on the design variables.

The inner subproblem in the above statement solves the problem of finding $\mathbf{x_s}$, for a given set of design variables ρ, while the outer subproblem solves the problem of minimizing the power input. The sensitivity of the above functional with respect to the design-variable vector ρ is given by **?**

$$\frac{d\Pi}{d\rho_i} = \frac{\hat{\omega}^2}{2}\left[\mathbf{x_s} \cdot \frac{\partial \mathbf{C}}{\partial \rho_\mathbf{i}}\mathbf{x_s} - \mathbf{x_r} \cdot \frac{\partial \mathbf{C}}{\partial \rho_\mathbf{i}}\mathbf{x_r}\right] - \hat{\omega}\mathbf{x_r} \cdot \left(\frac{\partial \underline{\mathbf{K}}}{\partial \rho_\mathbf{i}} - \hat{\omega}^2\frac{\partial \mathbf{M}}{\partial \rho_\mathbf{i}}\right)\mathbf{x_s}. \quad (2)$$

Once the analysis has been carried out, and $\mathbf{x_r}$ and $\mathbf{x_s}$ have been found using Eqn.(1), the incremental cost of computing the sensitivity expressions given above is quite trivial. The sensitivity of the constraint (which will typically be an explicit function of the design variables) is also easily computed. Note that no assumption about the nature of the structure (plate, shell etc.) has been made in deriving any of the above expressions, and hence they are valid for any structure that can be approximated by Eqns.(1). In case the loading is a superposition of several harmonics, then it can be shown that the total input power is simply the sum of the power due to each individual harmonic. From this it follows that the sensitivity of the total input power is the sum of the individual sensitivities as given by Eqn.(2).

Although, we have treated only the case of distinct excitation frequencies, oftentimes in practice, the excitation frequency varies over a band (see discussions in, e.g., [16],[18]. This broadband excitation with possibly many 'bands' as shown in Fig.1, may or may not include natural frequencies of the structure. Since the main contribution to the power is due to the resonant frequencies in each band, our main interest is in moving the natural frequencies which are in the bands of excitation outside these bands, while simultaneously ensuring that natural frequencies outside the bands do not enter them as a result of the redesign. Within the context of the proposed methodology, this can be achieved by assuming an *artificial* loading function with discrete frequency components placed in each band of excitation. For example, in the situation shown in Fig. 1,

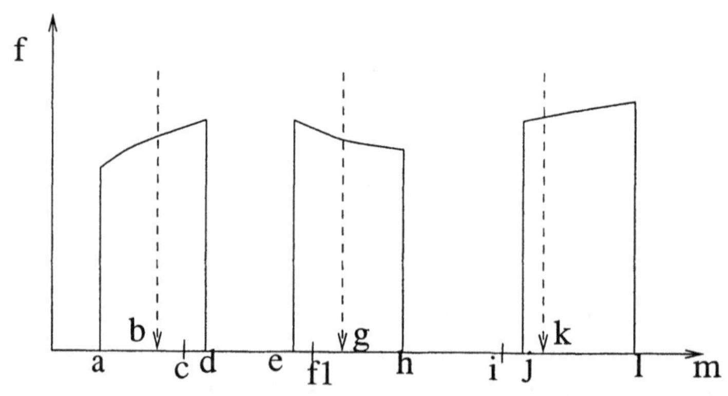

Figure 1. Placement of artificial loads in the case of broadband excitation.

the artificial loads are placed in each of the bands at frequencies $\hat{\omega}_1$, $\hat{\omega}_2$ and $\hat{\omega}_3$. Since the natural frequency ω_1 lies close to the right end of the first band, ω_{1u}, and since ω_2 lies close to the left end of the second band, ω_{2l}, we have chosen $\hat{\omega}_1$ and $\hat{\omega}_2$ to be slightly less than and slightly greater than ω_1 and ω_2, respectively, so that moving these natural frequencies out of these bands is easier. At the same time, note that we have placed an artificial load in the third band near to ω_{3l} to prevent the natural frequency ω_3 from entering the third band due to the influence of $\hat{\omega}_2$.

Of course, a little bit of trial and error may be required to judge the frequencies of the artificial loads, but since, as we shall see in the examples section, the natural frequencies move away from the driving frequencies by significant amounts during the redesign process, the above strategy should prove to be effective.

The solution methodology that we follow is an optimality criterion method first proposed by Ma et al. [14]; the reader is referred to [9] for further details. We assume a uniform thickness h_i for each element i, and use the normalized quantities $\rho_i = h_i/h_{max}$, where h_{max} is the maximum allowable thickness, as the set of design variables. However, we note that in the formulation there is no restriction on the choice of design variables. Other possible choices of design variables could be the thickness and spacing of beam stiffeners attached to the structure, orientation of fibers in a fiber composite, and so on. The constraint that we impose is the volume constraint, i.e., $V = \int_A h\, dA = \bar{V}$, where $\bar{V}$ is the specified volume.

The acoustic power radiated by a baffled plate is calculated using the expression developed by Naghshineh et al. [15], namely

$$P_o = \frac{1}{4\pi} \int_{S'} \int_S \hat{\omega} \rho_f v_n(\mathbf{r_s}) \left(\frac{\sin(\mathbf{kR})}{\mathbf{R}} \right) \mathbf{v_n^*}(\mathbf{r'_s})\, \mathbf{dS}\, \mathbf{dS'}, \tag{3}$$

where $\hat{\omega}$ is the excitation frequency, ρ_f is the fluid density, $R = |\mathbf{r_s} - \mathbf{r'_s}|$, k is the acoustic wave number, $\mathbf{r_s}$ and $\mathbf{r'_s}$ are the position vectors of the source and receiver points located on the plate surface, v_n is the plate normal velocity and v_n^* is the complex conjugate of v_n. In case of a load vector which is a superposition of several harmonics, we simply sum up the power due to each frequency component to obtain the total acoustic power. The power in decibels is obtained from the power in Watts using the formula

$$P_{dB} = 10 \log_{10} \left(\frac{P_{Watts}}{10^{-12}} \right).$$

The sensitivity of P_o with respect to a design variable is obtained by first finding the sensitivity of v_n using Eqn.(1), and then substituting it in the sensitivity expression for P_o obtained by differentiating Eqn.(3). Note that the explicit expression for the power given by Eqn.(3) greatly simplifies the optimization process in the case of baffled plates. However, even in this case, the sensitivity calculation becomes very time-consuming as the number of design variables increases due to the need to solve multiple load-cases while finding the sensitivity of v_n with respect to each design variable. Needless to say, the problem of determining sensitivities gets further aggravated when the assumption of a baffled plate cannot be made, or when we deal with shell structures where a simple expression such as Eqn.(3) is not available for computing the radiated power.

3. Numerical Examples

In this section, we present numerical examples to show the correlation between the dynamic compliance and the radiated acoustic power. When the dynamic compliance (input power) is being minimized by the optimization strategy, we also compute the sound power radiated at each design iteration. Since the input power is equal to the sum of the power dissipated and the acoustic power radiated [1], it is not obvious that reducing the input power will automatically reduce the acoustic power–indeed there are some iterations in the later stages of the optimization process where the input power reduces but the acoustic power increases. However, these increases are very small, and for all practical purposes there is almost a perfect correlation between the input and radiated sound power as can be seen from the correlation graphs presented below. We also present topologies with sound power as the objective function to show that the reductions in radiated sound power obtained as an indirect benefit while minimizing the input power are almost as much as those obtained with sound power itself as the objective function.

Although, we have implemented both the MITC4 and MITC9 plate elements developed by Bathe and coworkers (see [1] and references therein), we shall present the results using the MITC9 elements due to the unstable nature of lower-order elements (see [8] for a detailed analysis). The acoustic power given by Eqn.(3) is computed using full integration. The Young modulus, Poisson ratio and density of the plate are taken to be 210 GPa, 0.25 and 7800 kg/m^3. For the fluid medium (air), the density is taken to be 1.2 kg/m^3, and the velocity of sound to be 341 m/s. Black regions in the density plots represent a value of ρ_{max} (equal to 1), white regions represent a value of ρ_{min} (which we have chosen to be 0.1), while grey regions represent intermediate densities. In the correlation graphs, the acoustic power radiated by the designs obtained during the input power minimization, and the input power itself, have been plotted as a function of the optimization iterations. The units used for both these quantities is Watts. The same model for damping is used

[1] In our case, since we are neglecting fluid loading, the input power is equal to the dissipated power.

as in [9], i.e., $\mathbf{C} = \alpha\mathbf{M} + \beta\mathbf{K}$, where

$$\alpha = \frac{2\omega_1\omega_2(\xi_1\omega_2 - \xi_2\omega_1)}{\omega_2^2 - \omega_1^2}, \beta = \frac{2(\xi_2\omega_2 - \xi_1\omega_1)}{\omega_2^2 - \omega_1^2}. \tag{4}$$

We choose ω_1 to be the first natural frequency and ω_2 to be the maximum frequency of interest in the loading or response. The modal damping parameters, ξ_1 and ξ_2, are taken to be $\xi_1 = \xi_2 = 0.01$.

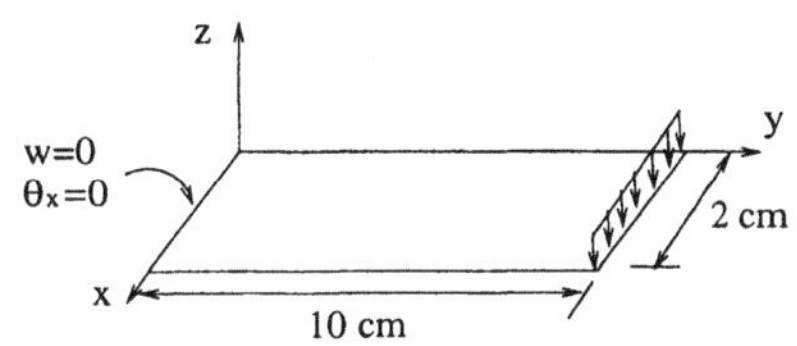

Figure 2. A cantilever beam subjected to end traction

The examples presented below have symmetric loading and geometries. We exploit this symmetry and conduct the analysis on only half or one-fourth of the structure depending on the type of symmetry. This not only reduces the computational cost, but also focuses on the symmetric (odd-odd) modes which are the most significant contributors to acoustic radiation. However, while presenting the final designs, we show the top view of the entire structure.

The effect of mesh refinement on the topologies will be presented elsewhere; we simply remark that based on our numerical experience, the dynamic compliance problem seems ill-posed just like its static counterpart, since, in some examples, we get topologies with increasingly finer members as the mesh is refined. Changing the volume fraction can also result in a change in the optimal design since the natural frequencies of the starting design are shifted in relation to the driving frequency.

Although we have not studied unsymmetric geometries and/or unsymmetric load cases in this work, we would like to reiterate that the methodology being general can be applied to such geometries or load cases also. Since minimizing the dynamic compliance simply moves the natural frequencies away from the driving frequency, the contribution of the unsymmetric modes to the vibration amplitudes would be reduced, notwithstanding the fact that they are not good radiators of sound. But since the contribution due to the symmetric modes is *also* reduced, we expect the sound power radiated to reduce, although, perhaps not as dramatically as in the examples presented below.

Example 1: Cantilever beam with a backing baffle

The candidate domain, loading and boundary conditions are shown in Fig.(2).

The maximum allowable thickness is $h_{max} = 0.5$. The distributed load at the free edge is given by $1.5 cos\hat{\omega}t\ N/cm$, where t is the time expressed in seconds. Since both the geometry and loading are symmetric, we consider only half the domain for the analysis. A mesh of 20×2 MITC9 elements is used. The move limit is taken to be 0.02. The specified volume fraction is 60%. Corresponding to this volume fraction, our starting design has a uniform thickness of 0.3 cm. The first three natural frequencies of this starting design obtained from the finite element model are $1589.8 rad/s$, $9918.1 rad/s$ and $27624.4 rad/s$. As in [9], we consider two cases:

(i) The driving frequency is less than the first natural frequency; we consider a value of $\hat{\omega} = 500 rad/s$.

(ii) The driving frequency is more than the first natural frequency but less than the second natural frequency; we consider a value of $\hat{\omega} = 2500 rad/s$.

In the first case, the topologies obtained with input power and acoustic power as objective functions are shown in Figs.3a. and 3b.

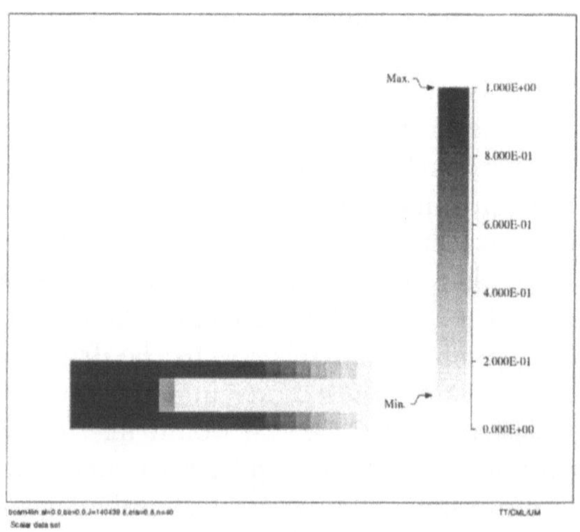

(3a)

The first three natural frequencies of the optimal topology shown in Fig.3a are 4416.7 $rads/s$, 17840 rad/s and 34602.7 rad/s, while those for the design in Fig.3b are 4183 rad/s, 16955 rad/s and 35766 rad/s. As can be seen from Table I, the sound power reduction is greater with the input power rather than sound power as objective function! This is an

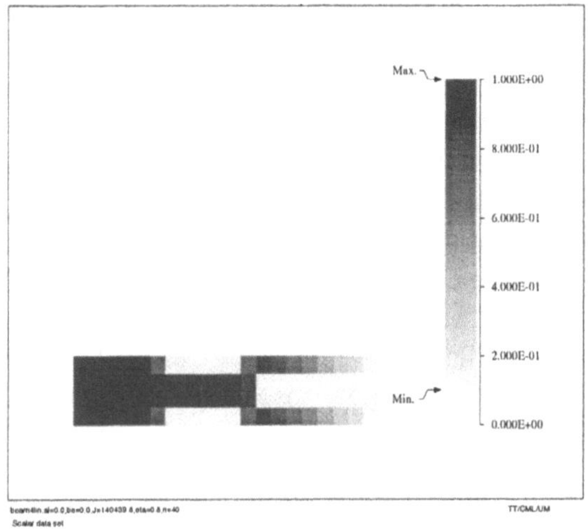

(3b)

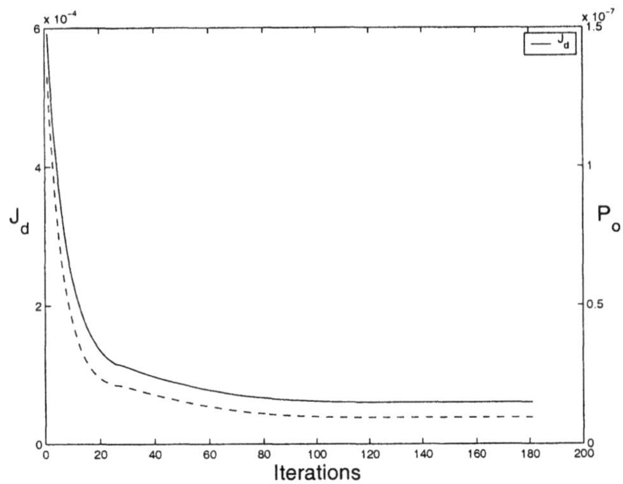

(c)

Figure 3. Density distribution for the cantilever beam example with $\hat{\omega} = 500\ rad/s$ (a) Input power minimization (b) Sound power minimization; (3c) correlation between input power and sound power radiated.

aberration, and occurs simply because in the latter case, the design gets trapped in the 'wrong' local optimum; as will be seen in all the other examples, directly minimizing the sound power will result in greater reduction of sound power than minimizing the input power, as expected.

When $\hat{\omega} = 2500\ rad/s$, the first three natural frequencies of the optimal topology shown in Fig.4a are 198 rad/s, 8656 rad/s and 26035 rad/s, while those for the design in Fig. 4b are 246.8 rad/s, 9156 rad/s and 31708 rad/s. The sound power optimization keeps the second natural frequency farther away from the driving frequency than the input power optimization strategy bringing about a greater reduction in sound power, but as can be seen from Table I, the difference is marginal. Since our strategies for input power and sound power minimization are both gradient-based, the choice of the starting design decides the local optimum design that we finally get. Since, from a static viewpoint, the designs in Figs 4a and 4b are flexible and since this may be undesirable, one can first solve an optimization problem in which one tries to raise the first natural frequency by either following the strategies in the works mentioned in the introduction, or more simply, by solving the static-compliance optimization problem–if a feasible design exists whose first natural frequency is above that of the actual excitation frequency, and if the optimization process is able to find this design, it can be used as a starting design to get an optimum design which will now be 'stiff' from both a static and dynamic viewpoint.

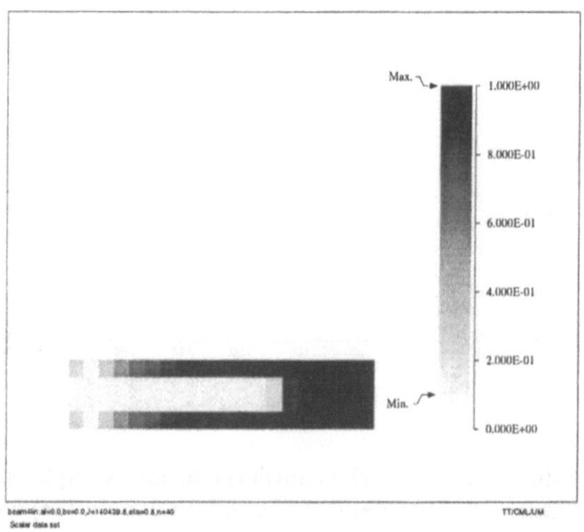

(4a)

For example, when $\hat{\omega} = 2500\ rad/s$, one first solves the static-compliance optimization problem which increases the first natural frequency; if this frequency is now greater than 2500 rad/s, we use the

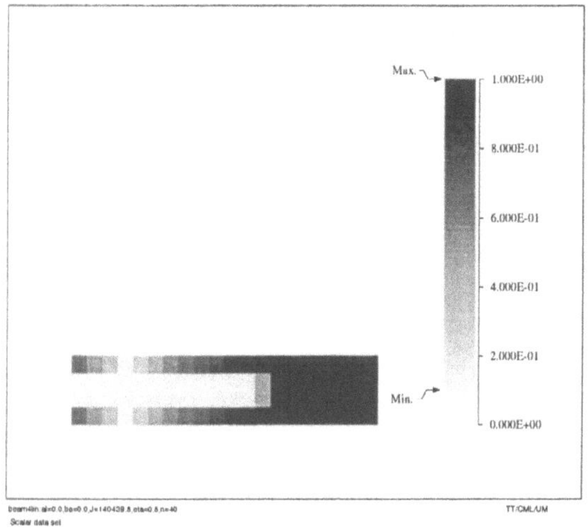

(4b)

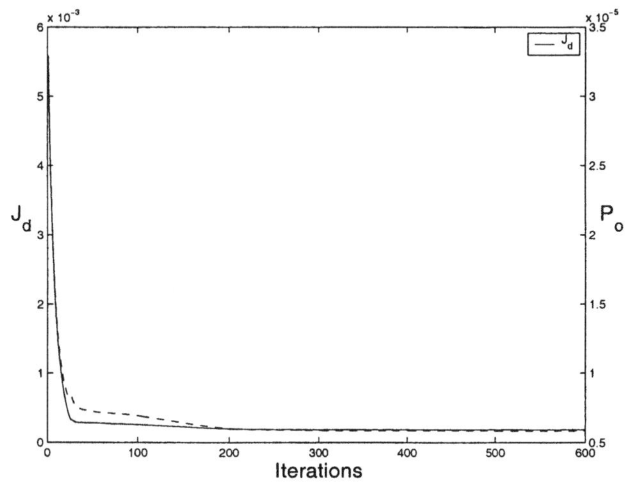

(4c)

Figure 4. Density distribution for the cantilever beam example with $\hat{\omega} = 2500\ rad/s$ (a) Input power minimization (b) Sound power minimization; (c) correlation between input power and sound power radiated.

static-compliance optimized design as a starting design for the actual excitation of $\hat{\omega} = 2500\ rad/s$.

It might seem that an alternate scheme for reducing vibration would be the minimization of the amplitude of tip displacement of the cantilever beam, which can be termed as a 'local' measure as opposed

Table 1. Summary of the optimization results with input power as objective function; figures in parenthesis are the corresponding results with sound power as the objective function.

Example	Driving frequency (rad/sec)	Initial sound power (dB)	Final sound power (dB)	Reduction (dB)
Cantilever	500	51.23	39.84 (40.3)	11.38 (10.9)
	2500	75.17	67.66 (67.11)	7.51 (8.06)
Plate	5000	92.38	88.42 (86.61)	3.96 (5.77)
	15000	100.4	85.6 (82.8)	14.8 (17.6)
	5000, 15000	101.0	91.0 (90.17)	10.0 (10.83)

to the dynamic compliance or sound power which are 'global' measures. However, although the local measure cuts down on the tip amplitude by a much larger amount than the global measures, it does so at the expense of greater amplitudes at other points in the structure, which is clearly not favorable from the viewpoint of sound power minimization. For the two cases which we have considered, namely, $\hat{\omega} = 500\ rad/s$ and $\hat{\omega} = 2500\ rad/s$, the acoustic powers of the optimal designs which minimize the tip displacement amplitudes are $40.85\ dB$ and $73.36\ dB$, respectively. As can be seen by comparing these values with those shown in Table I, these are not as good as the reductions obtained using input power as an objective function.

Example 2: Simply supported plate in an infinite baffle

The example that we now consider is that of a simply supported square plate of dimension $32\ cm$ and $h_{max} = 1\ cm$, and loaded at the center. The specified volume fraction is 60%. Corresponding to this fraction, the starting design is taken to be a uniform plate of thickness $0.6\ cm$. Due to the symmetry of the structure and loading, we consider only a quarter of the plate for the analysis. The first five natural frequencies are $1787.1 rad/s$, $8894.7 rad/s$, $8894.7 rad/s$, $15945.3 rad/s$ and $22924.7 rad/s$. We consider three cases with the load in each case given by

(i) $80 \cos 5000t\ N$.

(ii) $80 \cos 15000t\ N$.

(iii) $80(\cos 5000t + \cos 15000t)\ N$.

We discretize the domain using a 8×8 uniform mesh of MITC9 elements and choose the move limit to be 0.02. The optimal topologies for these three cases with input power and sound power minimization as objective functions along with the correlation graphs are shown in Figs.5-7.

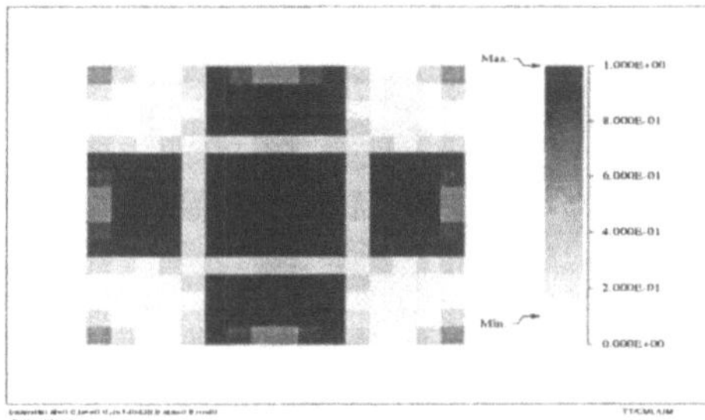

(5 a)

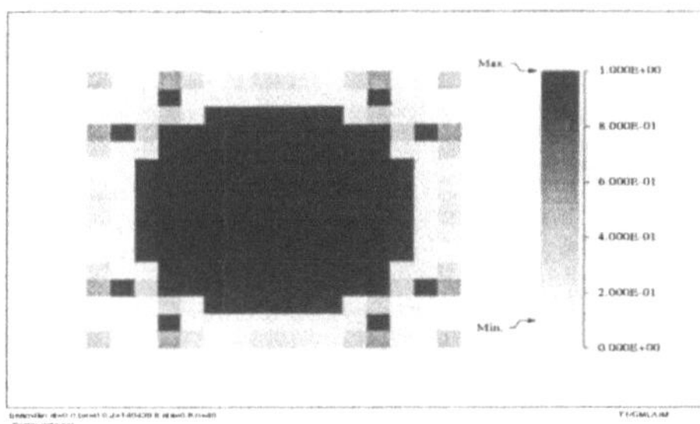

(5 b)

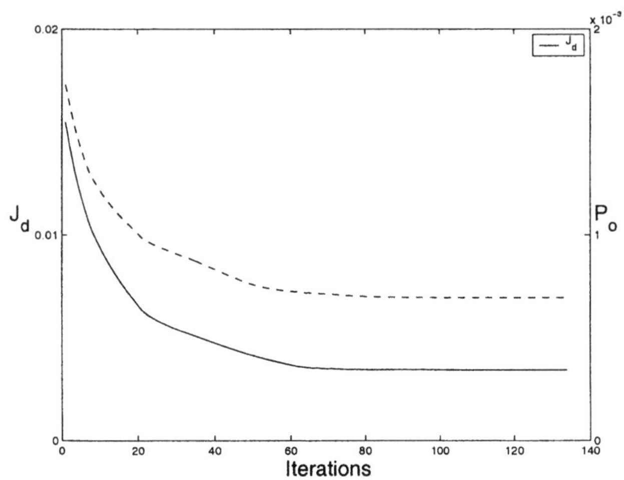

(5 c)

Figure 5. Density distribution for the plate example with $\hat{\omega} = 5000\ rad/s$ (a) Input power minimization (b) Sound power minimization; (c) correlation between input power and sound power radiated.

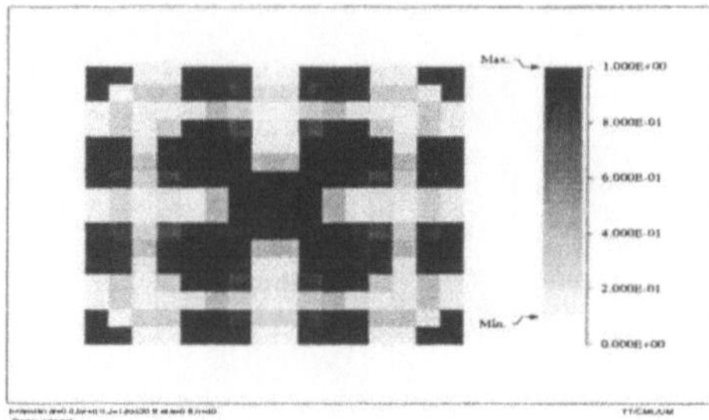

(6 a)

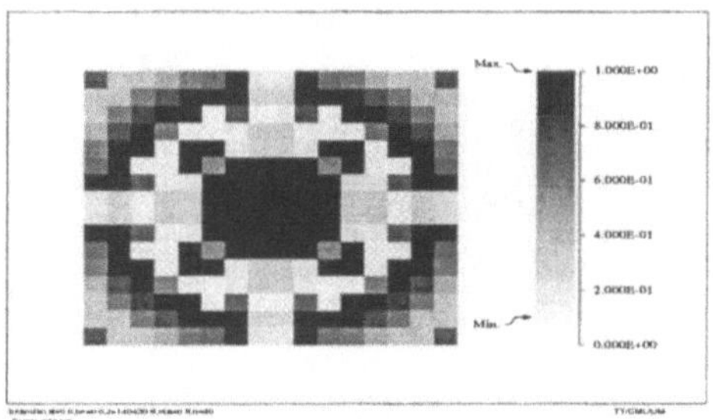

(6 b)

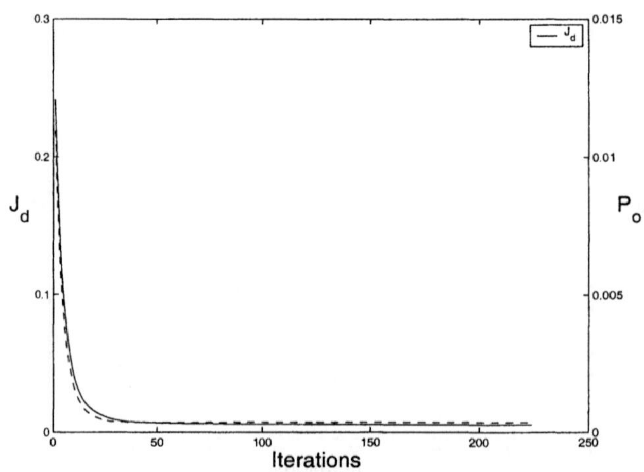

(6 c)

Figure 6. Density distribution for the plate example with $\hat{\omega} = 15000\ rad/s$ (a) Input power minimization (b) Sound power minimization; (c) correlation between input power and sound power radiated.

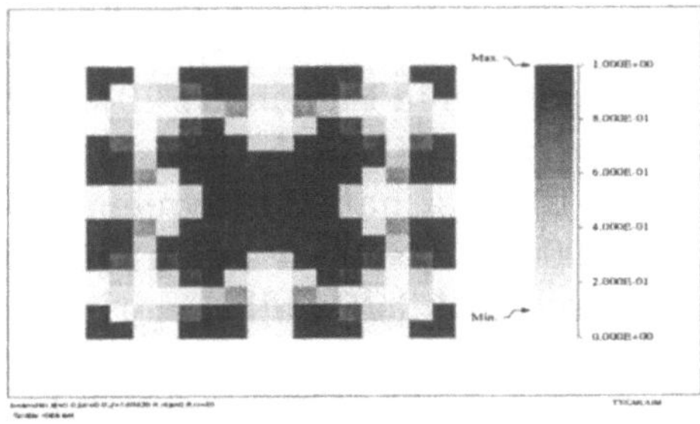

(7 a)

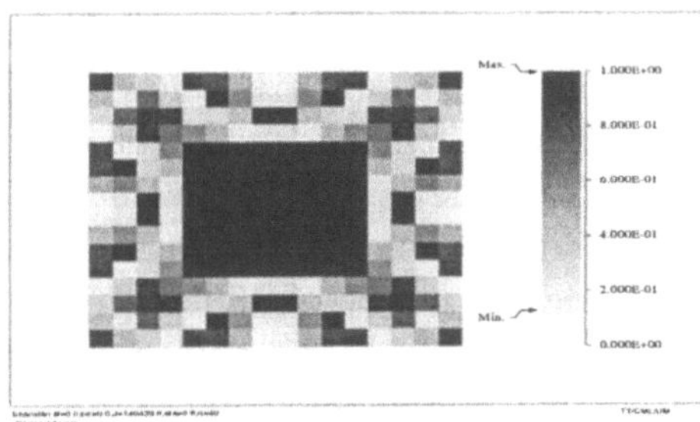

(7 b)

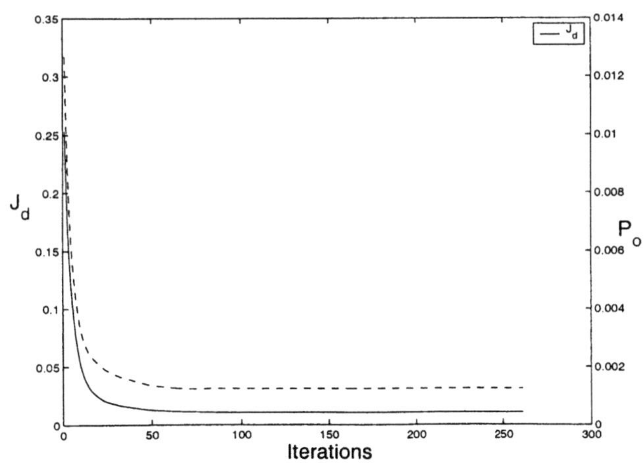

(7 c)

Figure 7. Density distribution for the plate example with $\hat{\omega}_1 = 5000\ rad/s$, $\hat{\omega}_2 = 15000\ rad/s$ (a) Input power minimization (b) Sound power minimization; (c) correlation between input power and sound power radiated.

The natural frequencies of the optimal topology shown in Fig. 5a are 893.4 rad/s, 9748.8 rad/s, 9758.4 rad/s, 12827.6 rad/s and 16524.6 rad/s, while those for the design in Fig. 5b are 797.2 rad/s, 5693.2 rad/s, 7734.2 rad/s, 14939.2 rad/s and 18226.5 rad/s. The sound power minimization strategy moves the first natural frequency away from the driving frequency of $\hat{\omega} = 5000$ rad/s by a larger amount than the input power minimization strategy, while for all the other natural frequencies, the reverse is true. As seen from Table I, the reduction in the sound power radiated achieved by the sound power minimization strategy is greater, indicating that the first mode is the dominant sound radiator. Nevertheless the reduction of almost 4 dB achieved by the input-power minimization strategy is quite substantial.

For the second case when $\hat{\omega} = 15000$ rad/sec, the first five natural frequencies of the optimal topologies shown in Fig.6a and 6b are 1322.8 rad/s, 7021.4 rad/s, 7026.3 rad/s, 23288 rad/s, 23311.2 rad/s, and 1758 rad/s, 7330 rad/s, 10425.6 rad/s, 16217.2 rad/s, 18716.5 rad/s, respectively. The natural frequencies in the former case are more widely separated from the driving frequency of 15000 rad/s, than in the latter case, and yet, as seen from Table I, the sound power reduction in the latter case is greater. This might seem anomalous; however, Eqn.(1) shows that the response is decided not merely by the separation of the natural frequencies from the driving frequency, but also the damping. The sound power minimization strategy maintains relatively high values of α and β (given by Eqn.(4) compared to the input power minimization strategy, thereby increasing the damping, decreasing the response amplitudes, and hence reducing the sound power radiated.

For the third case, the first five natural frequencies of the optimal topologies shown in Fig.7a and 7b are 1347 rad/s, 8095 rad/s, 8513 rad/s, 22462 rad/s, 22490 rad/s, and 1284.5 rad/s, 7751.1 rad/s, 9053.9 rad/s, 15235 rad/s, 18691.4 rad/s, respectively. From Table I, we observe that the acoustic power reduction using the former strategy (10 dB) is almost as much as that with the latter (10.83 dB). The results again indicate that the first mode is the dominant mode for sound power emission since the first natural frequency of the optimal design in the latter case is slightly lower than the one in the former case. Although the fourth natural frequency in the latter case is quite close to the driving frequency component of 15000 rad/s, this does not seem to have an effect on the sound power radiated.

4. Conclusions

We have shown that minimizing the input power using techniques of topology optimization results in a substantial reduction in the vibration, and hence the noise radiated, for a light fluid-loaded structure. The numerical examples presented show that the reductions are almost as good as the ones obtained with sound power as an objective function, but are obtained at a significantly lower cost. We note that the formulation presented is completely general (valid for any continuum structure), and there is no need to solve the acoustic equations. The associated sensitivity expressions are easy to evaluate, and can be obtained at a marginal cost once the structural analysis has been conducted. In contrast, if acoustic power is directly used as an objective function, the sensitivity analysis can be quite cumbersome, and, even in the rather simple case of baffled plates, is computationally intensive.

Acknowledgements

The author gratefully acknowledges many helpful discussions with Dr. V. R. Sonti, Mechanical Engg. Dept, IISc. This work has been made possible largely due to the financial support from the Department of Science and Technology of the Government of India through FRITA.

References

Bathe K. J. (1997) *Finite Element Procedures*, Prentice-Hall, New Jersey.

Bendsøe M. P. and Diaz A. R. (1994) Optimization of material properties for improved frequency response, *Structural Optimization* **7**, 138–140.

Bendsøe M. P. (1995) *Optimization of Structural Topology, Shape and Material.*, Springer Verlag, Berlin.

Christensen S. T., Sorokin S. V., and Olhoff N. (1998) On analysis and optimization in structural acoustics–Part I: Problem formulation and solution techniques, *Struct. Optim.* **16**, 83–95.

Christensen S. T., Sorokin S. V., and Olhoff N. (1998) On analysis and optimization in structural acoustics–Part II: Exemplifications for axisymmetric structures, *Struct. Optim.* **16**, 96–107.

Constans E. W., Belegundu A. D., and Koopmann G. H. (1998) Design approach for minimizing sound power from vibrating shell structures, *AIAA J.* **36(2)**, 134–139.

Diaz A. R. and Kikuchi N. (1992) Solutions to shape and topology eigenvalue optimization problems using a homogenization method, *Inter. J. Numer. Methods in Engrg.* **35**, 1487–1502.

Jog C. S. and Haber R. B. (1996) Stability of finite element models for distributed-parameter optimization and topology design, *Comput. Meth. Appl. Mech. Engrg.* **130**, 203–226.

Jog C. S. (2001) Topology design of structures subjected to periodic loading, *Journal of Sound and Vibration*, to appear.

Koopmann G. H. and Fahnline J. B. (1997) *Designing Quiet Structures*, Academic Press, London.

Krog L. A. and Olhoff N. (1999) Optimum topology and reinforcement design of disk and plate structures with multiple stiffness and eigenfrequency objectives, *Computers and Structures* **72**, 535–563.

Lamancusa, J. S. (1993) Numerical optimization techniques for structural-acoustic design of rectangular plates, *Computers and Structures* **48**, 661–675.

Ma Z. D., Kikuchi N., Cheng H. C., and Hagiwara (1995) Topological optimization technique for free vibration problems, *Trans. of the ASME, J. of Applied Mechanics* **62**, 200–207.

Ma Z. D., Kikuchi N., and Cheng H. C. (1995) Topological design for vibrating structures, *Comput. Meth. Appl. Mech. Engrg.* **121**, 259–280.

Naghshineh, K., Koopmann G. H., and Belegundu, A. D. (1992) Material tailoring of structures to achieve a minimum radiation condition, *J. Acoust. Soc. Am.* **92(2)**, 841–855.

Salagame, R. R., Belegundu, A. D., and Koopmann G. H. (1995) Analytical sensitivity of acoustic power radiated from plates, *J. Vibration and Acoustics* **117**, 43–48.

St. Pierre R. L. and Koopmann G. H. (1995) A design method for minimizing the sound power radiated from plates by adding optimally sized, discrete masses, *J. Vibration and Acoustics* **117**, 243–251.

Tinnsten M., Esping B., and Jonsson M. (1999) Optimization of acoustic response, *Struct. Optim.* **18**, 36–47.

Tinnsten M. (2000) Optimization of acoustic response–a numerical and experimental comparison, *Struct. Optim.* **19**, 122–129.

COMPRESSOR RELATED NOISE CONTROL IN AIR-CONDITIONERS AND REFRIGERATORS

S. MANIVASAGAM and J. SENTHILNATHAN
Product Engineering Department
Kirloskar Copeland Ltd.
Karad-415 110, India
manivasagam@kircop.com

Abstract Various noise reduction techniques for developing quiet compressors and appliances are briefly explained in this paper. In Air-conditioners and Refrigerators, the compressor is known to be one of the major noise sources. Therefore it is highly essential for the designer of the compressor to understand the noise mechanisms, including the sources and the transfer paths, so that low noise compressors can be effectively developed. It is also equally important for the domestic appliance manufacturers to understand the interaction effects between the compressor and the components of the appliance so that effective tuning of the system can be carried out. Hermetically sealed reciprocating compressors are considered here even though most of the noise reduction techniques are common across the various types of compressors. The techniques include handling both the source as well as the noise path for noise reduction. A few recommendations are given towards developing quiet appliances.

1. Introduction

As the affordability of the people of the developing counties is rapidly improving, more and more consumers have started purchasing Refrigerators and Air-conditioners for their domestic requirements. They are also gradually getting educated on the noise related environmental issues and started demanding quite appliances. In Air-conditioners and Refrigerators, the compressor is known to be one of the major noise sources. Therefore it is highly essential for the designer of the compressor to understand the noise mechanisms, including the sources and the transfer paths, so that low noise compressors can be effectively developed. It is also equally important for the domestic appliance man-

M.L. Munjal (ed.), IUTAM Symposium on Designing of Quietness, 237–246.

ufacturers to understand the interaction effects between the compressor and the components of the appliance so that effective tuning of the system can be carried out.

This paper discusses a few popular aspects of noise control techniques with applications to hermetically sealed compressors as well as a few guidelines for the appliance manufacturers to target overall quiet appliance development.

2. Noise Mechanism in Hermetic Compressors

A hermetically sealed reciprocating compressor consists of an induction motor and a reciprocating pump mechanism. This mechanism is basically an assembly of crankcase, crankshaft, connecting rod, piston, valve plate and valves, gaskets and cylinder head. The inner pump assembly is mounted on relatively soft spring supports, which are attached to the outer housing which is nothing but an elliptical closed shell. A cut section of a typical hermetically sealed reciprocating compressor is shown Figure1. The Compressor is mounted on rubber grommets, which are firmly connected to the appliance base plate. For lubrication purposes a prescribed quantity of oil is poured inside the shell cavity, which in few applications cools the motor also. The complete inner assembly, including the motor, forms the source of noise for the compressor. The noise is transferred to the outer shell through the acoustic cavity path as well as through the structural path like springs and tube connections.

Figure 2 shows the schematic representation of the noise generation and transmission mechanism in a reciprocating compressor [1].

3. Compressor Noise Control Techniques

The general approach for controlling the noise at the source and by modifying the transmission path involves the following activities.

3.1 Balance of Rotating and Reciprocating Masses.

It is strongly recommended that the rotating assembly should be dynamically balanced for force as well as for moment. Any imbalance of considerable value will lead to heavy vibration of the sprung assembly, which may get transmitted to the appliance base-plate [2]. Figure 3 shows the balancing mass added to the rotor ring.

3.2 Attenuation of Gas Pulsating by Using Efficient Mufflers

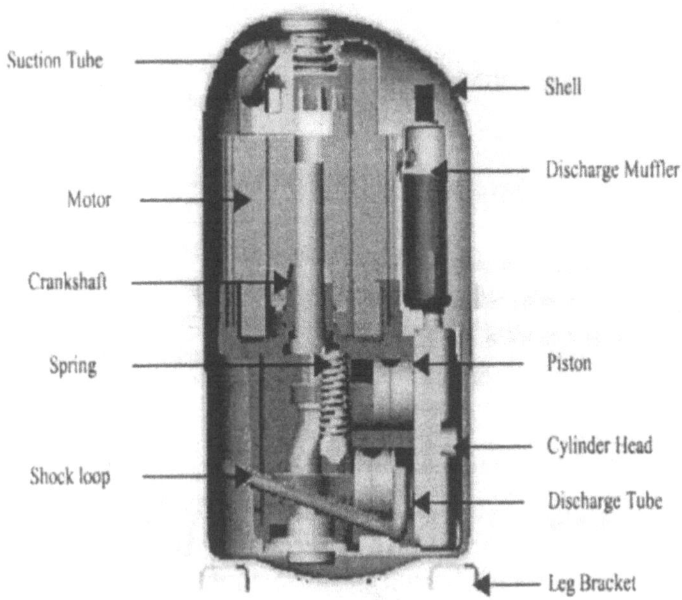

Figure 1. Cut view of a typical hermetic compressor

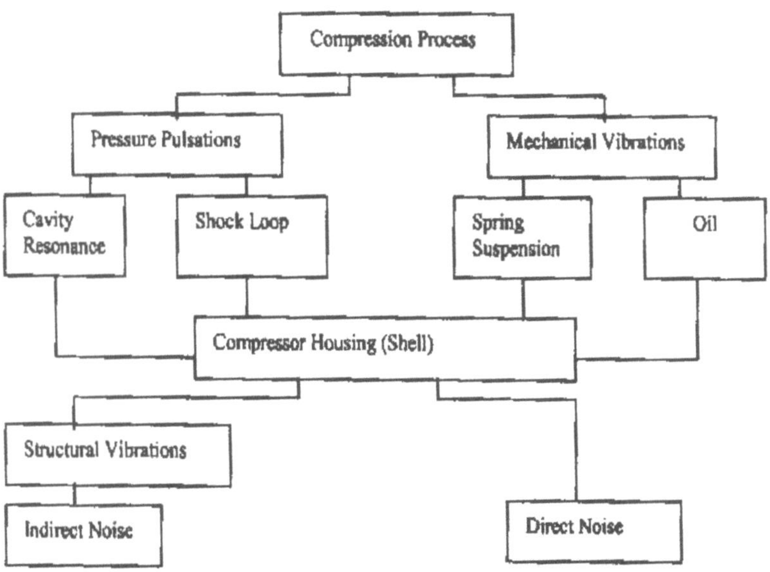

Figure 2. Noise generation and transmission mechanism

Gas pulsation not only leads to cavity resonance, but also excites the compressor as well as the appliance tubing connections and result

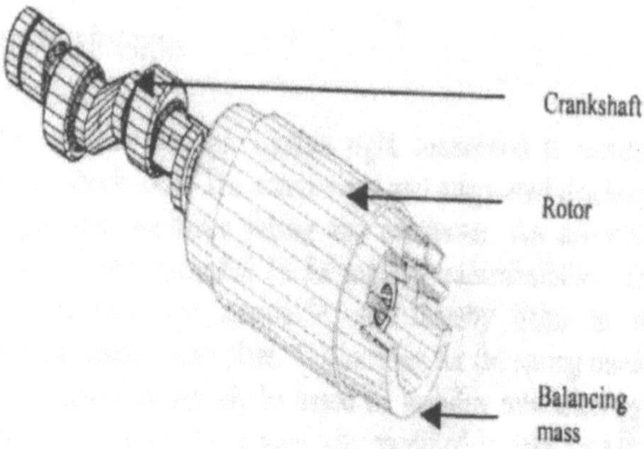

Figure 3. Shaft and rotor assembly with rotor balancing mass

in appliance noise and vibration. Therefore it is highly important to attenuate the suction and discharge pulsations by properly selecting or designing the mufflers. It is also important to take care the back pressure increase due to restrictive muffling activities.

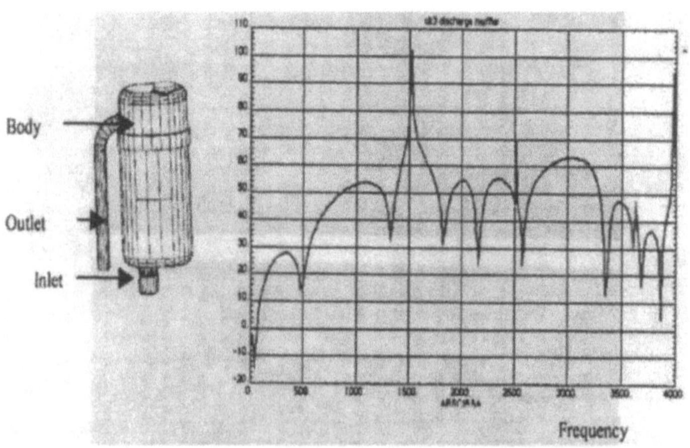

Figure 4. Discharge muffler and its Transmission Loss curve

Perforated type mufflers [3] may be recommended where backpressure considerations are very serious. Figure 4 shows the configuration of a typical discharge muffler of an air-conditioner compressor and its transmission loss characteristics. The transmission loss calculations have been carried out by using SYSNOISE Vibroacousic software. It is always recommended to have wider attenuation characteristic mufflers, 500 Hz to 2000 Hz for the discharge side.

3.3 Isolation/Suspension

In addition to the mounting springs, another rigid connection is present inside the compressor known as shock loop. The compressed and attenuated discharge gas flows through this and gets into appliance tubing and condenser. An assembly frequency below 10 hertz is always recommended for having low transmissibility. This helps the housing to vibrate at very low amplitude, and thereby helps in reducing the transmissibility to the appliance base plate. The location for the spring mounting as well as for the shock loop should preferably be based on mobility mismatch conditions [4]. Figure 5 shows how an existing shock loop was modified to take care transportation shock absorption, start stop transients and resonance at the operating frequencies.

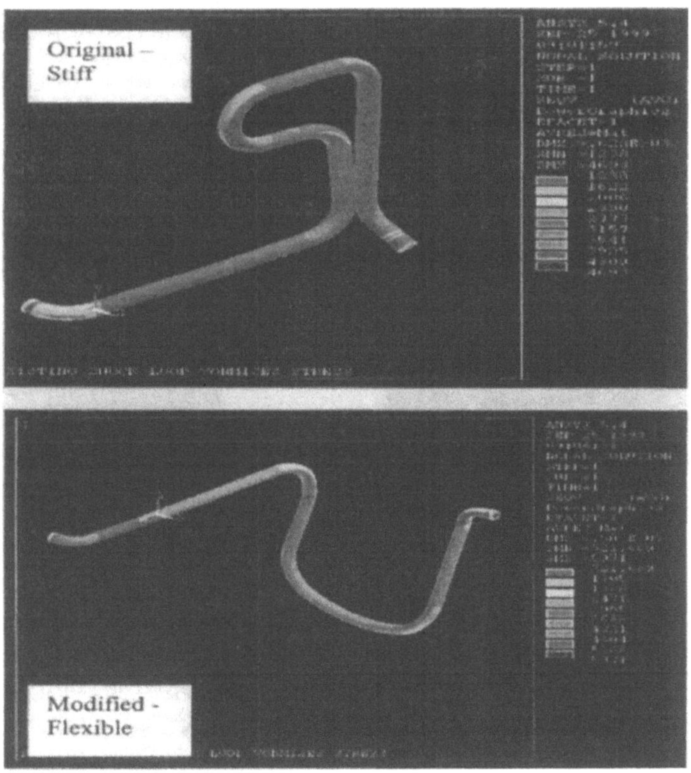

Figure 5. Original stiff shock Loop and modified shock loop with added flexibility.

3.4 Noise Control by Shell Redesign

The other very popular approach for noise reduction in compressors is attacking the transmission path namely the shell [5]. This can be achieved in two ways; the thickness of the shell can be marginally increased or the stiffness of the shell can be increased by properly contouring it. The later approach involves no additional material cost increase and therefore being targeted globally by compressor designers. The best possible shell option would be a perfect sphere, which because of assembly and size considerations is not feasible. Therefore, designers always try to give larger fillets between the top and sides of the compressor shell and thereby coupling the modes as much as possible. Figure 6 shows how the top mode and body mode shift towards higher side by changing the curvature of the upper and lower portions of the shell. More often, it can be observed that the shift in top mode is around 1000Hz. A 15 to 20 percent reduction in the modal density of the shell can also be observed. This technique has been proved to be a cost-effective way of reducing the shell radiated noise.

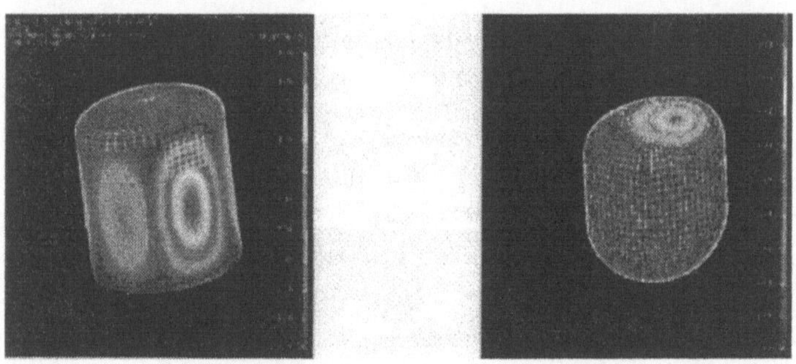

Figure 6. Shell stiffening by contouring.

Some times due to some practical difficulties with respect to manufacturing or assembly, it may not be possible to opt for shell contouring. Also, frequencies lying in the mass controlled regionsmay still radiate noise as forced response. Under these conditions, a proven way of handling noise issue is by increasing the shell thickness marginally, with little impact on the cost. Figure 7 shows the SYSNOISE response of a node for increased thickness of about 10 percent. Figure 8 shows the experimental Sound Power Spectrumcomparison for 10 percent increase in the shell thickness.

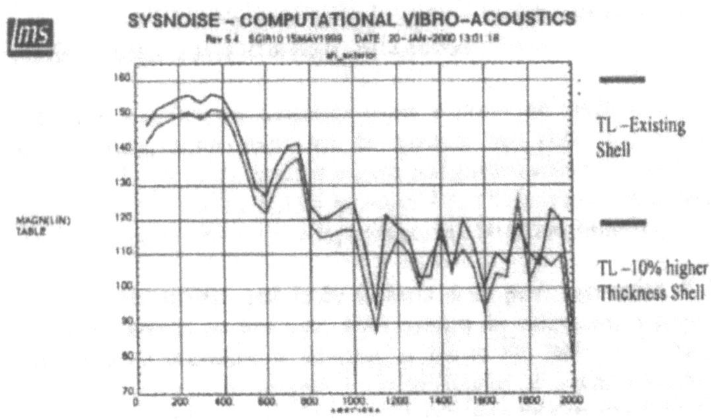

Figure 7. Transmission Loss response of a typical node

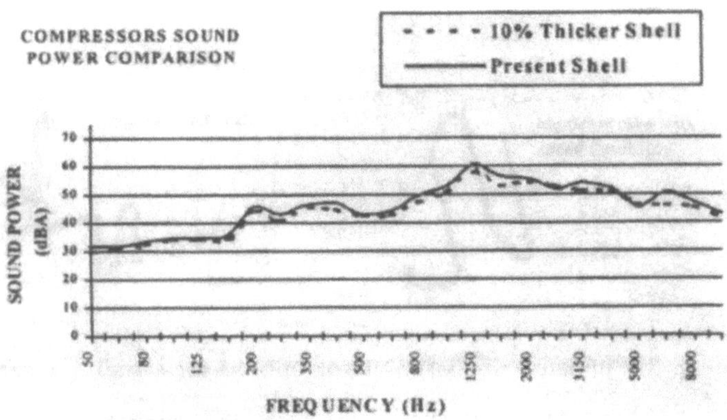

Figure 8. Experimental compressor sound power comparison

4. Appliance Noise Control

As it was pointed out in section 4.2, very frequently, noise reduction activities lead to performance deterioration. This may be due to restrictive gas flow passages or stiff valve reads etc. Even though people are still working on high efficiency compressors with low noise, ultra high efficiency compressors, whose Energy Efficiency Ratio(BTU output divided by wattage power input) is above 12, may be noisier than the normal compressors. This is also due to the fact that high efficiency compressors should also be available at the affordable price, which prevents cost addition towards noise reduction. As most of the developed countries are coming out with strict Energy Efficiency Norms to counter the ever-expanding power needs, the appliance manufacturers/designers role in controlling appliance noise is becoming very significant.

The noise path from the compressor source is mainly the condenser and the evaporator tubing, rubber mount and the appliance base plate. It is strongly recommended that the tubing should be properly configured to avoid resonance due to operating frequencies and their first few harmonics. Figure 9 shows how a modification in the tube configuration helps in avoiding resonance with the exciting frequency.

A major noise radiation path in the appliance is the panel construction of the appliance. Particularly, the base plate, which supports the compressor, is prone to resonate at lower frequencies in the range of 180 to 220 Hertz. It is strongly recommended to the appliance design-

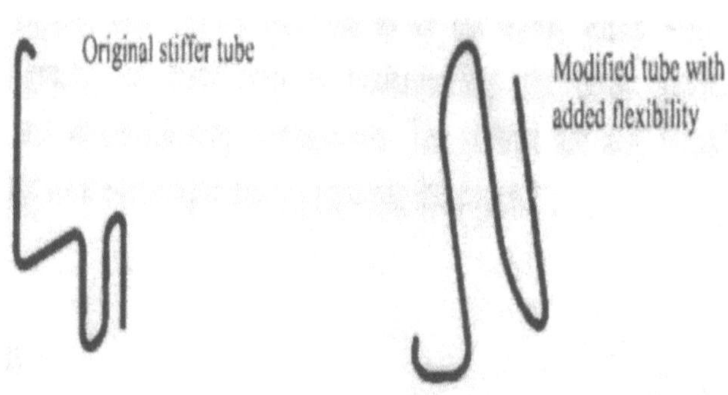

Figure 9. Appliance discharge tubing options for avoiding resonance

ers that before investing on expensive tooling, to ensure that the base vibration natural frequency is away from the compressor operating frequency harmonics; for example 200 Hertz. Tools like FEM are of great help in predicting the natural frequencies of the panels [6,7]. Figure 10 shows a comparison between an existing base plate of an appliance and a proposed base plate considering resonance behavior.

5. Conclusion

Various noise reduction techniques for developing quiet compressors and appliances are briefly explained. The techniques include handling both the source as well as the noise path for noise reduction. A few recommendations are given towards developing quiet appliances. Many compressor and appliance manufacturing industries including ours, have gained reasonably good benefits in the noise control assignments.

Figure 10. Appliance discharge base plate options for avoiding resonance

6. Acknowledgements

The authors express their sincere gratitude to all the teams, which work on noise reduction programs, for their help in understanding the noise mechanisms in refrigeration and air-conditioning compressors. The authors are also thankful to the management for their encouragement in preparing this paper.

References

Hamilton, J. F. 1988 *Measurement and Control of Compressor Noise*, Purdue University Publication, West Lafayette, US.

Gatecliff, G.W.: 1974 Analysis of free vibration of spring mass of a reciprocating compressor including comparison with experiments., *Proceedings of the International Compressor Conference* Purdue University , **221-229**

Munjal, M.L.: *Acoustics of Ducts and Mufflers with Application to Exhaust and Ventilation System Design,* John Wiley and Sons, 1987.

Kjeldson, K. and Madsen, P.: Reduction of compressor vibrations by optimizing the location of counter weight and internal springs, *Proceedings of the International Compressor Conference,* Purdue University (1978), 55-59.

Lowery, D.C.: An improved shape of compressor housing, *Proceedings of the International Compressor Conference,* Purdue University (1984), 285-290.

Zhang, K., Yang, J. and Walt, M.: Acoustic analysis of a vehicle ribbed floor, *Proceedings of the Noise and Vibration Conference,* Society of Automobile Engineers Publication (1977), Paper No. 971945.

Onsay, T., Akanda, A., and Goetchis, G.: Vibroacoustic analysis of a bead stiffened flat panels-FEA, SEA, Experimental analysis, *Proceedings of the Noise and Vibration Conference,* Society of Automobile Engineers Publication (1999), Paper No. 1999-01-1698.

AN EXPERIMENTAL VERIFICATION OF STRUCTURAL-ACOUSTIC MODELING AND DESIGN OPTIMIZATION

H.-J. BEER, J. GIER,
H.-J. HARDTKE, S. MARBURG
Institut für Festköpermechanik, Technische Universität
01062 Dresden, Germany

F. PERRET
Ecole Supérieure de Mécanique de Marseille,
IMT - Technopole de Chateau-Gombert,
13451 Marseille Cedex 20, France
(on leave at Institut für Festkärpermechanik)

R. RENNERT
Institut für Materialforschung und Anwendungstechnik GmbH
01101 Dresden, Germany

Abstract A number of papers have been published on the simulation of structural–acoustic design optimization. However, extensive work is required to verify these results in practical applications. Herein, a steel box of 1.0×1.1×1.5 meters with an external beam structure welded on three surface plates was investigated. This investigation included experimental modal analysis and experimental measurements of certain noise transfer functions (sound pressure at points inside the box due to force excitation at beam structure). Using these experimental data, the finite element model of the structure was tuned to provide similar results. With a first structural mode at less than 20 Hertz the reliable frequency range was identified up to about 60 Hertz. Obviously, the finite element model could not be further improved only by mesh refinement. The tuning process will be explained in detail since there was a number of changes that helped to improve the structure. Other changes did not improve the structure. Although this model of the box could be expected as a rather simple structure it can be considered to be a complex structure for simulation purposes. A defined modification of the physical model verified the simulation model. In a final step, the opti-

M.L. Munjal (ed.), IUTAM Symposium on Designing of Quietness, 247–278.

mal location of stiffening beam structures was predicted by simulation. Their effect on the noise transfer function was experimentally verified. This paper critically discusses modeling techniques that are applied for structural–acoustic simulation of sedan bodies.

1. The Problem

Several papers that deal with structural–acoustic optimization indicate significant improvements of the acoustic properties inside or outside the structure. Noise reductions of 10 to 50 dB have been reported. Most of them will be discussed in what follows.

Hambric [9,10] reported the improvement of a ripped cylindrical shell with hemispherical ends by about 10 dB using location of ribs, shell thickness and loss factors as parameters. As the objective function, he used the radiated sound power as well as Lamancusa [16,17] who reported 8 to 15 dB reduction in a frequency range (100 to 1000 Hz) only by optimizing the local distribution of plate thickness at a constant mass. In an older paper, Lamancusa [15] discussed minimization of noise transmission through an intake system of internal combustion engines. While optimally designing diameters and lengths of ducts, he managed to reduce the average noise transmission up to 20 dB over the frequency range of 50 to 250 Hz. Experimental verification was provided.

Hibinger [11] discussed optimization problems of two plates perpendicularly oriented to each other. First, he verified the calculation model with an existing physical model. Then he optimized structure–borne noise in these plates by a locally distributed plate thickness and by ribs. He presented similar investigations at a box structure. It was mentioned that more than 10 dB average reduction over a large frequency range were achieved at real structures. Experimental measurements at optimized structures were not reported. However, once a reliable simulation model is available this model should be reliable for the modified structure as well.

Kozukue et al.[14] and Pal and Hagiwara [25,26] presented an example where they have observed an improvement of about 50 dB in a box only by optimizing the plate thickness of flat panels. Tinnsten et al. [28] reported similar reduction of noise inside a box structure. They applied locally distributed plate thicknesses and the fibre orientation as optimization parameters. In most of the papers that optimize plate thickness distributions at least one constraint concerning an upper limit of the mass was necessary and, therefore, included.

Crane et al. [6], Cunefare et al. [7] and Engelstad et al. [8] described the minimization of sound transmission through a clamped cylinder. Application of twenty optimization variables [6] that represented twenty circumferential bands of constant thickness allowed them to reduce the transmission level by about 20 dB while keeping the mass constant. With respect to these results, they emphasized the particular value of structural acoustic optimization in passive noise control. In [7] and [8], they presented results of optimal locations and cross section shapes of frames and stringers. 8.6 dB reduction were gained mainly by optimizing the positions and distributions of the stiffening structures over the cylinder. Cross sections of the stiffeners were of minor importance in the optimization process.

Belegundu and Koopmann and several different co-workers have published a number of papers on structural-acoustic optimization, see for instance Naghshineh et al. [24], Belegundu et al.[3], St. Pierre et al. [27] and Constans et al. [5]. There, they have reduced the radiated sound power of plates and shells. In one of these papers [3], a plate of optimized stepped thickness distribution confirmed that even a weaker structure can radiate less sound than a stiffer structure. In that example, the first eigenfrequency of an engine cover plate was decreased while the radiated noise level reduced by about 12 dB. Similar investigations [24] were reported for piecewise constant material data like YOUNG's modulus or density. Reductions between 2 and 23 dB were achieved. Optimally sized discrete masses lead to an improvement of 30 dB in the simulation and, still, 22 dB in the experiment. Although, this can be considered as a good prediction the authors discussed the differences between theoretical and experimental results [27]. In the special cases of a clamped plate and a semicylindrical shell[5], reductions of 33 dB for the clamped plate and 9.6 dB for the shell were gained by adding lumped masses. Both, the position of the masses and their size were considered in their investigation.

There are few papers dealing with shape modification of the shell curvature. In their work on optimization of the directivity pattern of a loudspeaker, Christensen and Olhoff [4] managed to change the original directivity pattern at different frequencies simultaneously by 10 to 20 dB. In that paper, eight keypoint positions and locally distributed shell thicknesses occurred as design parameters. The papers by Marburg and Hardtke et. al.[21,22,23] indicate that significant reduction of noise emission is possible only by changing the curvature of a shell structure. If this modification remains small with respect to a characteristic dimension of the shell the mass remains nearly constant. Hence, a mass constraint is negligible.

Considering very complex structures in general and vehicle bodies in particular the most challenging problem arises in creating a reliable calculation model. Obviously, this model should be a compromise. It should be simple enough to be handled easily and to avoid unnecessary expensive computation especially in the optimization process and it should be detailed enough to represent all or at least most of the desired effects being under consideration. Modeling of the body structure of a car usually is so complex that existing local effects of vibration cannot reliably be validated in the calculation model. Moreover, methods that proposed changes of the geometry of vehicle panels [21.22] cannot be verified at these structures. The possibility of substituting the panel by another panel of the new geometry would include a destruction of the complex structure and even the identical panel would provide a new vibroacoustic behaviour. An alternative would be to produce two body structures, one with the original and one with the optimized panel. However, it is well-known that in the case of complex structures we will hardly find two identical compositions. Thus, the improvement of complex structures can only show an average effect for a large set of examples.

For that reason, a simpler structure is chosen to verify improvements of structural optimization experimentally. This construction consists of two beam frames and six more or less flat panels. Together they form a box with an external beam construct. In general, the components are welded together to have a vehicle body like assembly. To avoid destruction of the box, feasible design modifications only include addition of stiffening beam structures.

In this paper, the authors will discuss the process of finding a reliable calculation model in a reasonable period of time. The simulation model is tuned with respect to modes found in the experimental modal analysis. Further, experimentally measured structural and noise transfer functions give an impression of the reliability of the simulation model. This frequency range of reliability is confirmed by an added mass block that was used to imply a defined modification. Finally, the model will be modified by four added beams of optimized length and position. All model modifications will be experimentally verified.

2. Model Description

2.1 Physical Model

The physical model was designed and constructed for experimental purposes. The assembly had been carried out similarly to sedan body assembly. This was confirmed by the automotive company that constructed the model. However, exactness of the assembly was somewhat rougher than those of real cars.

The box–like model consists of an internal beam–frame, an external frame structure and six surface panels. It is made of steel. The internal frame and the surface metal sheets are spot–welded together whereas the external frame is fixed at the surface panels at certain welding joints. Owing to welding points, welding joints and additional prestrain effects, it can be easily observed that the surface panels are not as plane as it could be expected for a simple structure.

Figure 1. Experimental setup of box model

To provide some measures: The box is 1.5 m long, 1.1 m wide and 1.0 m high. We define the end of the external frame as the front end. It shall be mentioned that the internal frame does not only support the edges of the box. There is an additional frame belt at about 85 cm from the front end. The upper and the lower adjacent surface panels have a gap along this frame belt. Both parts, are detached from each other but spot–welded at the frame belt. The side panels are continuous from the front to the rear. The rear panel can be dismounted since it is screwed to the base–frame. The external beam structure stiffens the lower part of the front panel and the front part of the floor panel. It is continued about 60 cm in front of the box.

The panels are of different thicknesses between 1.2 and 3 mm while the beam structure has a 20 mm square cross section of 2 mm thickness.

The bearing conditions can be considered as free–free bearing. Hence, we should expect six rigid body modes. These modes can actually be identified. The maximum eigenfrequency of a rigid body mode is found at about 4 Hz and the lowest elastic mode is observed above 19 Hz. Furthermore, the bearings are fixed at the upper corners. These points are particularly stiff so that the bearing conditions should hardly influence the elastic vibration modes.

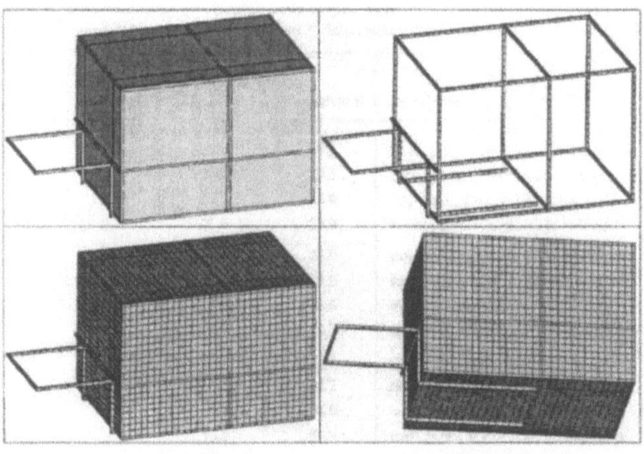

Figure 2. Visualization of structural model, area plot (upper left), beam frame elements(upper right) and model of element size $\leq 5cm$(lower subfigures)

The force excitation is applied at the left corner of the external beam frame in front of the box. This ensures that the force excitation does not directly act on a noise emitting panel. This shall be a similarity to engine excitation of cars. Moreover, the external beam frame transmits the structural vibrations on to the front and the lower panel. The remaining four panels are excited by these two.

Three holes have been bored into the upper panel. These holes are required for sound pressure measurements inside the box. Their diameter is about 30 mm, just sufficient to conduct one hanging microphone through.

2.2 Simulation Model

For simulation purposes, a geometry based model has been created for the structure. This model consists of 583 keypoints. Connection of these keypoints are realized using 1134 spline–interpolated lines while these lines form 572 interpolated areas.

Table 1. Eigenfrequencies of elastic mode shapes, experimental and simulation data, mode numbering based on simulation I (element size $\leq$ 5cm), simulation II (element size $\leq$ 2.5cm)

Mode	Eigenfrequencies [Hz] Experiment	Simulation I	Simulation II	Description
1	19.3	19.3	19.3	lower panel, rear part, first mode
2	20.4	20.2	20.3	upper panel, front part, first mode
3	23.6	23.7	24.0	rear panel, global, first mode
4	23.3	27.8	27.9	lower panel, rear part, second mode
5	28.1	28.1	28.0	external beam frame, first mode
6	29.2	28.6	28.8	lower panel front, upper panel rear
7	30.5	29.8	30.0	almost same as mode 6
8	—	35.5	35.7	upper panel, front part, second mode
9	37.1	36.9	37.2	side panels and others
10	39.3	38.2	38.7	front panel, upper part, first mode
22	55.7	54.2	54.6	external beam frame, second mode
24	57.1	56.6	57.3	front panel, upper part, second mode

Table 2. Number of eigenfrequencies of elastic mode shapes (rigid body modes excluded) below certain frequency limits for simulation I (element size $\leq$ 5cm), simulation II (element size $\leq$ 2.5cm) and the optimized structure (discretization as in simulation I)

type	Number of eigenfrequencies below 50 Hz	100 Hz	150 Hz	200 Hz	250 Hz	300 Hz
simulation I	20	61	108	159	217	275
simulation II	19	61	108	156	212	268
optimized	19	58	104	153	211	266

The panel surfaces are modeled with variable curvature to include the uneven characteristics of these areas. All panels and even the beam structures are included as areas although it could be assumed that modeling of beams as lines would be sufficient for our purposes. However, especially the spatially distributed cross sections of the beams and, henceforth, the chance of spatially distributed and selective coupling between beams and surface panels appeared to be a particular sensitive issue in the process of model adjustment.

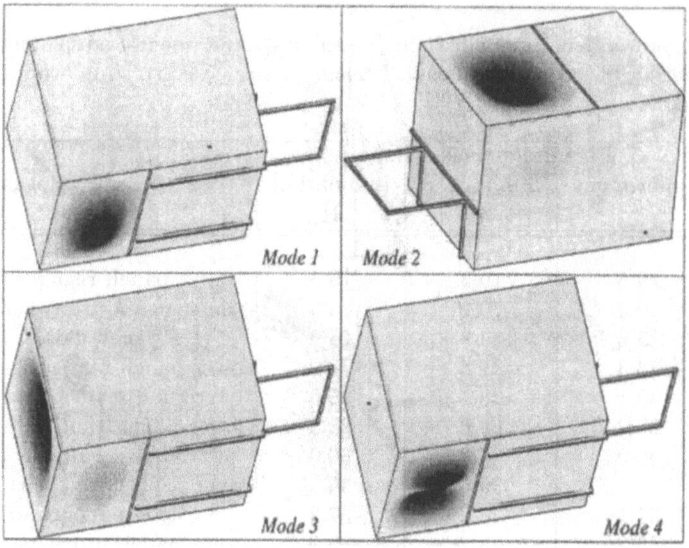

Figure 3. Mode 1 to 4 of the box model, frequencies of experimental and simulated modal analysis in Table 1

One of the advantages of a geometry based model is the possibility of variable meshes. Checking different mesh sizes we finally end up with a compromise of a maximum shell element size of 5 cm, i.e. there are many smaller elements. Altogether, the finite element model of the structure consists of 6577 (linear) shell elements and about 6604 nodes. The constraints between the three different parts of the structure are modeled as very short and light but stiff beams. The commercial finite element code ANSYS Rev. 5.4 [2] that is used for structural analysis offers usage of constraint equations as well. However, in this case they provide a ten times greater wavefront in the solution process. The simulation model contains 1235 of these short beam elements.

To briefly discuss a refined mesh of the structure that will be referred to in the modal analysis section we mention that this model consists of 18773 linear shell elements of size $\leq$ 2.5cm while the number of nodes reaches 18836. Compared to the coarser model the wavefront in the solution process increases from 1830 to 2370.

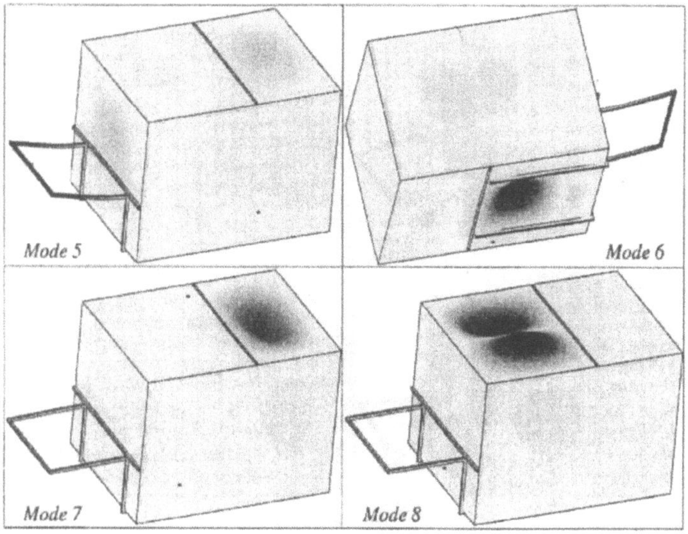

Figure 4. Mode 5 to 8 of the box model, frequencies of experimental and simulated modal analysis in Table 1

The fluid model is much coarser than the model of the structure. It consists of 850 linear boundary elements and 852 nodes. That is equivalent to an element size of 10 cm. With the material data of air being 340 m/s as the speed of sound and 1.3 kg/m^3 as the fluid's density and the upper frequency limit of 100 Hz we have 34 elements per wavelength. Even with the knowledge of a bad convergence of linear acoustic elements, cf.[12,18], we can expect very small numerical errors for the fluid analysis. Generally, the boundary conditions for the fluid are assumed rigidly. One test will be described where the fluid's boundary conditions include complex boundary admittances.

The (non–commercial) boundary element based computer program AKUSTA is used for the acoustic calculations. AKUSTA allows the computation of acoustic influence coefficients that can be efficiently used for optimization purposes, cf. Marburg et al. [22] and Marburg [19].

It shall be mentioned that for this type of multifield calculation a one–way fluid structure interaction is assumed. This implies that we first compute the structural vibrations in vacuum and, then, take the harmonic particle velocity distribution as a boundary condition for an acoustic analysis.

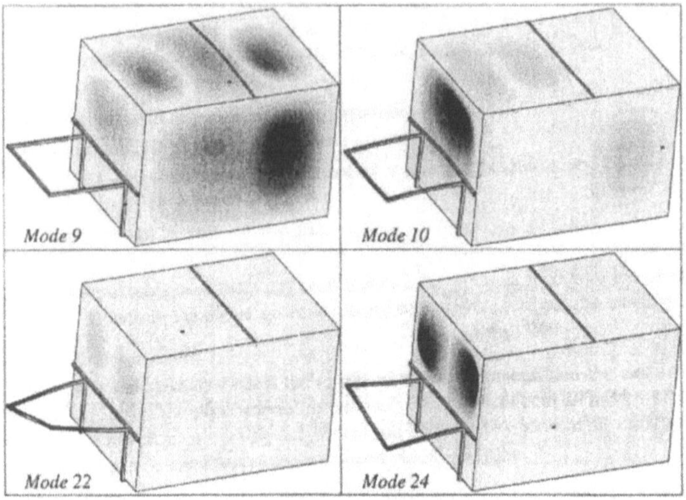

Figure 5. Mode 9,10, 22 and 24 of the box model, frequencies of experimental and simulated modal analysis in Table 1

3. Experimental and Simulated Modal Analysis

It is well-known that special simulation models initially show many differences in their vibration behaviour when compared to the real model. This dilemma increases for complex models. Moreover, it seems beyond a reasonable solution if large numbers of modes should be considered. Although initially a simple model was looked for we obviously have to expect numerous eigenfrequencies in the frequency range up to about 100 Hertz.

The comparatively stiff edges of the box can likely be assumed as one of the major advantages of this model for low frequencies. Hence, it is possible to consider most of the surfaces separately. This is a reasonable simplification for the first one or two modes of each panel. However, hardly any useful success may be reported in some cases even for low frequencies.

The experimental modal analysis of the structure considers data of 70 piezoelectric accelerometers of the negligible mass of about 1 g. Major focus is put on the vibrations of the front panel and the external beam frame. A more detailed modal analysis especially using a laser scanning vibrometer would have been useful for this investigation especially for the vibrations of the flat shell panels.

Several different measures are taken to match modes and frequencies of simulation and experimental modal analysis. Most successful changes include:

- Optimization of panel curvature, especially front panel
- Prestrain effects at welding joints
- Carefully incorporated coincidence of welding spots/joints and beam constraints
- Adjustment of panel thicknesses

A number of additional trials has been realized, for example incorporation of self weight. However, the above given items are best suited to adjust the calculation model.

Now we would like to discuss the results of the experimental and the simulated modal analysis. The experimental modal analysis provides about 20 modes. Obviously, the lowest six modes are subject to rigid body modes. Most of the remaining 14 modes can be identified in the simulated modal analysis.

Twelve mode shapes are visualized in Figures 3 to 5. Eigenfrequencies in the order of occurrence in the simulation can be found in Table I. We observe an excellent agreement for most of the eigenfrequencies. Worst deviation occurred for the Mode 4, the second mode of the rear floor panel. It is assumed that this is subject to welding joints between this panel and the internal base frame. Modes 2 and 3 are calculated very close to the experimentally determined frequencies. Though, modes 6 and 7 look very similar and can even be found at neighboring frequencies they are both confirmed by the experimental data. Mode 8 can not be verified in the experimental modal analysis. This may happen because the number of accelerometers in this domain is chosen too small. However, there are two modes below 40 Hz in the experiment that can not be identified. So this problem can also be originated by an inconsistency of the experimental data. For more explanations on this subject, we refer to Auweraer et al. [29]. Another typical mode shape is mode 9. While for lower frequencies usually only one panel accounts for a single vibration mode shape higher modes are characterized by global eigenvector shapes. These global modes are much more difficult to adjust.

A good and important mode matching can be reported for the external frame structure and the front panel. Adjustment of beam modes 5 and 22 requires exact positioning of the welding joints and correct modeling of the free length of the beam frame. Although the beam structure on the front panel is significantly stiffer than the panel itself an important interaction can be observed. This interaction is expressed by modes 10 and 24. These modes show vibrations of the upper part of the front panel and vibrations of the external frame. The frequencies match very well.

It is important to mention in connection with mode 24 that the idealized shape being provided by finite element analysis is not found in the experiment. This mode can be identified clearly, though. However, the displacements in the left part of the panel (looking from rear to front) are about twice of those in the right part.

Few more modes can be identified in the experiment and the simulation. One example is the rear panel where both of the second mode shapes are clearly observed. Since these modes hardly effect the acoustic noise transfer function as being described later we dispense with their detailed description.

Only few remarks are given for the fluid modes inside the box. Assuming hard reflecting walls, the lowest fluid mode can be expected at 113 Hz. Successively, higher modes can be found. It is mentioned at this point that – hard reflecting walls provided – an additional rigid body mode at 0 Hz occurs. However, this mode can exist only if a slowly increasing pressure inside the model does not escape. Gaps between welding spots, joints and bore holes definitely destroy this assumption. We are going to discuss this matter later with the noise transfer function.

Finally in this section, we would like to briefly discuss issues of the numerical reliability of the structural model with respect to its discretization. As mentioned in the previous section, two finite element models of the structure are considered. One consists of elements up to 5 cm length, the other one of elements up to 2.5 cm length. Comparison of the eigenfrequencies of these two meshes should suffice for such a check. In general, we observe about one percent deviation between eigenfrequencies of matching modes up to a frequency of 150 Hz. Table II presents the number of eigenfrequencies up to certain frequency bounds. These numbers coincide well in the range up to 150 Hz. Moreover, these counts provide an impression of the modal density of our complex model.

4. Structural and Noise Transfer Functions

4.1 Structural Transfer Function

Structural transfer functions are measured at 22 locations using accelerometers. The harmonic accelerations are then integrated twice to find the structural transfer functions, i. e. the displacements due to a unit force excitation at the left corner of the external beam frame in front of the box. Measured transfer function values are available in steps of 0.25 Hz.

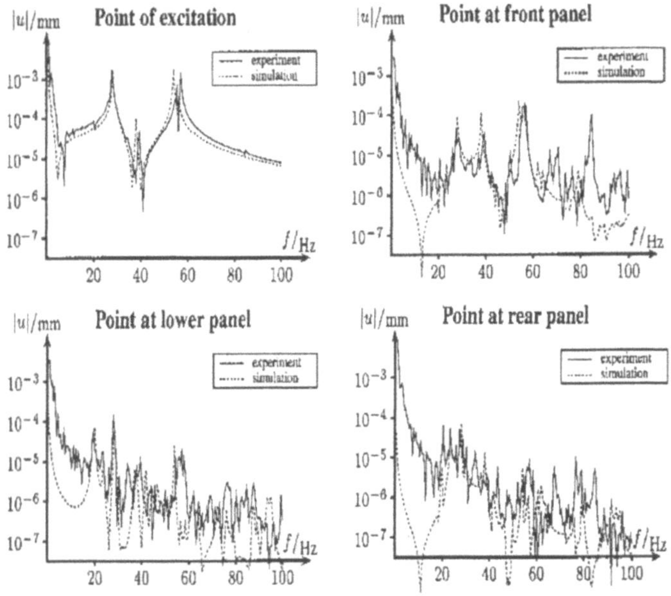

Figure 6. Structural transfer functions measured and calculated for four diferent points, excitation at the external base-frame, averaging not included

A corresponding configuration is applied to the simulation model. A finite element approach is used. Further, we assume a harmonic time dependence of the force excitation as $\hat{f}(\vec{x},t)=f(\vec{x})\,e^{i\omega t}$. Vector $\vec{x}$ represents the position in space, ω is the circular frequency and t and i denote time and imaginary unit respectively. Now, we calculate the structural transfer function by solving the linear harmonic system of equations

$$A(\omega)u(\omega) = f(\omega) \tag{1}$$

for the harmonic displacements assembled in the column matrix u. The other column matrix f contains the nodal excitation force vectors. A is the global system matrix of finite elements more commonly known as the dynamic stiffness matrix given by

$$A(\omega) = K + i\omega B - \omega^2 M \tag{2}$$

It consists of the (static) stiffness matrix K, the damping matrix B and the mass matrix M. Apparently, the harmonic displacement of a single node appears as a vector. However, in the current example we have clearly defined major displacement directions that coincide with the directions of the Cartesian coordinate axes. These are the vertical

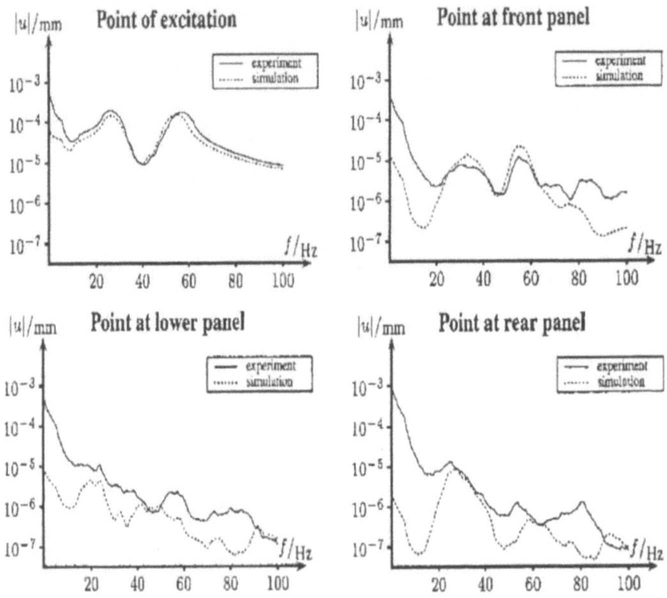

Figure 7. Structural transfer functions measured and calculated for four diferent points,excitation at the external base-frame,10-Hertz-averaging included

direction for the excitation point and a point at the lower panel, the horizontal one for the front and the rear panel. Figure 6 gives an impression of the structural transfer functions for points at these four parts of the construction.

Best coincidence between experiment and simulation is observed for the excitation position. For the point at the front panel, both functions show a very similar behaviour in the range between 20 and 60 Hz. However, matching gets worse the longer the transfer path becomes. Examples for the lower and the rear panel are shown in Figure 6.

While for short transfer paths a comparison between experiment and simulation appears to be simple or at least possible, this comparison is difficult or even impossible for longer transfer paths. A useful strategy that is often applied for complex structures is a suitable averaging. This average displacement $\bar{u}$ is evaluated in a certain frequency window of $2\Delta\omega$ size (reaching from $\omega - \Delta\omega$ to $\omega + \Delta\omega$) using the formula

$$\bar{u} = \frac{1}{2\Delta\omega} \int_{\omega-\Delta\omega}^{\omega+\Delta\omega} u(\omega), d\omega \tag{3}$$

Herein, in all applications a 10 Hz window ($\Delta\omega = 5\ Hz$) is used. The average intervals are suitably adjusted at the lower and the upper bounds of the complete frequency range. The resulting transfer functions that have been discussed above are plotted again in Figure 7.

Although the general information of Figure 6 is the same as in Figure 7, the latter is much easier to survey. Especially for model adjustments an average is advantageous and highly recommended.

In the process of modeling some additional measures and tests are included to achieve a better adjustment of experimental and calculated transfer functions. More than the mode shapes, the structural transfer function proves to be highly sensitive in terms of the coupling conditions between both frames, in particular the location and the number of coupling points.

As in many other applications of modeling, the structural damping accounts for another major challenge. Owing to practical reasons, especially simplicity of the simulation model, we decide to use mode specific damping rates. Hence, each mode is supplied with its own damping rate. However, the experimental modal analysis indicates very low modal damping rates. For the lower vibration modes being identified, rates of 0.1 to 0.2 % are found. The experimentally discovered modes between 50 and 60 Hz, cf. Figure 5, provide values of approximately 0.7 %. In general, modal damping rates are estimates less than one percent rather tending to zero than to greater values.

Note that the structural transfer functions of the simulation underestimate the measured ones. Most likely this is a result of the high sensitivities of the welding joints and welding spots. More realistic values for the coupling elasticity are unknown but highly desirable.

4.2 Noise Transfer Function

In what follows, the sound pressure level at certain position inside the box due to a unit force excitation at the external frame is called the noise transfer function.

Noise transfer functions are measured at three locations inside the box. Their position can be described as distances from the front panel, from the lower panel and from the left panel. For these points the distances are

1 30 cm from front, 35 cm from left and 62 cm from lower panel

2 89 cm from front, 20 cm from left and 20 cm from lower panel

3 75 cm from front, 55 cm from left and 50 cm from lower panel.

The first two positions are arbitrarily chosen, the third is located in the centre of the box. As mentioned previously, the microphones are hung through bored openings in the upper panel.

The calculation procedure of the noise transfer function is identical with the one proposed in [19]. To resume briefly, we start with the structural transfer function, i.e. harmonic displacements of the nodes of the finite element mesh assembled in the column matrix u. For consideration of noise emission inside the box we split the number of nodes into noise emitting and non–emitting ones. For the set of noise emitting nodes, we add the factor of $i\omega$ to calculate the harmonic velocity. Scalar multiplication of each nodal velocity vector with its normal vector (directing outwardly) provides the normal components of the particle velocity of the structure. Assembling the components of all normal vectors in a matrix N and extending equation 1 allows us to formulate

$$v_s(\omega) = i\omega N A^{-1}(\omega) f(\omega) \tag{4}$$

where v_s denotes the (normal) particle velocity of the structure. The sound pressure at certain positions inside the structure can be calculated simply by a scalar multiplication of a column matrix of influence coefficients b and that of the nodal particle velocities v_s. We can write for the sound pressure p_i

$$p_i(\omega) = b^T(\omega) v_s(\omega) = \omega b^T(\omega) N\, A^{-1}(\omega) f(\omega) \tag{5}$$

These influence coefficients allow an explicit mapping of the particle velocity of the structure to the sound pressure, i. e. they represent the solution of the fluid's boundary value problem. Hence, they depend on the geometry of the fluid's domain, the frequency, the boundary admittance or impedance respectively and, obviously, on the position of the internal point. However, if the geometry does not change significantly, if the admittance remains constant and if the sound pressure is required at few positions only, acoustic influence coefficients account for an efficient tool for structural acoustic optimization since they have to be calculated only once in the entire optimization process. Examples for their application can be found in Marburg et al. [22, 23] and Marburg and Hardtke [21]. The influence coefficients are calculated using the non–commercial boundary element–based code AKUSTA. A reference that confirms reliability of this code can be provided [18].

Another problem that arises in the context of modeling fluid domains for acoustic calculation is the choice of suitable boundary conditions. Often, this question is evaded and hard, i.e. fully reflecting, walls are assumed. This is equivalent to a vanishing boundary admittance value. Further, we can expect standing wave phenomena if viscous damping can be neglected. In that case, however, we should also expect infinite sound pressure values at eigenfrequencies. Since there is an eigenfrequency at zero Hertz for closed domains and hard reflecting walls this assumption requires that when uniformly and slowly increasing the static pressure inside the domain the fluid must not escape. While for a well sealed sedan cabin this requirement is fulfilled, cf. [20], it does not seem realistic here.

Since realistic values for the boundary admittance in this case are not available, a comparison between hard reflecting walls and an (arbitrarily chosen) value is presented. For that, we use the boundary admittance of an infinite plate being

$$Y_0(\omega) = \frac{1}{\rho_f c - i\omega m_s'} = \frac{1}{\rho_f c - i\omega\rho_p h} \tag{6}$$

The fluid density and speed of sound are represented by ρ_f and c respectively, m_s' denotes the specific mass of the plate being equal to the product of the plates density ρ_p and its thickness h. Clearly, this choice is an arbitrary one. Results for this case are presented to show some possible effects and to emphasize the significance of reliable values of the boundary condition.

Figure 8 contains noise transfer functions for the three positions, measured und simulated data. The need of an averaging is confirmed when looking to the upper left subfigure. It shows the non–averaged noise transfer functions for the first position. Hardly any substantial information can be taken from this picture.

Comparison between experiment and simulation shows that the transfer functions of simulation and experiment generally behave similarly. For low frequencies ($\leq 15\ Hz$), the experimentally determined noise transfer function is found between the simulation results. Choice of hard reflecting walls overestimates, a boundary admittance as given above underestimates the noise transfer. This frequency range of the noise transfer function should not overvalued, though. Experimental data below 15 Hz can be expected to contain substantial uncertainties. For the first peak, the situation is the other way around as it is for very low frequencies. Use of rigid walls underestimates, use of the non–zero admittance overestimates the measured data. However, we can realize for frequencies above 10 Hertz that hard reflecting walls generally lead to noise transfer functions that either look similar to the experimental one or underestimates these data as the simulated structural transfer function did before.

The calculated functions for non–zero boundary admittances show a considerable deviation from the zero condition. This deviation becomes small above 60 Hz. In Figure 7, we observe an overestimation of the vibration amplitudes of the front panel, an underestimation for

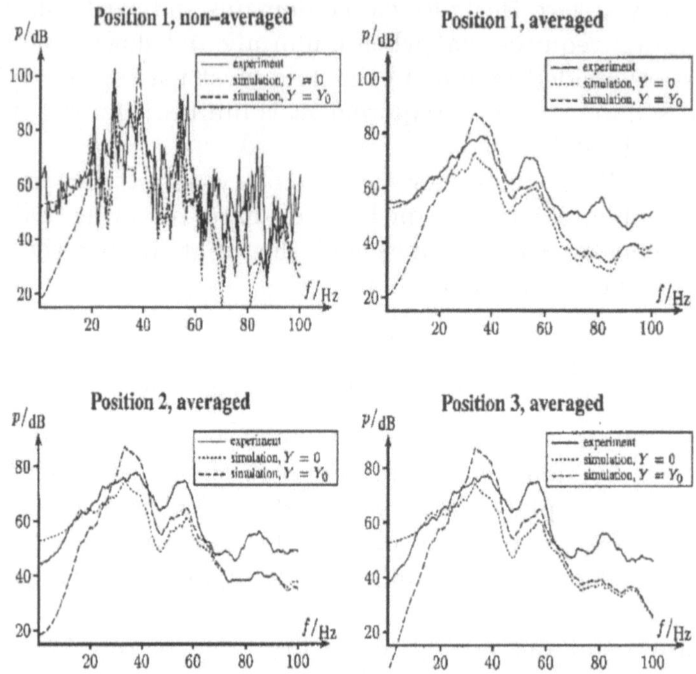

Figure 8. Noise transfer functions at three positions inside the box, experimentally determined functions versus simulation, two boundary admittance values applied as fluid boundary condition

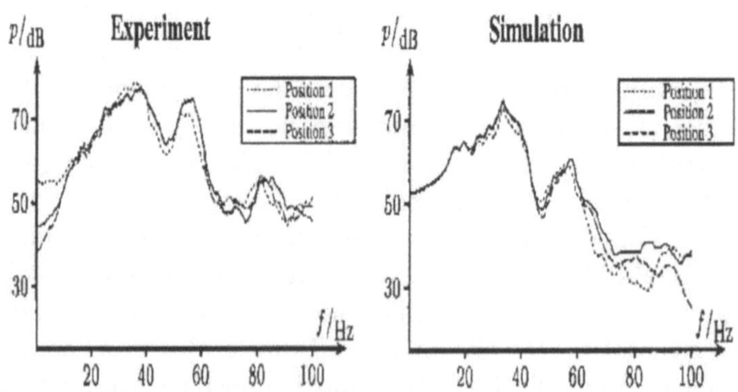

Figure 9. Noise transfer functions at three positions inside the box, comparision of three measured functions (left) and three computed for hard reflecting walls (right)

the lower panel and good matching for the rear panel in the frequency range between 25 and 40 Hertz. Both side panels agree as well. A panel contribution analysis will be provided in the optimization section but it should be remarked here that for the chosen frequency of 38 Hz the front panel, both side panels and the rear one contribute most to the sound pressure at the internal point 2. Hence, an underestimated noise transfer function does not seem very realistic. Suitable values for the boundary condition are desirable.

Moreover, Figure 9 indicates that hard reflecting walls as boundary condition for the fluid may be insufficient for acoustic simulation. This assumption becomes due when comparing the measured noise transfer functions for three positions. We observe that the peak between 50 and 60 Hz is several decibel lower at position 1 compared to the other two positions. Such a behaviour cannot be explained for hard reflecting walls in a frequency range of about half of the first acoustic eigenfrequency (except the rigid body mode at 0 Hz). A comparison between the corresponding simulation data shows that only much smaller differences are expected between these functions. Note that position 1 is the closest position from the front panel that is likely responsible for this peak, cf. Fig. 5, mode 22 and 24.

It can be observed that the peak between 50 and 60 Hertz is much smaller in the simulation, cf. Figure 7. It is assumed that this is subject to the non–perfect second mode of the front panel that has been discussed in the previous section. The simulation uses a more idealized mode of two loops. These loops look like a dipole, cf. Figure 5 (lower left subfigure), and erase each other. This assumption fails in reality since both loops have significantly different amplitudes. Several trials to model this behaviour failed because either unrealistic geometry would have been required or the eigenfrequency was shifted away from the desired range. At the end, it is found satisfying that the simulation can supply this peak but not as high as the experiment.

Finally, we can resume that in general a good agreement of experiment and simulation is achieved up to a frequency of 60 Hertz. The upper frequency limit corresponds to the upper limit from structural transfer functions especially for the front panel. The simulation model is too stiff for higher frequencies. This is likely due to idealized coupling conditions between the beam frames and the panels.

5. Verification of the Simulation Model

After building the calculation model it must be verified to experience its reliability. On this purpose, a mass is added to the metal sheet. This mass consists of a compact steel block that is mounted to the panel by just one screw. It weighs 10 kg. The front panel is chosen since this part of the box seems to contribute most to the noise inside. A heavy additional structure should significantly change the modal properties of the front panel.

Solid three–dimensional brick elements are used to model this block. Coupling of this new structure to the front panel is realized by coupling four single nodes. Hence, there is a more distributed coupling in the calculation model. Mainly, this is subject to numerical considerations. A concentrated coupling at just one node of the front panel may cause unexpected singularities. With respect to the physical model we can even argue that the screw and washers on both sides of the panel rather represent a distributed coupling than a concentrated one. The position of the added structure is shown in Figure 10.

Position and size of the added mass allow to predict that especially mode 10 and 24 are changed. This actually happens. Modification of mode 10 reduces the noise transfer function, see for example for position 2 in Figure 11. However, modification of mode 24 in the experiment reduces vibration of the right loop while the left remains. Mode 24 vanishes in the simulation after modification. This is likely due to idealization of modeling as discussed in the previous sections.

Generally, we can conclude that the simulation model represents the modification in the noise transfer function. This claim holds for frequencies up to 50 Hz. Furthermore, it can be explained what the reason for different behaviour of simulation and experiment for the range between 50 and 60 Hertz is. Hence, the range of reliability of the simulation model reaches up to 60 Hz if we are careful with the peculiarities of the front panel.

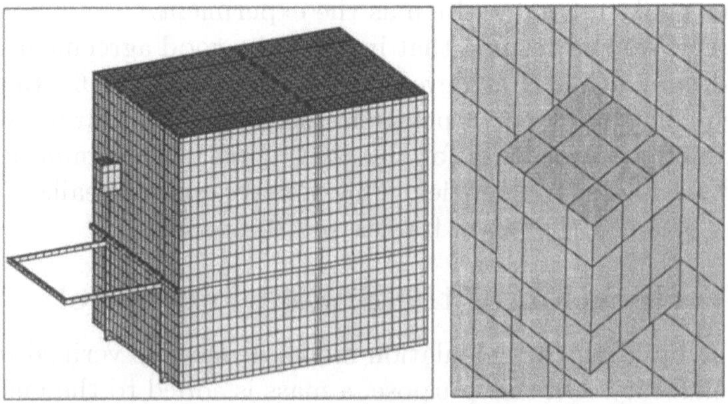

Figure 10. Added mass structure, global view (left), close-up view (right)

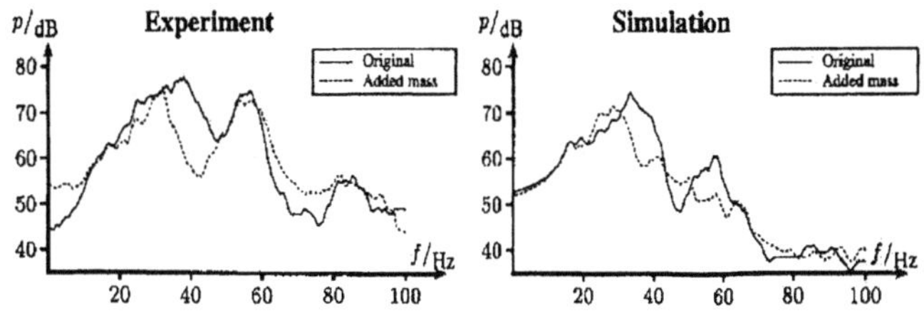

Figure 11. Noise transfer functions for added mass

6. Optimization of the structure

6.1 Feasible Modifications

It is one idea of this project to provide evidence that a design optimization of a complex structure can actually lead to an improved physical construction. As explained in the first section this evidence presupposes a non–destructive modification of the model. For that reason, the decision is made that two panels are to be selected for modification. Initially, two beams per panel should be stuck up. However, this technique proved to be unsuited from the production–engineering point of view since both panel are too uneven. Thus, the beams are to be mounted while washers are used to bridge the local distances between straight beams and distorted metal sheets. This technique should ensure that the beams are coupled tightly but do not modify the uneven geometry of the metal plate. It is emphasized that welding of the beams would likely change the geometry of the panel. Moreover, the chance of dismantling would have been spoiled.

The optimization concentrates on the noise transfer function of position 2. This position is chosen because the corresponding noise transfer functions of simulation and experiment matched better than for position 1. The other one is not selected because of its special position in the centre of the box.

A panel contribution analysis is made to select two surfaces for modification in the optimization process. The approach is similar to the one originally proposed by Ishiyama et al. [13] and improved by Adey et al.[1]. Following their approach or returning to equation 5, we can calculate the sound pressure p_i at a specified position inside the box as

$$p_i = \sum_{j=1}^{N_p} \sum_{k=1}^{N_j} b_k v_{s_k} \tag{7}$$

N_p an N_j denote the number of panels and the number of nodes per panel, respectively. As explained above, b_k and v_{sk} represent nodal values of the influence coefficients and the particle velocity of the structure, respectively. Summing up all nodal contributions of one panel supplies the panel contribution p_c

$$p_{c_j} = \sum_{k=1}^{N_j} b_k v_s \tag{8}$$

The value p_{c_j} describes the sound pressure at the internal point if only the j–th panel contributes to the noise inside.

Apparently, panel contributions depend on the frequency and on the position. However, Figure 9 proves that for frequencies below 60 Hertz hardly any dependence on the location of the internal point can be observed. Usually, decrease of high peaks in the noise transfer function is a primary target for constructions. For that reason, the frequencies of 38 and 56 Hertz are chosen for the panel contribution analysis. These frequencies correspond to maxima in the noise transfer function of position 2.

Table III contains the contributions of all panels for the above chosen frequencies. Unquestionably, we identify the front panel and the lower panel as those that contribute most to the interior noise. However, remarkable low contribution of the front panel to the noise at 56 Hz has the question of the second front panel mode arisen again. As discussed previously, both vibration loop erase each other. Hence, the part of the surface that shows the greatest displacements emits negligible noise to the inside.

Table 3. Panel contributions to sound pressure at position 2 at two frequencies

	Contribution of panel in [dB]						
frequency	front	lower	upper	right	left	rear	all panels
38 Hz	91.4	43.5	54.8	74.4	71.8	62.9	93.7
56 Hz	46.8	65.9	43.8	50.1	44.9	38.6	69.3

6.2 Design Parameters and Objective Function

In the context of feasible modifications we have mentioned that two beams can be added to both, the front and lower panel. Obviously, these beams must not cross each other. For numerical reasons that will be discussed later in this subsection we require that they all are parallel. Hence, two parameters per added beam have to be controlled, one for the position and one for the length. For simplicity, we require that every beam is mounted symmetrically (presupposing an approximately symmetric construction). Consequently, for two beams per panel and two design parameters per beam the model can be modified by eight parameters.

A numerical problem that occurs when fixing these beams to the surface panels consists in coupling both structures. As can be seen in Figure 2 the area model is prepared to connect virtually coincident nodes. Hence, the distances between two nodes are very short and it is clarified which node is to be coupled with another. The situation for a continuously shifted beam is more complicated. If the beam coincides with a line of nodes of the panel elements automated mapping for coupling is simple. However, this beam is just located between two lines of nodes this mapping is not as clear. Furthermore, jump of the coupling from one line of nodes to another causes certain discrete properties. So it was decided that beam positioning is allowed only along a line of nodes and the parameter domain in between is interpolated.

The objective function has been discussed in [19]. Recalling this function, we can write

$$F = \frac{1}{\omega_{max} - \omega_{min}} \int_{\omega_{min}}^{\omega_{max}} \Phi\left\{p_L(\omega)\right\} d\omega \tag{9}$$

The operator $\Phi\{\}$ applied to the sound pressure level p_L represents a kind of a weighting function. An example for this weighting function is

$$\Phi\left\{p_{\mathrm{L}}\right\} = \begin{cases} \left(p_{\mathrm{L}} - p_{\mathrm{Ref}}\right)^n & \text{for} \quad p_{\mathrm{L}} > p_{\mathrm{Ref}} \\ 0 & \text{for} \quad p_{\mathrm{L}} \leq p_{\mathrm{Ref}} \,. \end{cases} \tag{10}$$

The exponent n controls the type of an average. For $n = 1$ equation (9) leads to the mean value where only values higher than a certain reference level p_{Ref} are taken into account. Similarly, for $n = 2$ this form provides the squared mean value. The major advantage of the $n = 2$ form is that high level peaks are higher valued than low level parts of the function. This helps to reduce these high level peaks during an optimization procedure and avoids deep valleys as compensation of high peaks. This effect is increased by using a reference level higher than the lowest values in the noise transfer function.

Herein, the optimization process is directed to minimize the objective function, equation (9) and (10), using the exponent $n = 2$ and the reference sound pressure level $p_{\text{Ref}} = 55$ dB.

The finite element code ANSYS that is applied for structural analysis is used for optimization. Due to the above sketched numerical problem a gradient based method fails. A gradient based method could work well in this case if the partially discrete character of this problem is considered. A random iteration is very competitive for this problem. It is simple to adjust this method to the stepped design modifications. Additional trials using a scan algorithm can be reported. However, best results are found by random iteration.

6.3 Optimized Design and its Mode Shapes

One reason for successful application of a random iteration method is the unique and easy–to–understand solution for most of the optimization variables. So, in all cases the optimal length of the beams is equal to their maximum length. Moreover, the beam in the upper part of the front panel is to be placed to calm down the first and the second vibration mode of this panel, in total being modes 10 and 24. Hence, only three variables remain to be optimally controlled.

A total of about 500 computations of the objective function – single evaluation of the objective function takes about 30 minutes CPU time on an SGI/Origin 2000 – provided the optimized design, cf. Figure 12. The locations of the stiffeners on the front panel look quite reasonable and, most likely, an engineer would design this part in a similar way. The solution for the lower panel appears somewhat more surprisingly. While one of the stiffeners is found in the center of the lower panel, the other one is shifted to the front edge. Due to the internal frame construction this part of the plate is very stiff in the direction of the added beam. Therefore, we assume that further stiffening of the lower panel would possibly increase the interior noise. The results of modal analysis show that the vibrations of the lower panel do not account for a complete mode shape. Starting with mode 6 and 7 of the original structure, there is at least one more panel vibrating together with the lower one at the same natural frequency.

Staying with these two modes (6,7) of the original simulation model, cf. Figure 4, we realize that these two modes reduce to one in the optimized design. This remaining mode consists of vibrations of the upper panel only. It is assumed that the positioning of the stiffener was mainly focused on destruction of this mode shape.

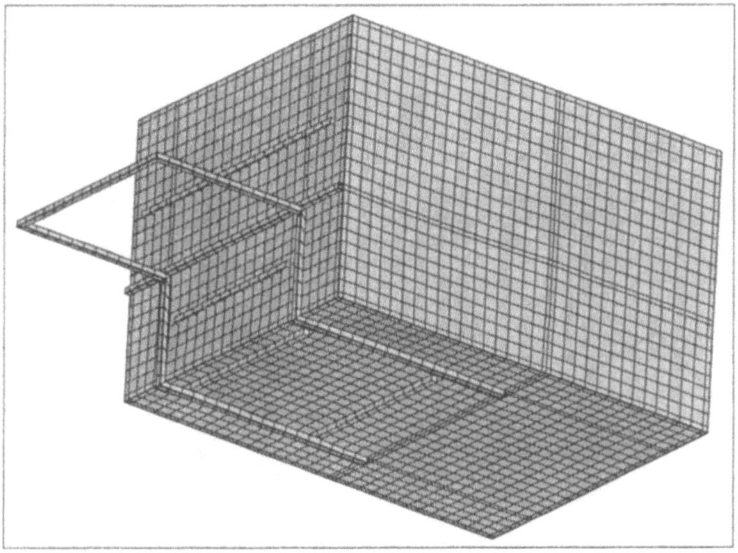

Figure 12. Optimized model: Four additional beams on front and lower panel

Mode 10 of the original model, cf. Figure 5, increases by about ten percent up to 42 Hz as the eleventh mode in the optimized one, cf. Figure 13. It can be seen that, compared to the original, the contribution of other than the front panel increases. A typical vibration mode shape of the lower panel is found at 45.9 Hz. Mode 22 of the original becomes mode 21 in the optimized model. In the original model, this mode is slightly coupled with the second mode of the upper front panel, cf. Figure 13. In the optimized model, the external frame is the only vibrating part at this frequency, cf. Figure 13. While mode 24 of the original does not occur in the optimized design, we would like to show one more mode of high mobility of the front panel. It is mode 32 at 64.8 Hz. One can clearly recognize that the stiffener in the upper part of the front panel helps to separate vibration loops of the shell and, apparently, it controls the corresponding eigenfrequency.

It is widely assumed that stiffening of a structure reduces the noise that is emitted. It appears much more difficult to measure the stiffness, in particular for complex structures. Defined as force per displacement, stiffness cannot be considered as a local size. More globally oriented measures for stiffness are the lowest eigenfrequency or the number of modes up to a certain frequency limit. It was shown in [23], that static stiffeners do not necessarily ensure better acoustics. Moreover, it was shown there that even the decrease of the first natural frequency by

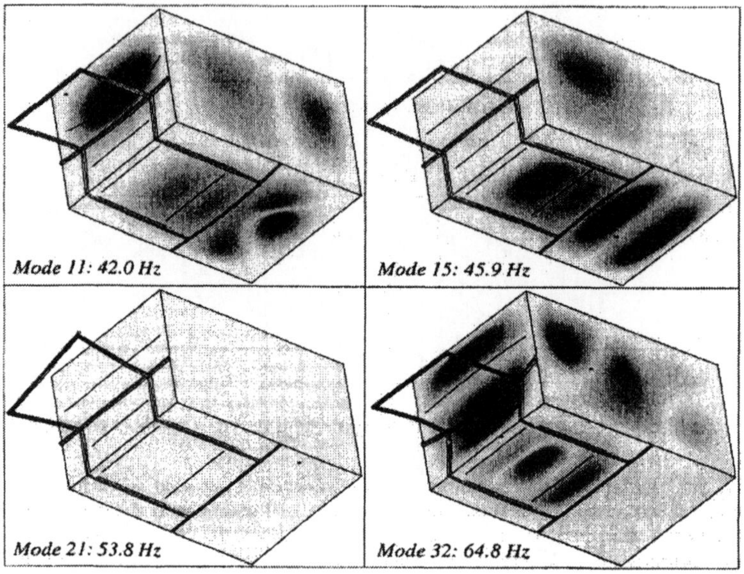

Figure 13. Selected mode shapes of optimised model

about 35% can remarkably decrease the modal density and improve the acoustic behaviour of the structure. It was proposed in that paper to measure the stiffness for acoustic purposes in the number of modes up to a certain frequency.

These modal densities of the original and the optimum are given in Table II. We realize that the number of modes up to certain frequency limits decreases very little for the optimized design. The new shape mainly takes advantage of a selective stiffening that destroys certain modes being unfavourable for acoustics.

6.4 Noise Transfer and Objective Functions of Optimized Design

A comparison between experiment and simulation of the original and the optimized models at positions 1 to 3 are shown in Figure 14. Clearly, the first wide peak from 20 to 40 Hz is significantly decreased. The second peak between 55 and 60 Hz is still observed in the experiment. Simulation provides a similar peak at somewhat lower frequencies but this one likely corresponds to another mode shape than the one observed in the experiment. It is in good agreement with the model verification using the added mass that the second peak at about 55 Hz does not vanish although predicted. Note, that in the frequency

Table 4. Difference of objective function between original and optimized model, results for different parameters of the objective function, equations (9) and (10), comparison of simulation and experiment (Optimization process provided minimum for $n = 2$ and $p_{\mathrm{Ref}} = 55$ dB in the frequency range between 0 and 100 Hz)

		Decrease of objective function in [dB]					
	p_{Ref}	0 – 100 Hz		0 – 60 Hz		10 – 50 Hz	
n	[dB]	Simulat.	Experim.	Simulat.	Experim.	Simulat.	Experim.
1	0	3.4	1.3	4.5	3.3	4.9	6.3
1	55	2.4	2.2	4.0	4.3	5.4	6.3
2	0	3.6	1.8	4.8	3.8	5.2	6.4
2	55	*4.1*	*3.3*	5.2	4.7	6.3	6.7

range up to 60 Hz hardly any differences between the data at different positions can be observed.

Table IV contains the values of the objective function for different parameters, cf. equations (9) and (10), for simulation and experiment. While the optimization process has been controlled by the parameters $n = 2$ and $p_{\mathrm{Ref}} = 55$ dB in the frequency range between 0 and 100 Hz these data are compared with the objective function values for $n = 1$ and/or $p_{\mathrm{Ref}} = 0$ dB. In the optimization process, the objective function is decreased by 4.1 dB while the experiment still shows 3.3 dB improvement.

However, in the case of simulation the noise transfer function does not exceed the reference level of $p_{\mathrm{Ref}} = 55$ dB for frequencies above 60 Hz. For that reason, the data for a reduced frequency range are also shown. Now, the simulation predicts 5.2 dB improvement whereas the experimental test validates 4.7 dB.

With respect to the experiment, the reliable frequency range starts between 10 and 20 dB while the reliable frequency range for the structural model ends at about 50 dB. Thus, the objective function in this frequency domain is computed as well. Since major peaks and improvements occur in this part of the noise transfer function the greatest decrease can be reported. Simulation predicts 6.3 dB and the experiment even supplies a 6.7 dB lower objective function.

Generally, trends that have been predicted by simulation can be proved in the experiment. But this statement reduces to the reliable frequency range. More than 6 dB decrease of the objective function can be reported for this range of reliability.

7. Final Remarks

Finally, we can resume that the entire process from model creation and adjustment, model verification and optimization has been presented in this paper. It could be shown that relatively great decreases of noise transfer that are predicted by simulation and optimization, are also found in the experiment. This verification, however, is impossible if the simulation model cannot predict the major effects that are responsible for noise. In the current application, these effects are well included for frequencies up to 50 Hz. For frequencies above 50 Hz the simulation model is likely modeled too stiff which is assumed to be due to an insufficiently detailed modeling at edges and welding joints. In general, a more reliable model requires more careful modeling techniques for these parts of the structure. Further, a detailed scan of the surfaces seems necessary. This scan should provide the geometry of the panels. Their exact incorporation accounts for another crucial point in the modeling process. It shall be mentioned that yet no word has been spent on the phase angles. Phase angles have been considered in the calculation model. But their values were not considered in the modeling process. Obviously, more detailed modeling should include the adjustment of phase angles as well. Apparently, these extensions would significantly increase requirements of manual work. However, doing such a work in a finite period of time remains one of the key questions in engineering.

The optimization process included positioning and length of additional beams. A case study of the addition of beams to the shell structure of a vehicle roof [23 proved that only little improvements are gained compared to modification of the shell curvature. However, experimental verification of such a case appears difficult and very costly. In general, one should rely on the simulation model once this model is verified.

Acknowledgement

We kindly acknowledge the financial support by the FORD–Werke AG Kln (Germany) where the physical model of the box was constructed.

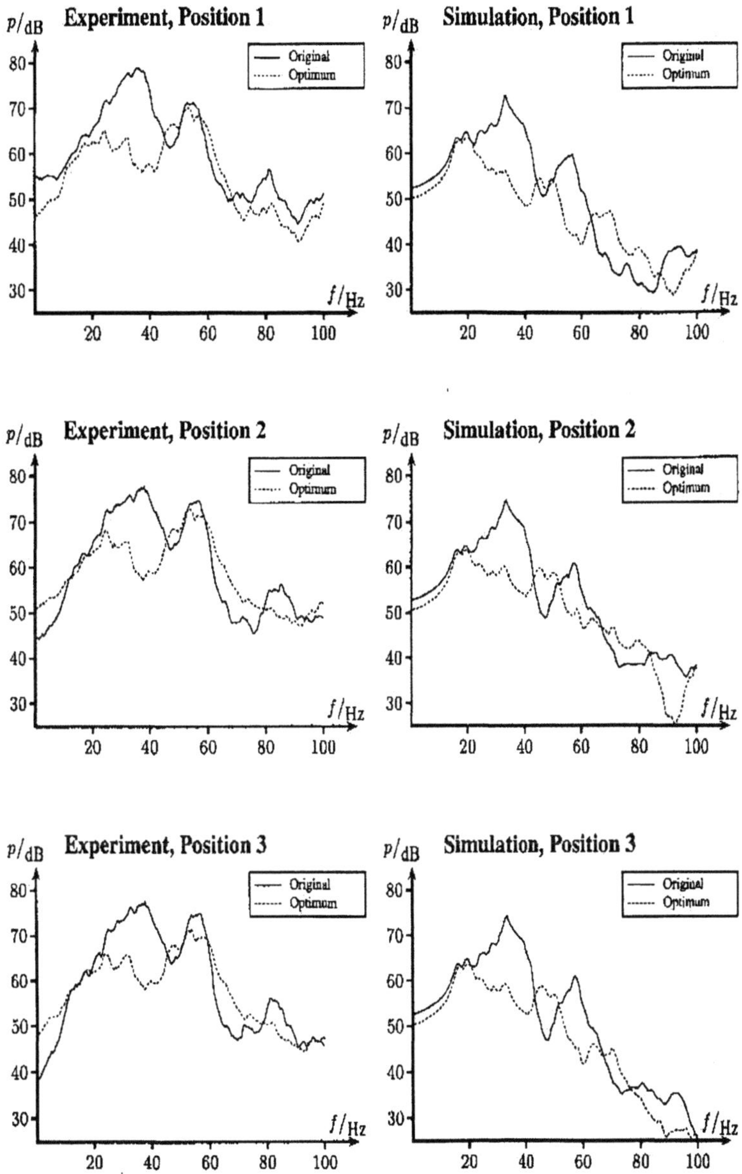

Figure 14. Noise transfer functions of original and optimized structures

References

Adey, R. A., S. M. Niku, J. Baynham, and P. Burns: 1995, 'Predicting acoustic contributions and sensitivity. Application to vehicle structures'. In: C. A. Brebbia (ed.): *Computational Acoustics and its Environmental Applications*. pp. 181–188, Computational Mechanics Publications.

ANSYS: 1997, 'ANSYS GUI Help Manual, ANSYS Release 5.4'. Swanson Analysis System Inc., Houston.

Belegundu, A. D., R. R. Salagame, and G. H. Koopmann: 1994, 'A general optimization strategy for sound power minimization'. *Structural Optimization* **8**(2–3), 113–119.

Christensen, S. T. and N. Olhoff: 1998, 'Shape optimization of a loudspeaker diaphragm with respect to sound directivity properties'. *Control and Cybernetics* **27**(2), 177–198.

Constans, E. W., A. D. Belegundu, and G. H. Koopmann: 1998, 'Design Approach for Minimizing Sound Power from Vibrating Shell Structures'. *AIAA Journal* **36**(2), 134–139.

Crane, S. P., K. A. Cunefare, S. P. Engelstad, and E. A. Powell: 1997, 'Comparison of design optimization formulations for minimization of noise transmission in a cylinder'. *Journal of Aircraft* **34**(2), 236–243.

Cunefare, K. A., S. P. Crane, S. P. Engelstad, and E. A. Powell: 1999, 'Design Minimization of Noise in Stiffened Cylinders Due to Tonal External Excitation'. *Journal of Aircraft* **36**(3), 563–570.

Engelstad, S. P., K. A. Cunefare, E. A. Powell, and V. Biesel: 2000, 'Stiffener Shape Design to Minimize Interior Noise'. *Journal of Aircraft* **37**(1), 165–171.

Hambric, S. A.: 1995, 'Approximation techniques for broad-band acoustic radiated noise design optimization problems'. *Journal of Vibration and Acoustics* **117**(1), 136–144.

Hambric, S. A.: 1996, 'Sensitivity calculations for broad-band acoustic radiated noise design optimization problems'. *Journal of Vibration and Acoustics* **118**(7), 529–532.

Hibinger, F.: 1998, 'Numerische Strukturoptimierung in der Maschinenakustik'. Dissertation, Technische Universität Darmstadt.

Ihlenburg, F.: 1998, *Finite Element Analysis of Acoustic Scattering*, Vol. 132 of *Applied Mathematical Sciences*. Berlin Heidelberg New York: Springer Verlag.

Ishiyama, S.-I., M. Imai, S.-I. Maruyama, H. Ido, N. Sugiura, and S. Suzuki: 1988, 'The application of ACOUST/BOOM – A noise level prediction and reduction code'. *SAE-paper 880910* pp. 195–205.

Kozukue, W., C. Pal, and I. Hagiwara: 1992, 'Optimization of noise level Reduction by Truncated Model Coupled Structural–Acoustic Sensitivity Analysis'. *Computers in Engineering (ASME)* **2**, 15–22.

Lamancusa, J. S.: 1988, 'Geometric Optimization of Internal Combustion Engine Induction Systems for Minimum Noise Transmission'. *Journal of Sound and Vibration* **127**(2), 303–318.

Lamancusa, J. S.: 1993, 'Numerical Optimization Techniques for Structural–Acoustic Design of Rectangular Panels'. *Computers and Structures* **48**(4), 661–675.

Lamancusa, J. S. and H. A. Eschenauer: 1994, 'Design Optimization Methods for Rectangular Panels with minimal sound radiation'. *AIAA Journal* **32**(3), 472–479.

Marburg, S.: 2000, 'Six elements per wavelength. Is that enough?'. *Journal of Computational Acoustics.* accepted for publication.

Marburg, S.: 2001, 'Efficient optimization of a noise transfer function by modification of a shell structure geometry. Part I: Theory'. *Structural and Multidisciplinary Optimization.* accepted for publication.

Marburg, S. and H.-J. Hardtke: 1999, 'A Study on the Acoustic Boundary Admittance. Determination, Results and Consequences'. *Engineering Analysis with Boundary Elements* **23**, 737–744.

Marburg, S. and H.-J. Hardtke: 2001, 'Efficient optimization of a noise transfer function by modification of a shell structure geometry. Part II: Application to a vehicle dashboard'. *Structural and Multidisciplinary Optimization.* accepted for publication.

Marburg, S., H.-J. Hardtke, R. Schmidt, and D. Pawandenat: 1997a, 'An Application of the Concept of Acoustic Influence Coefficients for the Optimization of a Vehicle Roof'. *Engineering Analysis with Boundary Elements* **20**(4), 305–310.

Marburg, S., H.-J. Hardtke, R. Schmidt, and D. Pawandenat: 1997b, 'Design optimization of a vehicle panel with respect to cabin noise problems'. In: *Proceedings of the NAFEMS World–Congress.* Stuttgart, pp. 885–896.

Naghshineh, K., G. H. Koopmann, and A. D. Belegundu: 1992, 'Material Tailoring of Structures to Achieve a Minimum Radiation Condition'. *Journal of the Acoustical Society of America* **92**(2), 841–855.

Pal, C. and I. Hagiwara: 1993, 'Dynamic Analysis of a Coupled Structural–Acoustic Problem. Simultaneous Multi–Modal Reduction of Vehicle Interior Noise Level by Combined Optimization'. *Finite Elements in Analysis and Design* **14**, 225–234.

Pal, C. and I. Hagiwara: 1994, 'Optimization of Noise Level Reduction by truncated modal coupled Structural–Acoustic Sensitivity Analysis'. *JSME International Journal, Series C* **37**(2), 246–251.

St. Pierre Jr., R. L. and G. H. Koopmann: 1995, 'A Design Method for Minimizing the Sound Power Radiated from Plates by Adding Optimally Sized, Discrete Masses'. *Journal of Mechanical Design* **117**, 243–251.

Tinnsten, M., B. Esping, and M. Jonsson: 1999, 'Optimization of Acoustic Response'. *Structural Optimization* **18**(1), 36–47.

van der Auweraer, H., W. Leurs, P. Mas, and L. Hermans: 2000, 'Modal Parameter Estimation from Inconsistent Data Sets'. In: *Proceedings of IMAC-XVIII: A Conference on Structural Dynamics.* San Antonio, Texas, pp. 763 – 771, Society for Experimental Mechanics Inc.

AN OPTIMIZATION TECHNIQUE IN STRUCTURAL-ACOUSTIC DESIGN OF SEDAN BODY PANELS

STEFFEN MARBURG and HANS–*JÜ*RGEN HARDTKE
Institut für Festkörpermechanik, Technische Universität,
01062 Dresden, Germany

Abstract Optimization techniques have so far been rarely used for structural–acoustic purposes. However, they are excellently suited for noise reduction of simple and large scale simulation models. The paper describes a general approach of utilizing optimization procedures in vehicle noise, vibration and harshness problems. We start with a review of structural–acoustic optimization. Then, the noise transfer function including structural harmonic analysis, acoustic influence coefficients and their coupling is described. This is followed by some remarks on the optimization process, discussing objective functions, useful design parameters and sensitivity analysis. Finally, two examples including a hat–shelf and a floor structure will be presented in the application section.

1. The problem

Structural–acoustic analysis of vehicle body structures using finite and boundary element methods has been addressed in a number of papers in the 1980's by Flangian and Borders[5], Nefske[32], Yashiro et al.[42], Ishiyama et al.[12],Yamazaki and Inoue [41], Suzuki et al. [39], and Kitamura et al. [14]. In these publications, the authors presented and discussed noise transfer functions of either fully coupled structure–fluid interaction models, e.g. Yashiro et al. [42] or noise transfer functions based on the one way structure–fluid coupling model that will be explained later in this section, e.g. Suzuki et al. [39]. In other related work, structural transfer functions were applied to quantify acoustic behaviour, e.g. Giebeler and Booz [7]. Analysis of fully coupled structural–acoustic modeling is derived by Freymann et al. [6]. This approach includes absorbing material and damping pads.

M.L. Munjal (ed.), IUTAM Symposium on Designing of Quietness, 279–297.

Vehicle interior noise simulation is usually accompanied by all challenges that very complex models supply. So, it is well known that large forces are transmitted through the structure and comparatively little energy is sufficient to cause much noise. However, the models have been improved and often the reliability of a simulation model is measured in terms of the highest frequency that can be reliably analyzed. Owing to a large number of publications on booming noise, the upper frequency limit of element based simulation models varies between 60 [5], and 250 Hz [14]. The upper limit is due to the structural model of the vehicle body. We can estimate that more than 200 vibration modes occur in the frequency range up to 200 Hz. Apparently, it will be difficult to create a reliable simulation model that considers all these modes properly.

The paper by Sielaff et al. [37] provides a detailed survey of engine excitation including comparison of different engine order resonance frequencies. It is well–known that a four–piston–engine mainly excites the body by its second motor–order. Booming noise computations are usually focused on this order. So, if the engine is run between 900 and 6000 rotations per minute this will correspond to major excitation frequencies between 30 and 200 Hz. Roesems [35] discussed the process of noise, vibration and harshness engineering. He supplied the frequency ranges for finite element based simulation models. For body–in–white models the frequency range up to 200 Hz is covered by existing models. Trimmed body models are reliable up to 150 Hz and full vehicles are analyzed up to 80 Hz.

Usually, the fluid model can be meshed much coarser than the structural model. It was shown that boundary element models of about 1000 to 2000 unknowns are sufficient to analyze a sedan cabin up to 500 Hz [21]. When using a one way coupling between structure and fluid – meaning that the structural vibrations are computed first and are then used as boundary conditions for the fluid analysis, whereas the fluid vibrations do not effect the structure – the damping behaviour of the fluid is of particular interest. This damping is governed by absorption at the fluid surface. The absorption can be modeled by using an admittance boundary condition. One possibility of determining the boundary admittance has been presented and discussed by Marburg and Hardtke [24].

The literature is full of heuristic methods that may be used to improve noise characteristics including the decrease of vehicle interior noise. Most of them appear as trial and error methods, in some cases accompanied by experimental measurements. *Mühlmeier* et al. [31] succeeded in decreasing low frequency noise by stiffening the rear door of a station wagon. Kitamura et al. [14] improved the acoustic characteristics by increasing the windshield mass. They also applied combined simulation and experimental techniques. Sielaff et al.[37] compared different schemes of beads to de–couple structural and acoustic resonance modes.

Numerical optimization, that includes more than just comparing several variants has been presented in few, most of them recent papers. In this context, a paper that addressed just a structural vibration problem and its optimization should be cited. It was published by Yim and Lee [43] and focuses on the optimization of the geometry of body joints. It is known that the stiffness of joints in body structures is most important for the entire vibration behaviour. However, acoustics was not directly addressed.

Esping [4] investigated and optimized the front floor panel of a car. 300 design parameters were optimized in that process. This number corresponds to the number of nodes where additional stiffening material was to be fixed. Shell thickness and curvature remained constant in that case but three stages of 5 mm thickness were added. A maximized lowest eigenfrequency was achieved in that example. Unfortunately, neither details about support conditions of the model nor details about design variables and their handling were supplied. With respect to the results indicating that the lowest eigenfrequency could be increased from 67 to 117 Hz, it is presumed that either a simply supported or a clamped structure was considered. Several additional industrial applications were focused in that paper. It also included a discussion of dynamic and acoustic response.

Milsted et al. [30] described the acoustic optimization of engine structures to minimize the emitted sound power. They mainly used design variables that do not change the mesh topology but were able to perform discrete, manually controlled steps to add or remove structural parts. In that paper, the authors emphasize the compromise between noise emission, reduction, and mass increase. In their examples, they always compared decreases of noise emission without increasing the mass and with a numerically controlled mass increase of six percent. Hnle and Sielaff [10] showed some optimization results in the content of vibrational comfort. Minimization of maxima in the structural transfer function enabled a significant decrease of a major peak. Unfortunately, these impressive results were not accompanied by scales. Hence, it was not clarified whether logarithmic or absolute scaling was used. In their paper, la Civita and Sestieri [10] investigated the potential of improving acoustic behaviour inside the cabin by optimizing the engine mounting system. They optimized design variables like coordinates for engine support locations, nonlinear elastic characteristics of certain engine mounts, and angles to characterize the direction of these mounts. They managed to decrease the sound pressure level at the driver's ear by about 5 dB.

Hagiwara and his co–workers published a number of papers on vehicle interior noise and its decrease, e.g. Hagiwara et al. [9], Ma and Hagiwara [17,18,19], Kozukue et al. [15], Pal and Hagiwara [33, 34] and Shi et al. [36].Mainly, they analyzed fully coupled structural–acoustic models. A number of these papers discussed computation of modal sensitivities for modal analysis [17] and harmonic analysis [18], and influence of truncation error in mode superposition techniques for harmonic analysis [19]. In these papers and in Hagiwara et al. [9], vehicle interior noise was decreased based on specially directed sensitivity analysis. Actual optimization techniques were applied in the papers by Kozukue et al. [15] and Pal and Hagiwara [33, 34]. They mainly modified plate and shell thicknesses and decreased the sound pressure level at specified points for prescribed discrete frequencies that correspond to the second order of engine excitation. A more recent paper by Shi et al. [36] discussed the use of holographic neural networks for structural–acoustic optimization. Unfortunately, their very simple example is hardly suited for generalization purposes. Wang [40] used a coarse concept model and a detailed model for optimization of a vehicle body structure. Beside a detailed description of sensitivity analysis, he presented examples where cross–section parameters and shell thicknesses were optimized. The theoretical part discussed analytic sensitivity analysis that can be efficiently used for gradient determination.

The authors of this paper investigated and optimized several structural-acoustic systems, i.e. a sedan roof in Marburg et al. [28], a dashboard and a hat–shelf in Marburg and Hardtke [25, 27]. The paper by Marburg [22] discussed the entire process of optimization including formulation of the noise transfer function, objective function, design variables and assumptions, optimization strategy and sensitivity analysis.

In the case of the vehicle roof [28], a stiffly supported structure was investigated. A pressure excitation from above was assumed. Using six keypoint coordinates as design variables, the average sound pressure level at the driver's ear was decreased. In the cases of unlimited design modification, the roof became between 7 and almost 30 cm higher. It could be observed that the optimizer concentrated on the strategy of shifting the modes off the frequency range under consideration. The structure was stiffened by optimizing the shell curvature and the lowest eigenfrequency appeared as an indicator of the roof's stiffness. With the limitation of just one centimeter modification, this stiffening could be observed in the number of modes in the frequency range under consideration and the lowest eigenfrequency was decreased from 88 to 64 Hz, though.

The component model of a vehicle dashboard was explained and justified by Marburg and Hardtke [25]. For that, the component under consideration was modeled as a detailed shell structure whereas the remaining body was represented by a superelement. A similar approach has been proposed by Hermans and Brughmans [11] recently. It has been assumed for simplification that just the detailed shell structure emits noise into the cabin. The sound pressure at the driver's ear was then calculated using acoustic influence coefficients [22].There, the large number of 44 design variables was introduced and optimized. While the modal density remained more or less constant, the improvement was achieved by modification of the vibration mode shapes. They changed from big global modes in the original model to smaller local bulges that partially erased each other. A maximum modification of just one centimeter was reported. The objective function had been decreased by about 3.8 dB.

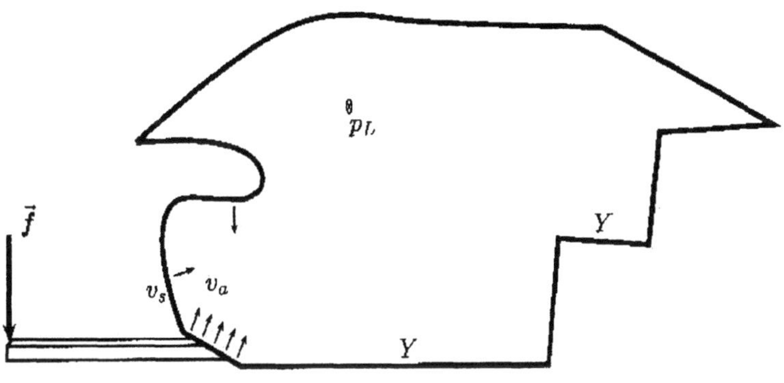

Figure 1. Vehicle body and cabin compartment: Noise transfer teansfer i.e. sound pressure level p_L at the drivers ear due to force excitation $f(\vec{x}, \omega)$, equation (20) and (21)

2. Noise transfer function

2.1 General considerations

In this article, the *noise transfer function* is calculated as the sound pressure level at a certain fluid field point due to a force excitation at the surrounding structure. This is illustrated for the special case of interior noise problems of a vehicle cabin in Figure 1. It is mentioned that this consideration mainly focuses on interior problems since for exterior problems in structural–acoustics it is more sensible to use the radiated sound power due to a force excitation as the noise transfer function, [3].

The calculation procedure itself consists of three parts, the structural harmonic analysis, the acoustic harmonic analysis and the coupling of both. A one–way–model for the structure–fluid–interaction is applied. This model contains the fluid excitation by the structure but structural excitation by the fluid is not included. This type of structure–fluid–interaction is reasonable for vehicle body applications since the structure consists of certain heavy material – often steel – and the fluid – air – is light. For application of a light structure or/and a heavy fluid a structure–fluid–interaction in both directions should be preferred.

Most interesting for the following considerations is the time-harmonic case. For that reason, we consider that all time-dependent functions $F(\vec{x},t)$ are time-harmonic functions where the spatial variable $\vec{x}$ and the time variable t are separated as

$$F(\vec{x},t) \quad = \quad f(\vec{x})\, e^{i\omega t} \tag{1}$$

In this equation ω is the circular frequency and i the imaginary unit where $i^2 = -1$

2.2 Structural analysis

Discretizing a shell structure into finite elements and introducing the above time dependency, the equations of motion yield the well–known system of linear complex equations of the harmonic analysis [1]

$$A(\omega) \;=\; K + i\omega B - \omega^2 M \;\; A(\omega)u(\omega) \;=\; f(\omega) \tag{2}$$

where u and f describe the column matrices containing the nodal displacement vectors and the nodal excitation force vectors. A is the global system matrix of finite elements more commonly known as the dynamic stiffness matrix given by

$$A(\omega) \;=\; K \;+\; i\omega B \;-\; \omega^2 M \tag{3}$$

In equation (3) the matrices K, B and M represent the (static) stiffness matrix, the damping matrix and the mass matrix.

A simple inversion of the dynamic stiffness matrix provides the nodal displacement vectors formally written as

$$u(\omega) \;=\; A^{-1}(\omega)\, f(\omega) \tag{4}$$

Time derivation of the nodal displacement vector $\vec{u}$ leads to the nodal velocity vector $\vec{v}$ as

$$\vec{v}(\omega) = i\,\omega\,\vec{u}(\omega) \tag{5}$$

The nodal particle velocity of the shell structure v_s is defined as the normal component of the nodal velocity vector given by

$$v_s(\omega) = \vec{n}\cdot\vec{v}(\omega) = i\,\omega\,\vec{n}\cdot\vec{u}(\omega) \tag{6}$$

Assembling the components of the nodal normal vectors $\vec{n}$ into a matrix N, the column matrix of the structural particle velocity is written in terms of the nodal displacements by

$$v_s(\omega) = i\,\omega\,N\,u(\omega) \tag{7}$$

and in terms of the force excitation and the system properties as

$$v_s(\omega) = i\,\omega\,N\,A^{-1}(\omega)\,f(\omega) \tag{8}$$

The nodal particle velocity is then used as a coefficient of the Robin boundary condition in acoustic analysis.

2.3 Acoustic analysis

The acoustic analysis is reduced to the calculation of discrete and continuous acoustic influence coefficients also known as sensitivities. These influence coefficients were first calculated by Ishiyama et al. [12]. More detailed discussion of their determination and their application are given by Marburg [20] and Marburg et al. [28]. One possible method to determine discrete and continuous acoustic influence coefficients, $b(\omega)$ and $\beta(\omega)$ is briefly reviewed in what follows.

To solve the boundary value problem for the harmonic wave equation – the Helmholtz equation – the surface of the considered fluid–filled cabin is discretized into boundary elements. Applying a boundary collocation method, a linear system of equations as

$$H(\omega)\,p(\omega) = G(\omega)\,v_f(\omega) \tag{9}$$

is found. The nodal values of the sound pressure p and these of the fluid's particle velocity v_f correspond to the Dirichlet and Neumann acoustic boundary data. The global system matrices G and H are well known from boundary element analysis. The sound pressure p_i at a point inside the cabin – in this article mostly called the interior point or the driver's ear – is calculated in terms of column matrices g and h and their scalar multiplication with the boundary data u and v_f. We find p_i as

$$p_i(\omega) \;=\; g^T(\omega)\, v_f(\omega) \;-\; h^T(\omega)\, p(\omega) \tag{10}$$

In a further step one has to incorporate the Robin boundary condition, the so–called admittance formulation [38] into equation (9) and (10).It is locally given by

$$v_f(\omega) - v_s(\omega) \;=\; Y(\omega)\, p(\omega) \tag{11}$$

The coefficient Y denotes the acoustic boundary admittance. The diagonal matrix Y contains the nodal values of the boundary admittance. Thus, equation (9) changes into

$$\Big[H(\omega) \;-\; G(\omega)Y(\omega)\Big]\, p(\omega) \;=\; G(\omega)\, v_s(\omega) \tag{12}$$

and for equation (10) we obtain

$$p_i(\omega) \;=\; g^T(\omega)\, v_s(\omega) \;-\; \Big[h^T(\omega) \;-\; g^T(\omega)Y(\omega)\Big]\; p(\omega) \tag{13}$$

Substitution of p in (13) by isolating p in equation (12) the sound pressure at the internal point is found to be (omitting ω dependence)

$$p_i \;=\; \Big\{g^T - \Big[h^T - g^T Y\Big]\, [H - GY]^{-1}\, G\Big\}\, v_s \tag{14}$$

which can be written in short form as

$$p_i(\omega) \;=\; b^T(\omega)\, v_s(\omega) \tag{15}$$

The discrete influence coefficients b_k represent the solution of the acoustic boundary value problem for just one internal point excluding the particle velocity as a boundary condition. Furthermore, they account for the sound pressure sensitivity at the internal point with respect to a structural vibration at a certain mesh point. The nodal product of $b_k v_k$ provides the contribution of the mesh point k to the sound pressure at the internal point.

As these influence coefficients are discrete they are not suited for visualization or for mesh interpolation and transformation. For that reason, it is useful to transform them into continuous influence coefficients β on behalf of the relation

$$p_i(\omega) \;=\; b^T(\omega)\, v_s(\omega) \;=\; \int_\Gamma \beta(\omega)\, v_s(\omega)\, d\Gamma \tag{16}$$

Γ represents the fluid's surface. Further, the continuous data on the right hand side of (16), β and v_s, are substituted by shape functions φ_j, i.e. piecewise formulated polynomials (Lagrangian polynomials), and nodal values. For the particle velocity, the column matrix v_s contains the same nodal values as on the left hand side of equation (16). The nodal values of the continuous influence coefficients are assembled in β. We can write

$$b^T(\omega)\, v_s(\omega) \;=\; \beta^T(\omega) \int_{\Gamma} \Phi\, \Phi^T\, d\Gamma\; v_s(\omega) \tag{17}$$

where the column matrix Φ contains the shape functions φ_j. Introducing a sparse symmetric matrix Θ as

$$\Theta_{ik} \;=\; \int_{\Gamma} \varphi_i(\vec{x}) \varphi_k(\vec{x}) d\Gamma(\vec{x}) \tag{18}$$

we are able to transform discrete into continuous influence coefficients and vice versa as

$$\beta(\omega) \;=\; \Theta^{-1}\, b(\omega) \qquad \text{and} \qquad b(\omega) \;=\; \Theta\, \beta(\omega) \tag{19}$$

As mentioned above, the calculation of acoustic influence coefficients is independent from the solution of the structural boundary value problem.

2.4 Coupling of structure and fluid model

The separate solution of structural and acoustic analysis enables us to use different meshes for both. Since we mainly consider vehicle interior noise problems, we deal with the problem that the fluid mesh may be much coarser then the mesh of the shell structure of the body. Applying equation (16) to the coarse boundary element mesh of the fluid continuous influence coefficients can be determined. Extracting the continuous coefficients at the nodes of the more detailed mesh of the structure one can find the discrete coefficients by multiplying them by Θ_s where Θ_s is now assembled for the fine mesh. Applying these discrete influence coefficients to the detailed mesh of the structure we can express the sound pressure at the internal point due to a specified force excitation by substituting v_s, equation (15) by equation (8). This yields

$$p_i(\omega) \;=\; i\,\omega\, b^T(\omega)\, N\, A^{-1}(\omega)\, f(\omega) \tag{20}$$

In this equation, b is the column matrix of the discrete acoustic influence coefficients of the fine structural mesh. The main objective of our investigation is the decrease of the noise inside a cabin. The perceived loundness of sound (in the required frequency domain $\leq 200Hz$) is virtually proportional to the logarithmic sound pressure. For that, the transformation of the noise transfer function into a sound pressure level p_L is useful. This leads to

$$p_L(\omega) \quad = \quad L\{p_i(\omega)\} \quad = \quad 20 \log_{10}\left(\frac{|p_i(\omega)|}{p_0}\right) \tag{21}$$

The reference sound pressure is given by $p_0 = 2 \cdot 10^{-5} Pa$. The unit of p_L is dB.

Herein, the sound pressure level at the internal point due to a force excitation, equation (21) with the substitution of p_i from equation (20), is called the *noise transfer function.* An illustration was given by Figure 1 for the special case of a vehicle cabin compartment.

3. Design parameters

A parametric description of the geometry requires the introduction of design parameters. The parameters used here are assembled into the vector ϑ.

Two concepts are known for controlling the shape of shell. The most straightforward approach, being utilized by Christensen and Olhoff [2] and by the authors of this paper, e.g. Marburg [22] and Marburg and Hardtke et al. [25, 28, 29] consists of creating a geometry based model that maybe meshed arbitrarily. An alternative approach has been presented in [23] and applied to body structures in Marburg and Hardtke [26, 27]. In that technique, the finite element shell mesh is modified directly.

Highly efficient sensitivity analysis and thus, a fast optimization process require some presumptions concerning the parameters. These are the following:

(*a*) It is presumed that all the modification of the geometry of the shell structure are small with respect to the fluid's wave length.

(*b*) It is presumed that all the modification of the geometry of the shell structure are small with respect to the shell structure size.

(*c*) It is presumed that the modification mainly shifts the surface shell in its normal direction.

(*d*) It is presumed that the modification does not effect the force excitation of the structure.

All these presuppositions and their conclusions simplify the sensitivity analysis and the optimization process.

The assumption (a) leads to the conclusion that the discrete acoustic influence coefficients do not depend on the parameter set

$$\frac{\partial b}{\partial \vartheta} = 0 \,. \tag{22}$$

That means, that the discrete influence coefficients are calculated only once. They can be used for each optimization step. However, this requires a constant topology of the mesh of the structure during the whole optimization process. If this cannot be ensured one can start with the continuous influence coefficients of the coarser fluid mesh and extract them at the new fine mesh. The new discrete influence coefficients are then determined by equation (19).

Consequence of the assumptions (b) and (c) is that the normal vectors do not depend on the parameter set

$$\frac{\partial N}{\partial \vartheta} = 0 \tag{23}$$

whereas the assumption (d) formulates the excitation force independence of the parameters as

$$\frac{\partial f}{\partial \vartheta} = 0 \tag{24}$$

Hence, the objective function F in terms of the parameter set ϑ can be written as

$$F(\vartheta) = \frac{1}{\omega_{max} - \omega_{min}} \int_{\omega_{min}}^{\omega_{max}} \Phi\left\{p_L(\omega, \vartheta)\right\} d\omega \tag{25}$$

with the substitution

$$p_L(\omega) = 20 \log_{10}\left(\frac{\left|i\,\omega\, b^T(\omega) N\, A^{-1}(\omega, \vartheta)\, f(\omega)\right|}{p_0}\right) \tag{26}$$

As the reader should now be familiar with the dependencies on ω it will not be explicitly mentioned in the following sections.

4. Objective function

As the sound pressure level still depends on the frequency an integral criterion over the whole frequency domain is required. A solution that meets both mathematical and technical interests is the following averaging

$$\tilde{F} = \frac{1}{\omega_{max} - \omega_{min}} \int_{\omega_{min}}^{\omega_{max}} \Phi \{p_L(\omega)\} \, d\omega \tag{27}$$

The operator $\Phi\{\}$ applied to p_L represents a kind of a weighting function. An example for this weighting function is

$$\Phi \{p_L\} = \begin{cases} (p_L - p_{Ref})^n & \text{for} \quad p_L > p_{Ref} \\ 0 & \text{for} \quad p_L \leq p_{Ref} \, . \end{cases} \tag{28}$$

The exponent n controls the type of average. For $n = 1$ equation (28) leads to the mean value where only values higher than a certain reference level p_{Ref} are taken into account. Similarly, for $n = 2$ this form provides the squared mean value. To enable a better comparison between mean values, we suggest to compute the n–th root of $\tilde{F}$ being

$$F = \tilde{F}^{\frac{1}{n}} \tag{29}$$

In the case of $n = 2$ this yields the root mean square value. The major advantage of the $n = 2$ form is that high level peaks are higher valued than low level parts of the function. This helps to reduce these high level peaks during an optimization procedure and avoids deep valleys as compensation of high peaks. The choice of the reference level p_{Ref} meets technical requirements as values below are not considered. So they cannot be overvalued. On the other hand, a reference level higher than the sound pressure level minimum destroys the C^1 continuity of function F in(27).

5. Sensitivity analysis

A crucial point for the optimization process is an efficient sensitivity analysis. It is recommended to avoid global finite difference calculations for sensitivity analysis. Moreover, it is highly recommended to apply kind of semi–analytic approach for determination of sensitivities of the harmonic displacements of the structure [8, 13]. This formulation is described in [22] and some results will be presented next.

Sensitivity analysis of the objective function given in equations (21)–(28) is to be separated first into the sensitivity of the objective function itself transferring the sensitivity problem to the sound pressure level. The sound pressure level sensitivity is then written in terms of the sensitivity of the structural displacements. If we use $\Re$ and $\Im$ for real and imaginary part of complex numbers and substitute as $\tilde{p} = b^T N u$ and $\psi = b^T N \partial u / \partial \vartheta$ we find

$$\frac{\partial \tilde{F}}{\partial \vartheta} = \frac{1}{\omega_{max} - \omega_{min}} \int_{\omega_{min}}^{\omega_{max}} \left\{ n \left[p_L - p_{Ref} \right]^{n-1} \frac{\partial p_L}{\partial \vartheta} \right\} d\omega,$$

$$\frac{\partial p_L}{\partial \vartheta} = \frac{20}{\ln 10} \frac{\Re \langle \tilde{p} \rangle \; \Re \langle \psi \rangle \; + \; \Im \langle \tilde{p} \rangle \; \Im \langle \psi \rangle}{\Re^2 \langle \tilde{p} \rangle \; + \; \Im^2 \langle \tilde{p} \rangle} \quad and \tag{30}$$

$$\frac{\partial u}{\partial \vartheta} = -A^{-1} \left[\frac{\partial K}{\partial \vartheta} - \omega^2 \frac{\partial M}{\partial \vartheta} \right] u$$

Sensitivities of the matrices K and M can be determined using finite differences. With respect to the highly nonlinear dependence of the objective function in terms of the parameters that control the shell structure geometry we recommend the use of central finite differences like (similar for M)

$$\frac{\partial K}{\partial \vartheta_k} \approx \frac{K(\vartheta_k + \Delta \vartheta_k) \; - \; K(\vartheta_k - \Delta \vartheta_k)}{2 \, \Delta \vartheta_k} \tag{31}$$

An alternative to the semi–analytic approach is the approach of global finite differences. This approach is usually realized in commercial optimization codes since these codes are expected to work for several different applications. However, for this multidisciplinary problem this approach is computationally too expensive and not efficient.

A more efficient sensitivity analysis may be achieved using analytic derivatives of the system matrices K and M. However, this requires basic manipulation of the finite element code. In the current approach, the structure was analyzed using the commercial code ANSYS.

6. Applications

Two examples will be presented briefly. Detailed descriptions of the models and the optimization process can be found in papers by Marburg and Hardtke for the hat–shelf [27] and for the floor structure [26]. In both of these applications, modification utilized a direct approach of to vary the geometry. A geometry base model was not created.

The hat–shelf structure has been considered on its own. Simply supported boundary conditions are assumed. Three different load cases were included. These load cases include three locations of force excitation in the hat–shelf itself. With respect to the force excitation in

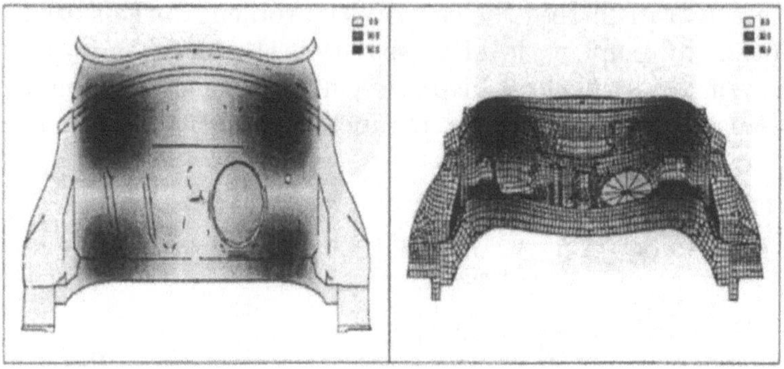

Figure 2. Optimized geometry as modification of original model,maximum modification: 60 mm

the component under consideration and, further, to the bearing conditions, we decided to maximize the lowest structural eigenfrequency. It was increased from 32 to 101 Hz, hence, shifted off the frequency range under consideration, being ≤100 Hz. Only four parameters were used in the optimization process. The new shape is six centimeters higher than the original design, cf. Fig. 2. That value was prescribed as the design limit for this component. It shall be mentioned that increasing the lowest eigenfrequency and, thus, the stiffness does not always provide quieter structures. This strategy usually requires force excitation in the considered component and stiffly supported boundary conditions. The objective function, mean value and root mean squared value, could be decreased between 4.4 and 13.9 dB, depending on load case, objective function and other parameters.

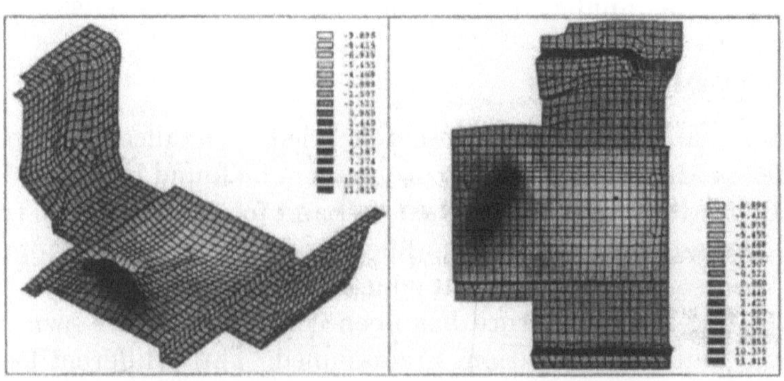

Figure 3. Vehicle floor structure:Optimized design as modification of original shape

Similar to the dashboard model [25], the floor panel has been modeled as a component embedded in a superelement environment. For such a component model the stiffening strategy does not necessarily provide better acoustic performance. It was shown by Marburg and Hardtke [25] that that particular model can be improved by increasing its stiffness. However, an alternative strategy controls the structural modes as shown in that paper. In the example of the floor panel, the optimization strategy utilizes both, stiffening and optimal control of modes [26]. 33 design variables have been used in an optimization process that applies an alternative concept for parametrization [23]. A decrease of about two decibel in the root mean square has been gained in this process. The modified geometry is shown in Figure 3

Acknowledgement

We kindly acknowledge the financial support of the AUDI AG Ingolstadt. We further thank Dr.-Ing. U. Knaust and Dipl.-Ing. Th. Adler of the AUDI AG for many interesting discussions in connection with these projects. They have provided all simulation models.

References

Bathe, K.-J.: 1982, *Finite Element Procedures in Engineering Analysis.* Prentice Hall, Englewood Cliffs.

Christensen, S. T. and N. Olhoff: 1998, 'Shape optimization of a loudspeaker diaphragm with respect to sound directivity properties'. *Control and Cybernetics* **27**(2), 177–198.

Christensen, S. T., S. V. Sorokin, and N. Olhoff: 1998, 'On analysis and optimization in structural acoustics – Part I: Problem formulation and solution techniques'. *Structural Optimization* **16**, 83–95.

Esping, B.: 1995, 'Design optimization as an engineering tool'. *Structural Optimization* **10**, 137–152.

Flanigan, D. L. and S. G. Borders: 1984, 'Application of Acoustic Modeling Methods for Vehicle Boom Analysis'. *SAE-paper 840744* pp. 207–217.

Freymann, R., R. Stryczek, and H. Spannheimer: 1995, 'Dynamic Response of Coupled Structural–Acoustic Systems'. *Journal of Low Frequency Noise and Vibration* **14**(1), 11–32.

Giebeler, W.-R. and G. Booz: 1994, 'Akustisches Verhalten als Zielvorgabe in der Karosserieentwicklung'. In: *Berechnung und Simulation im Fahrzeugbau.* pp. 237–252, VDI–Report 1153.

Haftka, R. T. and H. M. Adelman: 1989, 'Recent Developments in Structural Sensitivity Analysis'. *Structural Optimization* **1**, 137–151.

Hagiwara, I., Z.-D. Ma, A. Arai, and K. Nagabuchi: 1991, 'Reduction of Vehicle Interior Noise Using Structural–Acoustic Sensitivity Analysis Methods'. *SAE Technical Paper Series No. 910208.* 10 pages.

Hänle, U. and J. Sielaff: 1998, 'Eine Berechnungsstrategie zur Auslegung des komfortrelevanten Karosserie–Strukturverhaltens'. In: *Berechnung und Simulation im Fahrzeugbau.* pp. 733–750, VDI–Report 1411.

Hermans, L. and M. Brughmans: 2000, 'Enabling Vibro–Acoustic optimization in a superelement environment: A case study'. *Proceedings– SPIE The International Society for Optical Engineering, Issue 4062//PT2* pp. 1146–1152.

Ishiyama, S.-I., M. Imai, S.-I. Maruyama, H. Ido, N. Sugiura, and S. Suzuki: 1988, 'The application of ACOUST/BOOM – A noise level prediction and reduction code'. *SAE-paper 880910* pp. 195–205.

Kibsgaard, S.: 1992, 'Sensitivity Analysis – The basis for optimization'. *International Journal for Numerical Methods in Engineering* **34**, 901–932.

Kitamura, H., M. Sano, M. Fukushima, T. Yagi, and M. Furuyama: 1990, 'Use of Structural–Acoustic Coupling Analysis to Improve the Acoustic Characteristics of the Vehicle Body'. *SAE-paper 905216* pp. 741–746.

Koopmann, G. H. and J. B. Fahnline: 1997, *Designing Quiet Structures: A Sound Power Minimization Approach.* San Diego, London: Academic Press.

Kozukue, W., C. Pal, and I. Hagiwara: 1992, 'Optimization of noise level Reduction by Truncated Model Coupled Structural–Acoustic Sensitivity Analysis'. *Computers in Engineering (ASME)* **2**, 15–22.

la Civita, M. and A. Sestieri: 1999, 'Optimization of an Engine Mounting System for Vibro–Acoustic Comfort Improvement'. *Proceedings– SPIE The International Society for Optical Engineering, Issue 3727//PT2* pp. 1998–2004.

Ma, Z.-D. and I. Hagiwara: 1991a, 'Sensitivity Analysis Methods for Coupled Acoustic–Structural Systems Part I: Modal Sensitivities'. *AIAA Journal* **29**(11), 1787–1795.

Ma, Z.-D. and I. Hagiwara: 1991b, 'Sensitivity Analysis Methods for Coupled Acoustic–Structural Systems Part II: Direct Frequency Response and its Sensitivities'. *AIAA Journal* **29**(11), 1796–1801.

Ma, Z.-D. and I. Hagiwara: 1994, 'Development of New Mode–Superposition Technique for Truncating Lower and/or Higher–Frequency Modes (Application of Eigenmode Sensitivity Analysis for Systems with Repeated Eigenvalues)'. *JSME International Journal, Series C* **37**(1), 7–13.

Marburg, S.: 1996, 'Calculation and Visualization of Acoustic Influence Co–efficients in Vehicle Cabins using Mode Superposition Techniques'. In: C. A. Brebbia, J. B. Martins, et al. (eds.): *Boundary Elements XVIII.* Southampton Boston, pp. 13–22, Computational Mechanics Publications.

Marburg, S.: 1998, 'Einige Untersuchungen zur akustischen Wandadmittanz'. In: A. Sill (ed.): *Fortschritte in der Akustik – Publikationen zur 24. Jahrestagung f"ur Akustik.* Zürich, pp. 606–607.

Marburg, S.: 2000, 'Six elements per wavelength. Is that enough?'. *Journal of Computational Acoustics.* accepted for publication.

Marburg, S.: 2001a, 'Efficient optimization of a noise transfer function by modification of a shell structure geometry. Part I: Theory'. *Structural and Multidisciplinary Optimization.* accepted for publication.

Marburg, S.: 2001b, 'A general concept for design modification of shell meshes in structural–acoustic optimization. Part I: Formulation of the concept'. *Finite Elements in Analysis and Design.* accepted for publication.

Marburg, S. and H.-J. Hardtke: 1999, 'A Study on the Acoustic Boundary Admittance. Determination, Results and Consequences'. *Engineering Analysis with Boundary Elements* **23**, 737–744.

Marburg, S. and H.-J. Hardtke: 2001a, 'Efficient optimization of a noise transfer function by modification of a shell structure geometry. Part II: Application to a vehicle dashboard'. *Structural and Multidisciplinary Optimization.* accepted for publication.

Marburg, S. and H.-J. Hardtke: 2001b, 'A general concept for design modification of shell meshes in structural–acoustic optimization. Part II: Application to a vehicle floor panel.'. *Finite Elements in Analysis and Design.* accepted for publication.

Marburg, S. and H.-J. Hardtke: 2001c, 'Shape optimization of a vehicle hat–shelf: Improving acoustic properties for different load–cases by maximizing first eigenfrequency'. *Computers and Structures.* accepted for publication.

Marburg, S., H.-J. Hardtke, R. Schmidt, and D. Pawandenat: 1997a, 'An Application of the Concept of Acoustic Influence Coefficients for the Optimization of a Vehicle Roof'. *Engineering Analysis with Boundary Elements* **20**(4), 305–310.

Marburg, S., H.-J. Hardtke, R. Schmidt, and D. Pawandenat: 1997b, 'Design optimization of a vehicle panel with respect to cabin noise problems'. In: *Proceedings of the NAFEMS World-Congress.* Stuttgart, pp. 885-896.

Milsted, M. G., T. Zhang, and R. A. Hall: 1993, 'A numerical method for noise optimization of engine structures'. *Proceedings of the Institution of Mechanical Engineers / Part D: Journal of Automobile Engineering* **207**, 135–143.

Mühlmeier, M., T. Kumbein, and N. Vogler: 1994, 'Identifikation und Reduzierung der Flächenbeteiligungen am niederfrequenten Innenraumgeräusch'. In: *Berechnung und Simulation im Fahrzeugbau.* pp. 221–235, VDI–Report 1153.

Nefske, D. J.: 1985, 'Acoustic finite element analysis of the automobile passenger compartment with absorption materials'. In: *Proceedings of NOISE-CON 85.* pp. 33–36.

Pal, C. and I. Hagiwara: 1993, 'Dynamic Analysis of a Coupled Structural–Acoustic Problem. Simultaneous Multi–Modal Reduction of Vehicle Interior Noise Level by Combined Optimization'. *Finite Elements in Analysis and Design* **14**, 225–234.

Pal, C. and I. Hagiwara: 1994, 'Optimization of Noise Level Reduction by truncated modal coupled Structural–Acoustic Sensitivity Analysis'. *JSME International Journal, Series C* **37**(2), 246–251.

Roesems, D.: 1997, 'A new methodology to support an optimized NVH engineering process'. *Sound and Vibration* **31**(5), 36–45.

Shi, Q., I. Hagiwara, A. Azetsu, and T. Ichkawa: 1998, 'Holographic neural network approximations for acoustic optimization'. *JSAE Review* **19**, 361–363.

Sielaff, J., A. Kropp, A. Irrgang, and H. P. T. Trong: 1998, 'CAE–gestützte Auslegung der Karosserie am Beispiel der Innenraumakustik'. In: *Entwicklungen im Karosseriebau.* pp. 231–259, VDI–Report 1398.

Suzuki, S.: 1989, 'Boundary element analysis of structural-acoustic problems with complicated boundary conditions'. In: C. A. Brebbia and J. Connor (eds.): *Advances in Boundary Elements*, Vol. 2. pp. 341–353, Springer–Verlag.

Suzuki, S., S. Maruyama, and H. Ido: 1989, 'Boundary element analysis of cavity noise problems with complicated boundary conditions'. *Journal of Sound and Vibrations* **130**(1), 79–91.

Wang, S.: 1999, 'Design Sensitivity Analysis of Noise, Vibration, and Harshness of Vehicle Body Structure'. *Mechanics of Structures and Machines* **27**(3), 317–336.

Yamazaki, I. and T. Inoue: 1989, 'An Application of Structural–Acoustic Coupling Analysis to Boom Noise'. *SAE-paper 891966* pp. 1–9.

Yashiro, H., K. Suzuki, Y. Kajio, I. Hagiwara, and A. Arai: 1985, 'An application of structural–acoustic analysis to car body structures'. In: *Proceedings of SAE Surface Vehicle Noise and Vibration Conference.* Traverse City, Michigan.

Yim, H. J. and S. B. Lee: 1997, 'Design optimization of vehicle structures for idle shake vibration'. *Proceedings- SPIE The International Society for Optical Engineering, Issue 3089//PT1* pp. 432–437.

HOW NOISE CONTROL IMPROVES HEALTH

DEEPAK PRASHER
Deepak Prasher
Institute of Laryngology and Otology
University College London
London
UK

Abstract Noise, a pervasive form of pollution, is not only an issue of acoustics but of great importance to Human health. Noise is of prime concern to a growing number of people and certainly an increasing number are being exposed to hazardous levels of noise often in combination with other oto-toxic agents. Though noise levels are comparatively simple to monitor the responses to noise are far from simple. Occupational noise is the major cause of acquired hearing loss in the adult population and environmental noise is a major source of annoyance and stress which may in turn lead to physiological and psychological changes affecting Man's well being.

1. Scale of the Problem of Noise

It is estimated that noise pollution adversely affects around 800 million people worldwide. According to the European Environment Agency about 450 million people, 65% of European population are exposed to noise levels above 55 dBA, which may result in annoyance, aggressive behaviour and sleep disturbance. A further breakdown shows some 113 million exposed to levels above 65dBA, and around 10 million to noise levels above 75dBA, which can potentially result in increased cardiovascular risk, stress and hearing impairment. The figures for the developing world are not known but it is thought that the noise levels both in occupational and environmental settings may be much higher than in the West due to older machinery and transport systems and highly populated cities. The cost to society has been estimated to range from 0.2 to 2% of GDP in Europe, which at the lower figure of 0.2% equates to 12 billion Euros annually.

M.L. Munjal (ed.), IUTAM Symposium on Designing of Quietness, 299–312.

Noise induced hearing loss is the single most common preventable occupational health problem. The World Health Organisation (WHO) set up a programme for the prevention of deafness and hearing impairment (PDH) in 1985. Their 1995 resolution estimated that there were 120 million people with disabling hearing difficulties worldwide and urged member states to set up national programmes as much deafness and hearing impairment was considered avoidable or remediable. The WHO held a special experts meeting in 1997 to discuss the issue of noise induced hearing loss and has published guidelines and recommendations for action. We have been running a European Commission concerted action on Protection against Noise (PAN) from 1996-1999 and Noise Pollution Health Effects Reduction (NOPHER) since 2000 (Prasher and Luxon, 1998; Prasher et al 1998). There are at present 54 partners from 16 European countries involved in this project. The aims of these actions are to co-ordinate the research activity in this field and disseminate the knowledge as widely as possible. This has resulted in the launch of a new inter-disciplinary international Journal of Noise and Health (Prasher 1998a), which has been publishing research on this topic for the past three years.

2. Noise and Hearing

The Human ear is sensitive to sounds between 20 and 20,000 Hz and has a vast dynamic range in terms of detection and tolerance of sound intensity. Hearing damage is the result of sound dose, which is a product of the magnitude of sound and duration of exposure. Intense impulsive sounds such as that from gunfire or explosions may cause mechanical damage through rupture of membranes in the middle and inner ears but the lower levels of continuous noise exposure as in the occupational setting or in the environment may result in damage through chronic metabolic processes and vascular changes affecting the inner ear through over-stimulation. The effects of noise are pervasive affecting the nerve endings, sensory and supporting cells and vascular supply of the cochlea (Prasher and Canlon1998). As most industrial noise is broad-band the external ear canal transformation results in a bandpass noise centred around 3kHz, and hearing loss due to this noise is centred an octave above the centre frequency giving a characteristic 4- 6 kHz notch in the audiogram. The acoustic stapedial reflex acts to restrict the transmission of sound of high intensity through the middle ear thereby affording some protection from continuous high level exposure. Those with loss of the reflex, such as in Bell's palsy, have been shown to be more vulnerable, having greater temporary threshold shift. Di-

rect shearing forces result in damage to outer hair cells (OHCs) of the organ of Corti in the cochlea and the delicate displacement detecting stereocilia which are the weakest link in the sound transduction process. OHCs provide bio-mechanical feedback to enhance cochlear sensitivity and frequency selectivity whilst the inner hair cells (IHCs) relay the signals to the brain via the VIII nerve. Once mechanical damage has occurred , the possibility for repair becomes limited, whereas the potential to reverse or counter the metabolic effects by pharmacological intervention are much higher (Prasher 1998b).
Pathological changes can be observed in the mitochondria, sub-surface cisternae, smooth endoplasmic reticulum, nucleus and there are increases in lysosomes and Hensen bodies. Furthermore membrane alterations lead to changes in ionic composition and herniation of cell contents. Vascular changes in the stria vascularis and the spiral ligament have also been reported. Thus there are changes in cellular metabolism, damage to cellular architecture leading to alteration in function although no direct relationship can be defined between pathology and function. Normal thresholds have been reported with considerable OHC and IHC loss. Accurate prediction of threshold is not possible on the basis of the extent of haircell loss or conversely a prediction of pathology for a given threshold. However, noise characteristically causes a notch in the auditory sensitivity in the region of 2-8kHz leading to reduction in the dynamic range of hearing and poorer frequency selectivity. There may be pitch distortion, reduced speech intelligibility and the troublesome symptom of tinnitus.

3. Occupational Noise Exposure: Legal Thresholds

It is known that prolonged exposure to noise in excess of 85dBA in the workplace will result in hearing impairment, the higher the level of noise the greater is the risk of acquiring a hearing loss if the individual remains unprotected. At present the legal threshold for noise exposure is 85dBA in the workplace in the UK. It is estimated that 1.7 million workers in the UK are exposed to noise above this level (which is the first action level) and over a half million are exposed above 90dBA (the second action level). A draft EU directive proposes to reduce the threshold level to 75dBA and for exposures between 85-90dBA noise assessments and control strategies to become mandatory. At present if UK workers are exposed to first action level then formal noise assessments must be made and at 90dBA or above ear protectors must be provided and reduction of exposure reduced so far as reasonably practicable.

Table 1. Admissible rounds due to different criteria

Level of Exposure dBA Leq(8hr)	10 years exposure Number of persons affected per 100	Lifetime exposure Number of persons affected per 100
80	1	3
90	5	11
100	17	32

It can be seen from Table 1 above that if 100 people are exposed to a lifetime of exposure of 100dBA, 32 of the 100 people will be affected and furthermore a reduction in the exposure level from 100dBA to 90 dBA will result in a reduction in the number of people affected by 21 to 11 for every 100 exposed. Although there is increasing effect with increasing levels of exposure, what is perhaps more surprising is why all the people are not affected similarly. What is the possible reason that there are still 68 people in every 100 exposed to 100dBA noise level who do not have a 50 dB hearing loss? This raises the issue of susceptibility of individuals to noise induced hearing loss. Animal and Human studies have shown that there is a broad range individual differences in sensitivity to any given noise exposure, which makes it extremely difficult to predict hearing outcome resulting from a given noise exposure in any one individual. Several factors have been proposed to explain the differences in noise induced hearing loss among individuals and over time in the same individual.

4. Susceptibility to noise induced hearing loss

Although reduction of noise levels must be the prime target in reducing the number of people adversely affected by noise exposure, it is important to also consider the role of individual susceptibility to noise damage and the possible factors which may contribute to the susceptibility. This may allow a prediction to be made about the population who may be most at risk hence allowing protection of these individuals.

Transmission of sound to the inner ear: does an efficient system increase risk?
It is accepted that no lower safe limit of sound intensity can be given for any one individual due to the large variation in susceptibility. Generally it is considered that sound levels below 75dBA do not result in permanent hearing loss and those above 85dBA do. The total sound energy reaching the ear is considered responsible for the damage. Thus sound intensity and duration both play a part in the cumulative damage suf-

fered by the ear. Thus an unusually efficient acoustic transfer through the external and middle ear system can theoretically be associated with increased risk of damage due to increased energy coupled to the inner ear.

Vascular changes in the cochlea: does reduced microcirculation and oxygenation of the cochlea increase susceptibility?

One of the possible mechanisms in the development of noise induced hearing loss is the vascular pathology in the micro-circulation of the cochlea as demonstrated by histological evidence of reduced cochlear blood flow following noise exposure. A number of recent studies have shown that localised ischaemia, micro-vascular alterations may occur during noise exposure indicating that poor blood circulation in individuals may render them more susceptible.

Genetic pre-disposition to noise induced hearing loss: is assessment possible? Although it has been shown that in-bred mice and those pre-disposed to age relating hearing loss have increased susceptibility to noise induced threshold shifts and auditory degeneration, the isogenic nature of the inbred lines is very unlike the Human condition.

Melanin and magnesium: indicators of susceptibility
Melanin estimation with eye and skin colour show that black or dark skinned individuals have reduced threshold shifts in comparison with white people with blue eyes. A comparison of hearing loss among soldiers also revealed better thresholds among black soldiers compared to whites. Animal and Human studies have also shown that magnesium deficiency increases the risk of noise induced hearing loss in a given noise exposure, furthermore oral intake of magnesium supplement reduces level of threshold shift after noise exposure.

Increasing resistance to damage by pre-exposure sound conditioning
It has been shown that interrupted exposure causes less damage than continous exposure of equal energy. Recent studies have shown that the extent of temporary threshold shift is reduced by prior exposure to a non-traumatic stimulus. It is considered that prior exposure " trains", " conditions" or "toughens" the ear to subsequent damaging sound exposure. Considerable research is needed to determine the best parameters and mechanisms responsible for these phenomenon.

Changes within an individual: age, drugs and chemical exposure
Babies born to mothers working in noisy environments have an increased risk of high frequency hearing loss. In childhood, there may be a sensitive period in the development related to the efferent pathway development.

Advancing age has only been linked with decreasing susceptibility but this is confounded by presbyacusis. Ototoxic drugs and chemicals in industrial settings appear to exacerbate the damaging effects of noise exposure.

Prediction of probability of risk to an individual

Clearly susceptibility to noise induced hearing loss in any individual is a combination of many factors. An early indication of the extent of the risk may be estimated from a number of investigations examining not only the status of the outer hair cells but their efferent motor control capability, extent of protection afforded by sound conditioning, level of melanin and magnesium. The specific individuals under threat remain at present unidentifiable prior to exposure.

5. Environmental Noise and Effects

Environmental or community noise largely stems from the road, rail and air traffic systems and indoor from ventilation systems and household appliances and neighbourhood activity. In addition there is noise from leisure activities such as clubs, outdoor events, sports events etc. Considerable progress has been made in noise technology over the last 30 years but surveys reveal that average noise exposure over the UK as a whole has hardly changed (Flindell and Stallen 1999) This is largely because traffic levels have increased as noise levels from individual vehicles have decreased. The remedy appears to be a reduction in the amounts of noise generating activities in order to reduce the levels of exposure to individuals. There is agreement that noise is a stressor and that it leads to annoyance but the relationship between the level of exposure and other health effects remains open. It is suggested that noise leads to disturbance of activities and communication causing annoyance which in turn leads to stress responses which may lead to further deterioration in physical and mental well being. A number of non-auditory health effects of exposure to noise are considered in detail below.

6. Noise and Annoyance

The most documented reaction to noise in the community is annoyance. It is particularly important psychological reaction as it affects an individual's quality of life. Furthermore any deterioration in the quality of life has implications for the health of the individual. It is recognised that as noise levels rise the level of annoyance also increases across individuals but absolute levels of noise are not the only determinants of annoyance. A very low intensity sound can cause annoyance such as a dripping tap, which may interfere with concentration or may be perceived as interfering. The subjective response to noise may also include fear and mild anger. It is also seen as intrusive and its meaning for an individual may also determine its annoyance level.

In road, rail and aircraft noise studies, noise levels have been associated with annoyance in a dose-dependent fashion although aircraft annoy the most and rail the least (Miedema and Vos1998). High frequency noise has been found to be more annoying than low frequency noise (Bjork 1986). Community annoyance is affected by acoustic properties of noise and other non-acoustic factors such as social, psychological or economic in nature. Noise sensitivity has been shown to be the most important non-noise predictor of annoyance (Taylor 1984), others include fear of noise (Gunn et al 1981) and attitudes to the noise source such as predictability and controllability of noise and general dislike of the environment including the belief that the authorities can control the noise. In addition personal factors such as benefits, employment, compensation as result of the noise may be important in determining annoyance. Stallen (1999) has proposed a theoretical framework for environmental noise annoyance, which he views as a form of psychological stress dependent on a number of non-acoustical factors.

7. Noise and Sleep

There is sufficient evidence to indicate that exposure to noise during sleep may cause awakenings and changes in sleep stages (Ohrstrom and Bjorkman1988 and Lambert and Vallet 1994). Road traffic noise may also increase the time taken to fall asleep. Sleep disturbance may in turn lead to poor performance and changes in mood (for review see Ouis 1999). Individuals sensitive to noise and the elderly may be more vulnerable to sleep disturbance. Recently it has been reported that chronic traffic noise exposure at night time may lead to chronic increase of cortisol which may in the long term be detrimental to health (Born and Fehm 2000). It has been shown that reducing indoor noise level results in an increase in the amount of REM and slow wave sleep. There may be some adaptation to sleep but complete habituation does not occur. There may also be consequences of disturbed sleep on performance next day. Drivers who fall asleep at the wheel cause approximately 20of traffic accidents on German motorways (Billiard 1993). In Europe 10-26(55-143 million) of people have severe difficulties in falling or staying asleep. Chronic sleep disorders lead to loss of efficiency in the work place, and affects psychological well being, learning in childhood, social interaction and driving ability.

8. Noise and Performance

There is evidence from laboratory studies that noise exposure leads to impaired performance (Smith and Broadbent 1992). Those with limited performance strategies for example with high anxiety or multiple tasks may be more vulnerable to the effects of distracting effects of noise. It seems that noise raises arousal level but decreases attention. The type of noise may play a significant role in the degradation of performance as may the perception of the degree of control over noise. Social performance may be affected by the impaired communication and by the distraction of attention from relevant cues in the social environment and by change of mood. There is a suggestion that noise may increase aggression and reduce helping behaviour. It has also been shown that noise slows rehearsal in memory. Learning in schools is impaired due to increased noise levels outdoor or poor acoustic environments indoor or sometimes both. This is particularly so for acquisition of language or music as quality of sound is important.

9. Noise Stress and Hormones

It has been suggested that prolonged exposure to stress may eventually lead to the development of disease in vulnerable individuals. The new theoretical stress model, the allostatic load model assumes that any physical or psychological threat to the organism results in a stress response, which must be contained once the threat has subsided (Krishbaum and Hellhammer 1999). However over or under activity of the allostatic system may add to the "wear and tear" of the organism. A large number of physiological processes change in response to environmental stress, including the production of stress-related hormones which allows the body use more energy in times of high demand. The hypothalamus-pituitary-adrenal (HPA) axis plays a central role in this process. Cortisol is closely associated with stress related health problems. Accumulating evidence (Prasher 1998c) suggests that noise can affect the HPA function and hence cortisol levels acutely as well as chronically.Salivary cortisol levels have been shown to be enhanced in response to mental work performed under 90dBA noise and not under quiet conditions (Miki et al 1998). Industrial workers who were chronically exposed to noise levels greater than 85dBA do not show the expected circadian decline in cortisol level in the afternoon but did so after reduction in noise levels with the use of headphones (Melamed and Bruhis 1996). Clarke et al (1994) have shown that noise not only has an activating effect on the HPA axis but can also have an organising effect in that offspring of rhesus monkeys who were stressed by unpredictable noise during mid-late gestation , showed enhanced HPA responses to stressors later in life.

10. Noise and Cardiovascular Effects

Noise has the potential to cause stress reactions according to Ising et al (1999). Acute reaction to sudden exposure to noise includes an increase in heart rate, blood pressure and peripheral vasoconstriction but this habituates rapidly. Heart rate reactions to noise disturbance during sleep do not habituate (Ohrstrom, 2000). Chronic exposure to noise may lead to dysregulation of physiological systems with an increase in the release of stress hormones leading to an increase in cholesterol or triglycerides in the blood. If repeated increases occur chronically this may lead to arteriosclersosis. These "flight or fight" type physiological reactions particularly with noise disturbances in sleep may in modern times be detrimental to health rather than being of value in saving the individual from perceived attack resulting from the noise.Ising et al (1999) suggest that chronic noise induced stress accelerates the ageing of the myocardium thereby increasing the risk of myocardial infarction. The pathomechanism suggested is the interaction between endocrine reactions and intracellular calcium/magnesium shifts. Melamed et al (1999) have shown in a study of 6000 industrial workers an association between noise exposure and serum lipids in women and young men. The strongest indication of a dose-response relationship between noise exposure and hypertension has come from a report in Chinese textile mill workers adjusting for age, years of work, salt intake, and family history of hypertension. In a recent epidemiological review (Babisch 2000) of traffic noise and cardiovascular disease, Babisch states that the available literature provides no epidemiological evidence of a relationship between noise exposure and mean blood pressure in adults, however noise related increases in blood pressure are consistently seen in children. There is little evidence that exposure to traffic noise is associated with an increased risk of hypertension, however there is some evidence for an increased risk for ischaemic heart disease in subjects who live in noisy areas with outdoor noise levels greater than 65-70dBA. Cardiovascular effects of occupational noise are contradictory (Babisch 1998) and he suggests that subjective measures of noise exposure may be better predictors than noise levels.

11. Noise and Mental Health

Stansfeld et al (2000) in a recent review of the topic indicate that recent community based studies suggest high levels of environmental noise are associated with mental health symptoms such as depression and anxiety but not with impaired psychological functioning. Chronic aircraft noise in children impairs quality of life but does not lead to depression or anxiety. They suggest that further studies should consider confounding factors such as social class, noise sensitive individuals leaving the noisy areas, and studies should use standardised instruments to measure mental health outcomes, physiological outcomes such as hormonal measures and detailed measures of noise exposure.

12. Noise and Fatigue

Physical environment may contribute to work fatigue and performance may be impaired during and after work in noise. Noise may over-stimulate, or monotonous noise may have a sleep-provoking effect or it may make a task difficult to perform due to loss of acoustic cues. There is some evidence that complaints about fatigue are more common amongst noise-exposed workers. Melamed and Bruhis (1996) reported increased fatigue and irritability as well as higher cortisol levels after one week in noise compared to one week with hearing protectors. Kjellberg et al (1998) have shown in epidemiological survey of 50,000 workers and experimental field studies that fatigue and headache were more common amongst noise exposed workers and reaction times were prolonged after a week's work in high noise levels, increasing gradually during the week compared to controls which showed the opposite.

13. Noise Effects in Children

In children, aircraft noise exposure has been associated with difficulties of attention, reading ability, learning and memory (Evans et al 1995; Hygge et al 1996). Auditory central processing and language acquisition are affected in the critical periods of development and deficits in sustained visual attention, concentration, motivation, auditory perception and speech comprehension have been reported. The noise exposure may be from outdoor as in road traffic noise or aircraft and/or indoor from poor classroom acoustics. There is an increase in resting blood pressure and raised levels of catecholamines with long-term exposure to road and aircraft noise.

14. Noise Control and Health

As indicated above, noise exposure has a significant impact on health. As decreasing the noise level and duration of exposure reduces the effects on health, it is clear that noise control will result in improving the quality of life of individuals exposed to occupational and/or environmental noise. Hearing conservation programmes should include accurate monitoring of noise exposure and identification of individuals affected, implementing acoustic controls whenever possible, providing workers with hearing protection devices, measuring hearing regularly, educating workers and management to the benefits of worker protection from noise and its effects.

References

Babisch, W. (1998) Epidemiological studies of the cardiovascular effects of occupational noise-a critical appraisal,*Noise and Health*,**1**,p .24-39 .

Babisch, W. (2000) Traffic noise and Cardiovascular disease: epidemiological review and synthesis, *Noise and Health*,**8**, p .9-32 .

Billiard, M. (1993) Die Zukunft der Schlafmedizin in Europa. *In: K.Meier-Ewert; E.Ruther (Hrsg); Schlafmedzin; Gustav Fischer Verlag Jena.*

Bjork, EA.(1986) Laboratory annoyance and skin conductance to some natural sounds. *J Sound Vib*,**109**,p .339-345 .

Born J and Fehm HL(2000) The neuroendocrine recovery function of sleep,*Noise and Health***7**,p .25-38 .

Clarke, AS., Wittwer, DJ., Abbott, DH. and Schneide,r ML. (1994) Long-term effects of prenatal stress on HPA axis activity in juvenile rhesus monkeys. *Dev. Psychobiol***27**p .257-269 .

Evans, GW., Hygge, S. and Bullinger, M. (1995), Chronic Noise and Psychological Stress. *Psychol Sci.*,**6**,p .333-338 .

Flindell, IH. and Stallen, PJM. (1999) Non-acoustical factors in environmental noise *Noise and Health*,**3**,p .11-16 .

Gunn, WJ., Shigehisa, T., Fletcher, J., Shepherd, WT. (1981), Annoyance response to aircraft noise as a function of contextual effects and personality characteristics,*J Audit Res***21**,p .51-83 .

Hygge, S., Evans, GW. and Bullinger, M. (1996), The munich airport noise study: Cognitive effects in children from before to after the change over of airports,*In : Hill FA and Lawrence R (Eds) Proceedings of Internoise 96 (book5)*,**Liverpool, UK**,p . 2189-2192 .

Ising, H., Babisch, W. and Kruppa, B. (1999), Noise induced endocrine effects and cardiovascular risk *Noise and Health*,**4**,p .37-48 .

Kirschbaum, C. and Hellhamme,r DH.(1999), Noise and Stress-salivary cortisol as a non-invasive measure of allostatic load,*Noise and Health*,**4**,p .57-66 .

Kjellberg, A. , Muhr, P. and Skoldstrom, B. (1998), Fatigue after work in noise-an epidemiological survey study and three quasi-experimental field studies ,*Noise and Health***1**p .47-55 .

Lambert, J. and Vallet, M. (1994), *Study related to the preparation of a communication on a future EC noise policy (LRN Report No.9420), Bron Cedex, France, Institut National Recherche sur Les Transports et leur Securite.*

Melamed, S. and Bruhis, S. (1996), The effects of chronic industrial noise exposure on urinary cortisol, fatigue and irritability: a controlled field experiment,*J. Occup. Environ. Med*,**38**,p .252-256 .

Melamed, S., Kristal-Boneh, E. and Froom, P. (1999) Industrial noise exposure and risk factors for cardiovascular disease, Findings from the CORDIS study, *Noise and Health*,**4**,p .49-56 .

Miedema, HME. and Vos, H.(1998), Exposure-response relationships for transportation noise*Journal of the Acoustical Society of America* , **104(6)**,p .432-3445 .

Miki, K., Kawamorita, K., Araga, Y., Musha, T., and Sudo, A. (1998), Urinary and salivary stress hormone levels while performing arithmetic calculation in a noisy environment,it Ind Health,**36**,p .66-69 .

Ohrstrom, E. (2000), Sleep disturbance caused by road traffic noise-studies in laboratory and field ,*Noise and Health*,**8**,p .71-78.

Ohrstrom, E. and Bjorkman, M.(1988), Effects of noise disturbed sleep- a laboratory study on habituation and subjective noise sensitivity , it J. Sound Vib, p.122:277 .

Ouis, D.(1999) Exposure to nocturnal road traffic noise :Sleep disturbance and its after effects,*Noise and Health*,**4**,p .11-36 .

Prasher, DK. and Luxon, LM. (Eds)(1998), Biological effects of Noise, *Whurr Publishers, London*

Prasher, DK., Luxon, LM .and Pyykko, I.(Eds)(1998), Protection against Noise, *Whurr Publishers, London*

Prasher, DK. and Canlon, B. (Eds)(1998), Cochlear Pharmacology and Noise Trauma, *nRn Publications, London*

Prasher, DK. (1998a), Birth of a new Journal *Noise and Health* ,**1**,p .1

Prasher, D. (1998b), New strategies for prevention and treatment of noise induced hearing loss, *The Lancet*,**352:9136**, p .1240-1242.

Prasher, D .(1998c) Traffic Noise increases stress by driving up cortisol, *The Lancet*, **352;9135: 1201**

Smith, AP. and Broadbent, DE. (1992), Non-auditory effects of noise at work: a review of the literature (HSE Contract Research Report No.30),*HMSO London UK*

Stallen, PJM. (1999), A theoretical framework for environmental noise annoyance,*Noise and Health* ,**3**,p .69-80 .

Stansfeld, S. A. , Haines, MM., Burr, M., Berry, B., Lercher, P. (2000), A review of environmental noise and mental health,*Noise and Health* ,**8**,p .1-8 .

Taylor, SM. (1984), A path model of aircraft noise annoyance,*J. Sound Vib* ,**96**,p .243-260 .

TRANSMISSION AND ABSORPTION - PREDICTIONS AND PERFORMANCE

R. J. HOOKER
Department of Mechanical Engineering
The University of Queensland
Brisbane
Australia

Abstract

The paper is based on test results for the transmission behaviour of single and double glazed window assemblies and for the absorption behaviour of perforated facing/air gap/glasswool arrangements for use as roadside barriers. Prediction procedures for these circumstances are applied and the predictions compared with the experimental data. Results are mixed.

1. Introduction

Much acoustic design is based on the experience of the designer - or that of others, through books, journals, colleagues and files. When new or changed circumstances arise design guidance is not always readily available and design might involve some elements of intuition and faith.

This paper has as its basis experimental testing in two areas, on transmission loss of single and double glazed windows for a hospital and on absorption coefficient determination for combinations of perforated facing, air gap and inxabsorbent material intended for use in roadside barriers. In both cases the testing was conducted before predictions were made, but the opportunity of making "predictions" has been taken.

2. Transmission

This part is based on tests carried out to determine transmission loss behaviour of single and double glazed window assemblies. The tests were carried out in a conventional reverberant two-room facility. Lami-

M.L. Munjal (ed.), IUTAM Symposium on Designing of Quietness, 313–321.

nated glass was used in all cases. The window areas were 7.5 m^2 for the single glazing and 5 m^2 for the double glazing. The assemblies consisted of the glass arrangements mounted in aluminium frames.

2.1 Prediction

Prediction of transmission loss is by the simple plateau method [1]. In this method, a field incidence transmission loss based on mass law behaviour with 6 dB/octave slope is taken at low frequencies. A "plateau" is set for the relevant material. For glass (presumably non-laminated), the plateau level is quoted as 27 dB in [1]. The plateau defines the upper end of the mass law region. Plateau "width" is set in terms of a frequency ratio, i.e., the ratio of the high frequency end of the plateau to the low frequency end. The reference quotes a frequency ratio of 10 for glass. At frequencies above the high end of the plateau the method sets a transmission loss rising with frequency at 10 dB/octave. Predictions based on this method are compared with test results in the following section.

2.2 Performance

Experimental and calculated results for single glazed assemblies are given in Figure 1 and Table 1, for two glass thicknesses. Testing was carried out in one-third octave steps as shown in the Figure. In the Table the results are simplified by conversion to octave band values. Since the transmission loss curves are relatively smooth (presumably because of the damping present in laminated glass) the octave band results give a simple but satisfactory basis for comparison.

Table 1. Transmission Loss Single Glazing

	Frequency Hz					
Thickness	125	250	500	1000	2000	4000
6.38 mm Expt	21	25	30	33	32	37
Calc	19	25	27	27	27	30
10.38 mm Expt	23	28	33	33	34	43
Calc	23	27	27	27	27	27

The test results are typical. There is a dip at low frequency, 125 Hz, presumably a inxpanel resonance. In the mass law region the test results follow prediction. In the coincidence region, the predicted plateau level of 27 is too low and a value of around 33 is indicated. A possible

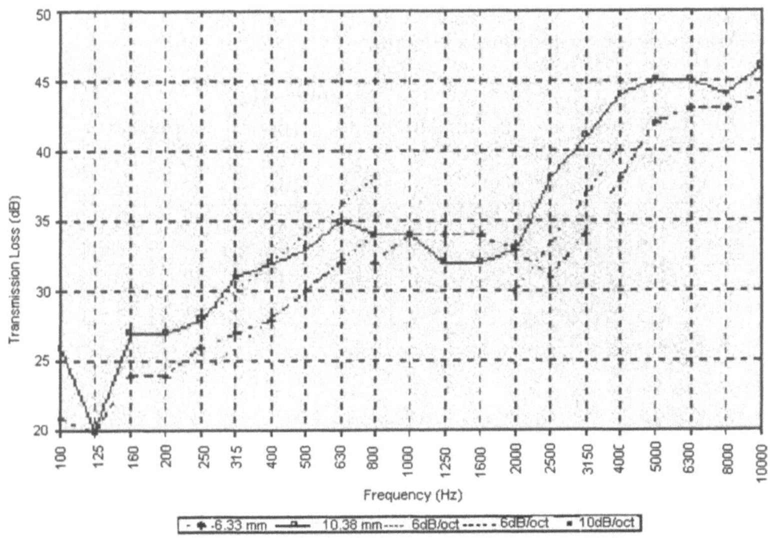

Figure 1. Transmission Loss, Single Glazing

explanation arises from the laminated construction. The laminated glass comprises two sheets of equal thickness with a central film 0.38 mm thick, i.e., the individual glass sheets are approximately half the total thickness.

For a sheet of half the thickness the coincidence effect would be expected at twice the frequency. The plateau will then intersect the mass law line at twice the frequency, which means at a level 6 dB higher. The flat plateau and the plateau width (considerably less than a frequency ratio 10) are consistent with the high damping due to the lamination layer. Results for one double glazed case are shown in Figure 2 and the octave band summaries for three cases are given in Table 2. The arrangements presented are:

Case 1. Glass 12.76 mm, gap 210 mm, glass 8.38 mm.
Case 2. Glass 8.38 mm, gap 85 mm, glass 6.38 mm.
Case 3. Glass 12.76 mm, gap 85 mm, glass 8.38 mm.

The arrangement of Case 1 was (as intended) the one of highest transmission loss. The two glass panels were mounted in independent frames and a small area of absorbent lining was exposed within the cavity. For Cases 2 and 3 the two panels were in a common frame. Test results for the single glazed thicknesses 8.38 and 12.76 mm are not available, so the calculated values are based on the plateau method and direct summation of the transmission loss for the two panels. This summation is based on independent behaviour of the two panels and no interaction arising

Table 2. Transmission Loss Double Glazing

	Frequency Hz					
Thickness	125	250	500	1000	2000	4000
Case 1 Expt	38	39	48	54	56	64
Calc	46	54	54	54	57	74
Case 2 Expt	27	33	42	47	47	49
Calc	40	52	54	54	54	64
Case 3 Expt	33	34	43	46	47	48
Calc	46	54	54	54	57	74

from the air cavity - an idealistic view. Independent behaviour assumes no flanking transmission through the frame. The half wavelength frequencies for the cavities, near which first air cavity resonances might be expected, are 820 Hz for the 210 mm gap and 2020 Hz for the 85 mm gap. The plateau has been retained at 27+27=54.

In Figure 2, the lower 6 dB/octave line is the mass law result for the total mass as a single panel. The double panel performance exceeds this. The upper 6 dB/octave line is the summation for the two panels. The performance on test does not reach this prediction. For Case 1, the results at the lowest and the highest frequencies are approaching the flanking limits of the test facility. At all frequencies for all cases, test item performance appears to be limited by flanking transmission through the window frame assemblies. This is particularly noticeable for Case 3, which has the same glass elements as Case 1 but a common frame, smaller air gap and much lesser performance.

3. Absorption

The absorption results relate to tests on proposals for roadside barriers with absorbent facing. Application to open-air noise testing facilities is also envisaged. The barrier arrangements being investigated were composed of a perforated metal sheet, air gap, absorbent material and rigid backing.

3.1 Prediction

Prediction of absorbent behaviour of acoustic materials and combinations has generally not been widespread. Most design information has been in the form of manufacturers' data determined by experiment. A recent significant contribution is the book, "Notes on Sound Absorption Technology" [2]. This material is now also available on CD-ROM [3].

Figure 2. Transmission Loss, Double Glazing, Case 2

These sources include computer programs for the calculation of absorption behaviour. One program gives a calculation of normal incidence and

Table 3. Absorption Coefficients Glasswool 100 mm

	Frequency Hz					
Condition	125	250	500	1000	2000	4000
Expt	0.57	0.99	1.	1.	0.99	0.99
Calc, R = 1.5	0.52	0.74	0.83	0.88	0.92	0.94
Calc, R = 2.0	0.54	0.71	0.79	0.85	0.90	0.93
Calc, R = 2.5	0.54	0.68	0.76	0.83	0.89	0.92
Calc, R = 3.0	0.53	0.64	0.73	0.81	0.88	0.92
Calc, R = 4.0	0.50	0.59	0.68	0.78	0.86	0.90

diffuse field absorption coefficient for multi-layer absorber combinations. The elements that can be included are air gaps, limp sheets, porous layers, perforated plates and wire-mesh screens. Parameters for each of the elements can be entered. The results now presented have been determined by this program. The porous layer element requires specification of the thickness of the layer and the normalised flow resistance. Flow resistance is not normally quoted by suppliers. During the experimental work a test was conducted on the glasswool acoustic absorbent used in the combination trials. Experimental results were compared with program predictions for a range of normalised flow resistance values R. Selected results are given in Table 3.

The experimental values quoted as 1. are adjusted - in the test a coefficient greater than 1.0 was obtained, as sometimes occurs in reverberation room testing. A coefficient greater than 1. is meaningless. Such results also suggest that experimental values approaching 1. are high. The comparison between calculation and experiment might perhaps be described as moderate. The discrepancy at 250 Hz suggests that some advantageous behaviour is occurring in the material. The effect of changing the value for R can be seen. For the study R values of 1.5 and 2.0 have been used. A set of calculated results is shown in Table 4.

Table 4. Absorption Coefficients
Effect of Open Area
Facing: 2.5 mm thick, holes 6mm dia, open area as given
Air gap: 25 mm
Absorbent: 100 mm thick, R = 1.5

	Frequency Hz					
Condition	125	250	500	1000	2000	4000
Open area 8%	0.56	0.85	0.97	0.63	0.23	0.10
Open area 13%	0.54	0.80	0.94	0.84	0.42	0.21
Open area 33%	0.53	0.76	0.88	0.93	0.80	0.60

This set of results demonstrates the possibility of establishing trends. In this case, the sometimes expressed view that reducing open area improves low frequency performance receives some slight support. The effect on high frequencies is dramatic.

3.2 Performance

Tests were performed in a reverberation chamber of volume 170 m3.The samples were placed on the floor and sample area was 10 m2.

Experimental results were determined from reverberation time tests using the Sabine equation, since this was a requirement of the specified standard. With respect to other approaches [1], it can be noted that for the circumstances of these tests, the Norris-Eyring equation gives similar values, and the Millington-Sette equation fails.

Results are available for several combinations. A set enabling comparisons and trends to be examined is given in Table 5 and Figures 3 and 4. Again the entries 1. correspond to cases where the experimental coefficients were greater than 1.0. A further set of combinations is set

Table 5. Absorption Coefficients
Effect of Open Area and Air Gap
Facing:2.5 mm thick, holes 6 mm.
Absorbent:100 mm,R = 2.0
Open area and gap as given

Conditions		Frequency Hz					
		125	250	500	1000	2000	4000
8%, 25 mm	Calc	0.58	0.81	0.97	0.68	0.24	0.10
	Expt	0.63	1.	1.	0.85	0.69	0.54
8%, 50 mm	Calc	0.59	0.83	0.95	0.53	0.23	0.14
	Expt	0.61	0.99	0.96	0.86	0.67	0.46
13%, 25 mm	Calc	0.57	0.77	0.92	0.88	0.43	0.21
	Expt	0.70	1.	1.	0.99	0.83	0.69

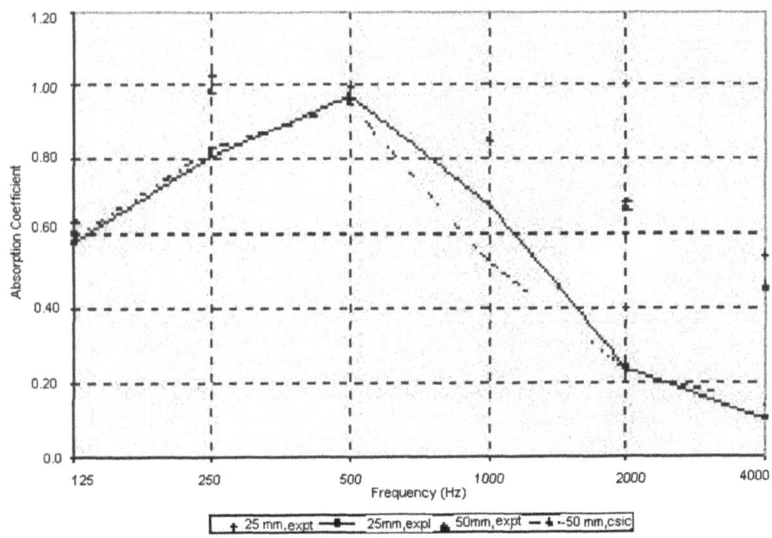

Figure 3. Absorption Coefficient, Effect of Air Gap
Open Area 8%, details in Table 5

out in Table 6. The "no air gap" comparison is shown in Figure 5. The calculated result for 2 mm gap (confirmed also by a calculation for zero gap) has not produced the substantial loss of performance as measured (and as might be expected).

From the results presented a moderate agreement between calculation and experiment might perhaps be claimed for the low frequency end, but there are major differences at the high frequencies. The best

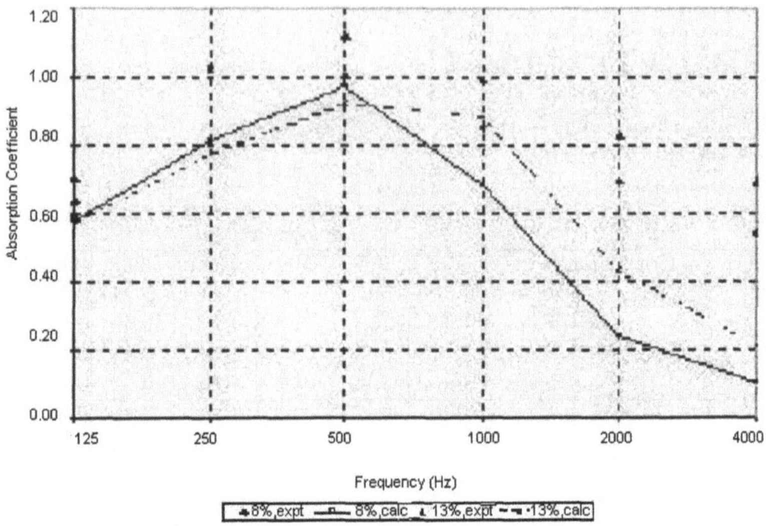

Figure 4. Absorption Coefficient, Effect of Open Area
Air Gap 25 mm, details in Table 5

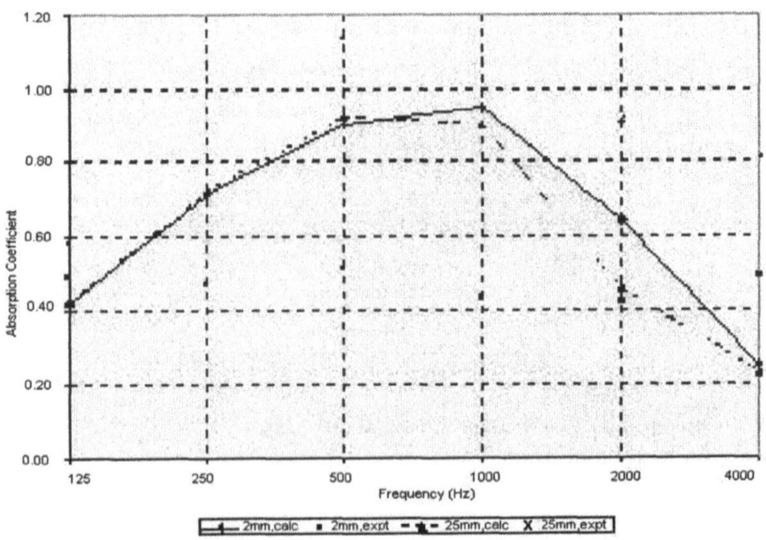

Figure 5. Absorption Coefficient, With and Without Air Gap
Open area 13%, details in Table6

that can be said is that the trends are consistent. It is apparent that the behavioural mechanism at high frequency is not well represented.

Table 6. Absorption Coefficients
Effect of Opne Area and Air Gap
Facing: 2.0 mm thick, holes 6 mm.
Absorbent: 75 mm, R = 2.0
Open area and gap as given

		Frequency Hz					
Conditions		125	250	500	1000	2000	4000
13%, 2 mm	Calc	0.42	0.71	0.90	0.95	0.64	0.25
	Expt	0.49	0.47	0.51	0.43	0.42	0.49
13%, 25 mm	Calc	0.42	0.72	0.92	0.90	0.46	0.23
	Expt	0.58	1.	1.	1.	0.91	0.81
13%, 25 mm	Calc	0.41	0.68	0.85	0.92	0.82	0.62
	Expt	0.56	1.	1.	1.	0.97	0.90

4. Conclusion

The results reported here show that the plateau method for transmission loss will give a good predictioni for single leaf laminated glass provided the plateau level is increased from 27 to 33. For double glazed assemblies, performance appears to be limited by flanking transmission through the window surround assembly. The absorption program can show a good prediction for absorbent material by selection of an appropriate flow resistance. The method can perhaps be claimed to give a moderate prediction at low frequency for the combinatio arrangements examined. At high frequencies the prediction was not good, although trends appear to be in the right directions.

References

Beranek, L. L.(1971) *Noise and Vibration Control. McGraw-Hill,*

Ingard, K.U.(1994) Notes on Sound Absorption Technology, *Noise Control Foundation,*

Ingard, K.U.(2000).

Lecture Notes and Computer Programs on CD-ROM, For details see *Noise-News Intnl,* **8 (2), June 2000, 86,**,*or http://www.ingard.com* ,

References

Beranek, L. L. (1971) *Noise and Vibration Control*. McGraw-Hill, [illegible]
[illegible] (1991) [illegible] [illegible]
[illegible]
[illegible] 2009 [illegible]
[illegible]
[illegible]

Index

Mechanics

SOLID MECHANICS AND ITS APPLICATIONS

Series Editor: G.M.L. Gladwell

49. J.R. Willis (ed.): *IUTAM Symposium on Nonlinear Analysis of Fracture*. Proceedings of the IUTAM Symposium held in Cambridge, U.K. 1997 ISBN 0-7923-4378-6
50. A. Preumont: *Vibration Control of Active Structures*. An Introduction. 1997 ISBN 0-7923-4392-1
51. G.P. Cherepanov: *Methods of Fracture Mechanics: Solid Matter Physics*. 1997 ISBN 0-7923-4408-1
52. D.H. van Campen (ed.): *IUTAM Symposium on Interaction between Dynamics and Control in Advanced Mechanical Systems*. Proceedings of the IUTAM Symposium held in Eindhoven, The Netherlands. 1997 ISBN 0-7923-4429-4
53. N.A. Fleck and A.C.F. Cocks (eds.): *IUTAM Symposium on Mechanics of Granular and Porous Materials*. Proceedings of the IUTAM Symposium held in Cambridge, U.K. 1997 ISBN 0-7923-4553-3
54. J. Roorda and N.K. Srivastava (eds.): *Trends in Structural Mechanics*. Theory, Practice, Education. 1997 ISBN 0-7923-4603-3
55. Yu.A. Mitropolskii and N. Van Dao: *Applied Asymptotic Methods in Nonlinear Oscillations*. 1997 ISBN 0-7923-4605-X
56. C. Guedes Soares (ed.): *Probabilistic Methods for Structural Design*. 1997 ISBN 0-7923-4670-X
57. D. François, A. Pineau and A. Zaoui: *Mechanical Behaviour of Materials*. Volume I: Elasticity and Plasticity. 1998 ISBN 0-7923-4894-X
58. D. François, A. Pineau and A. Zaoui: *Mechanical Behaviour of Materials*. Volume II: Viscoplasticity, Damage, Fracture and Contact Mechanics. 1998 ISBN 0-7923-4895-8
59. L.T. Tenek and J. Argyris: *Finite Element Analysis for Composite Structures*. 1998 ISBN 0-7923-4899-0
60. Y.A. Bahei-El-Din and G.J. Dvorak (eds.): *IUTAM Symposium on Transformation Problems in Composite and Active Materials*. Proceedings of the IUTAM Symposium held in Cairo, Egypt. 1998 ISBN 0-7923-5122-3
61. I.G. Goryacheva: *Contact Mechanics in Tribology*. 1998 ISBN 0-7923-5257-2
62. O.T. Bruhns and E. Stein (eds.): *IUTAM Symposium on Micro- and Macrostructural Aspects of Thermoplasticity*. Proceedings of the IUTAM Symposium held in Bochum, Germany. 1999 ISBN 0-7923-5265-3
63. F.C. Moon: *IUTAM Symposium on New Applications of Nonlinear and Chaotic Dynamics in Mechanics*. Proceedings of the IUTAM Symposium held in Ithaca, NY, USA. 1998 ISBN 0-7923-5276-9
64. R. Wang: *IUTAM Symposium on Rheology of Bodies with Defects*. Proceedings of the IUTAM Symposium held in Beijing, China. 1999 ISBN 0-7923-5297-1
65. Yu.I. Dimitrienko: *Thermomechanics of Composites under High Temperatures*. 1999 ISBN 0-7923-4899-0
66. P. Argoul, M. Frémond and Q.S. Nguyen (eds.): *IUTAM Symposium on Variations of Domains and Free-Boundary Problems in Solid Mechanics*. Proceedings of the IUTAM Symposium held in Paris, France. 1999 ISBN 0-7923-5450-8
67. F.J. Fahy and W.G. Price (eds.): *IUTAM Symposium on Statistical Energy Analysis*. Proceedings of the IUTAM Symposium held in Southampton, U.K. 1999 ISBN 0-7923-5457-5
68. H.A. Mang and F.G. Rammerstorfer (eds.): *IUTAM Symposium on Discretization Methods in Structural Mechanics*. Proceedings of the IUTAM Symposium held in Vienna, Austria. 1999 ISBN 0-7923-5591-1

Mechanics

SOLID MECHANICS AND ITS APPLICATIONS

Series Editor: G.M.L. Gladwell

69. P. Pedersen and M.P. Bendsøe (eds.): *IUTAM Symposium on Synthesis in Bio Solid Mechanics.* Proceedings of the IUTAM Symposium held in Copenhagen, Denmark. 1999 ISBN 0-7923-5615-2
70. S.K. Agrawal and B.C. Fabien: *Optimization of Dynamic Systems.* 1999 ISBN 0-7923-5681-0
71. A. Carpinteri: *Nonlinear Crack Models for Nonmetallic Materials.* 1999 ISBN 0-7923-5750-7
72. F. Pfeifer (ed.): *IUTAM Symposium on Unilateral Multibody Contacts.* Proceedings of the IUTAM Symposium held in Munich, Germany. 1999 ISBN 0-7923-6030-3
73. E. Lavendelis and M. Zakrzhevsky (eds.): *IUTAM/IFToMM Symposium on Synthesis of Nonlinear Dynamical Systems.* Proceedings of the IUTAM/IFToMM Symposium held in Riga, Latvia. 2000 ISBN 0-7923-6106-7
74. J.-P. Merlet: *Parallel Robots.* 2000 ISBN 0-7923-6308-6
75. J.T. Pindera: *Techniques of Tomographic Isodyne Stress Analysis.* 2000 ISBN 0-7923-6388-4
76. G.A. Maugin, R. Drouot and F. Sidoroff (eds.): *Continuum Thermomechanics.* The Art and Science of Modelling Material Behaviour. 2000 ISBN 0-7923-6407-4
77. N. Van Dao and E.J. Kreuzer (eds.): *IUTAM Symposium on Recent Developments in Non-linear Oscillations of Mechanical Systems.* 2000 ISBN 0-7923-6470-8
78. S.D. Akbarov and A.N. Guz: *Mechanics of Curved Composites.* 2000 ISBN 0-7923-6477-5
79. M.B. Rubin: *Cosserat Theories: Shells, Rods and Points.* 2000 ISBN 0-7923-6489-9
80. S. Pellegrino and S.D. Guest (eds.): *IUTAM-IASS Symposium on Deployable Structures: Theory and Applications.* Proceedings of the IUTAM-IASS Symposium held in Cambridge, U.K., 6–9 September 1998. 2000 ISBN 0-7923-6516-X
81. A.D. Rosato and D.L. Blackmore (eds.): *IUTAM Symposium on Segregation in Granular Flows.* Proceedings of the IUTAM Symposium held in Cape May, NJ, U.S.A., June 5–10, 1999. 2000 ISBN 0-7923-6547-X
82. A. Lagarde (ed.): *IUTAM Symposium on Advanced Optical Methods and Applications in Solid Mechanics.* Proceedings of the IUTAM Symposium held in Futuroscope, Poitiers, France, August 31–September 4, 1998. 2000 ISBN 0-7923-6604-2
83. D. Weichert and G. Maier (eds.): *Inelastic Analysis of Structures under Variable Loads.* Theory and Engineering Applications. 2000 ISBN 0-7923-6645-X
84. T.-J. Chuang and J.W. Rudnicki (eds.): *Multiscale Deformation and Fracture in Materials and Structures.* The James R. Rice 60th Anniversary Volume. 2001 ISBN 0-7923-6718-9
85. S. Narayanan and R.N. Iyengar (eds.): *IUTAM Symposium on Nonlinearity and Stochastic Structural Dynamics.* Proceedings of the IUTAM Symposium held in Madras, Chennai, India, 4–8 January 1999 ISBN 0-7923-6733-2
86. S. Murakami and N. Ohno (eds.): *IUTAM Symposium on Creep in Structures.* Proceedings of the IUTAM Symposium held in Nagoya, Japan, 3-7 April 2000. 2001 ISBN 0-7923-6737-5
87. W. Ehlers (ed.): *IUTAM Symposium on Theoretical and Numerical Methods in Continuum Mechanics of Porous Materials.* Proceedings of the IUTAM Symposium held at the University of Stuttgart, Germany, September 5-10, 1999. 2001 ISBN 0-7923-6766-9
88. D. Durban, D. Givoli and J.G. Simmonds (eds.): *Advances in the Mechanis of Plates and Shells The Avinoam Libai Anniversary Volume.* 2001 ISBN 0-7923-6785-5
89. U. Gabbert and H.-S. Tzou (eds.): *IUTAM Symposium on Smart Structures and Structonic Systems.* Proceedings of the IUTAM Symposium held in Magdeburg, Germany, 26–29 September 2000. 2001 ISBN 0-7923-6968-8

Mechanics

SOLID MECHANICS AND ITS APPLICATIONS

Series Editor: G.M.L. Gladwell

90. Y. Ivanov, V. Cheshkov and M. Natova: *Polymer Composite Materials – Interface Phenomena & Processes*. 2001 ISBN 0-7923-7008-2
91. R.C. McPhedran, L.C. Botten and N.A. Nicorovici (eds.): *IUTAM Symposium on Mechanical and Electromagnetic Waves in Structured Media*. Proceedings of the IUTAM Symposium held in Sydney, NSW, Australia, 18-22 Januari 1999. 2001 ISBN 0-7923-7038-4
92. D.A. Sotiropoulos (ed.): *IUTAM Symposium on Mechanical Waves for Composite Structures Characterization*. Proceedings of the IUTAM Symposium held in Chania, Crete, Greece, June 14-17, 2000. 2001 ISBN 0-7923-7164-X
93. V.M. Alexandrov and D.A. Pozharskii: *Three-Dimensional Contact Problems*. 2001 ISBN 0-7923-7165-8
94. J.P. Dempsey and H.H. Shen (eds.): *IUTAM Symposium on Scaling Laws in Ice Mechanics and Ice Dynamics*. Proceedings of the IUTAM Symposium held in Fairbanks, Alaska, U.S.A., 13-16 June 2000. 2001 ISBN 1-4020-0171-1
95. U. Kirsch: *Design-Oriented Analysis of Structures*. A Unified Approach. 2002 ISBN 1-4020-0443-5
96. A. Preumont: *Vibration Control of Active Structures*. An Introduction (2nd Edition). 2002 ISBN 1-4020-0496-6
97. B.L. Karihaloo (ed.): *IUTAM Symposium on Analytical and Computational Fracture Mechanics of Non-Homogeneous Materials*. Proceedings of the IUTAM Symposium held in Cardiff, U.K., 18-22 June 2001. 2002 ISBN 1-4020-0510-5
98. S.M. Han and H. Benaroya: *Nonlinear and Stochastic Dynamics of Compliant Offshore Structures*. 2002 ISBN 1-4020-0573-3
99. A.M. Linkov: *Boundary Integral Equations in Elasticity Theory*. 2002 ISBN 1-4020-0574-1
100. L.P. Lebedev, I.I. Vorovich and G.M.L. Gladwell: *Functional Analysis*. Applications in Mechanics and Inverse Problems (2nd Edition). 2002 ISBN 1-4020-0667-5; Pb 1-4020-0756-6

Kluwer Academic Publishers – Dordrecht / Boston / London

ICASE/LaRC Interdisciplinary Series in Science and Engineering

1. J. Buckmaster, T.L. Jackson and A. Kumar (eds.): *Combustion in High-Speed Flows.* 1994 ISBN 0-7923-2086-X
2. M.Y. Hussaini, T.B. Gatski and T.L. Jackson (eds.): *Transition, Turbulence and Combustion.* Volume I: Transition. 1994 ISBN 0-7923-3084-6; set 0-7923-3086-2
3. M.Y. Hussaini, T.B. Gatski and T.L. Jackson (eds.): *Transition, Turbulence and Combustion.* Volume II: Turbulence and Combustion. 1994 ISBN 0-7923-3085-4; set 0-7923-3086-2
4. D.E. Keyes, A. Sameh and V. Venkatakrishnan (eds): *Parallel Numerical Algorithms.* 1997 ISBN 0-7923-4282-8
5. T.G. Campbell, R.A. Nicolaides and M.D. Salas (eds.): *Computational Electromagnetics and Its Applications.* 1997 ISBN 0-7923-4733-1
6. V. Venkatakrishnan, M.D. Salas and S.R. Chakravarthy (eds.): *Barriers and Challenges in Computational Fluid Dynamics.* 1998 ISBN 0-7923-4855-9
7. M.D. Salas, J.N. Hefner and L. Sakell (eds.): *Modeling Complex Turbulent Flows.* 1999 ISBN 0-7923-5590-3

KLUWER ACADEMIC PUBLISHERS – DORDRECHT / BOSTON / LONDON

MIX
Papier aus verantwortungsvollen Quellen
Paper from responsible sources
FSC® C105338

If you have any concerns about our products,
you can contact us on
ProductSafety@springernature.com

In case Publisher is established outside the EU,
the EU authorized representative is:
Springer Nature Customer Service Center GmbH
Europaplatz 3, 69115 Heidelberg, Germany

Printed by Libri Plureos GmbH
in Hamburg, Germany